Mosaicism in Human Skin

Rudolf Happple

Mosaicism in Human Skin

Understanding Nevi, Nevoid Skin Disorders, and Cutaneous Neoplasia

Springer

Rudolf Happle
Department of Dermatology
Freiburg University Medical Center
Freiburg
Germany

ISBN 978-3-662-50205-1 ISBN 978-3-642-38765-4 (eBook)
DOI 10.1007/978-3-642-38765-4
Springer Heidelberg New York Dordrecht London

Softcover reprint of the hardcover 1st edition 2014

Printed on acid-free paper

Springer is part of Springer Science+Business Media (www.springer.com)

Preface

During my training in dermatology in the years 1968 through 1972, I had seen some linear skin lesions but never heard or read of Alfred Blaschko since, among dermatologists, his lines had fallen into oblivion. Shortly thereafter, when I worked at the University of Münster, the geneticist Rudolf Pfeiffer told me of Blaschko's atlas and how to find this book. I was fascinated by Blaschko's innovating ideas and clarity of writing. Some years later, I began to give lectures on possible genetic mechanisms giving rise to this system of lines. In part my proposed explanations were influenced by some seminal thoughts that I had found in the book *Medical Genetics* written by Widukind Lenz from Münster. Today, this field of medical science has extensively grown by recognition of other patterns and mechanisms. In the years to come, further molecular research will show which of the hypotheses presented here can be corroborated and which of them may turn out to be wrong.

Admittedly, this book deals mainly with rare mosaic disorders. Studying such phenotypes, however, has far-reaching practical implications. For example, if a type 2 segmental manifestation of tuberous sclerosis is mistaken as simple mosaicism, this will mislead genetic counseling. The theory of superimposed segmental manifestation of polygenic skin disorders will help elucidating further the molecular background of common diseases such as psoriasis or atopic dermatitis. And the concept of revertant mosaicism may help develop novel approaches of gene therapy for severe disorders such as epidermolysis bullosa.

Some readers may feel puzzled by neologisms such as rhodoid nevus, sash-like pattern, or livedo reticularis congenita. In defense of these new terms, let me refer to the Swedish scientist Carl von Linné who wrote during the eighteenth century: "Nomina si nescis, perit et cognitio rerum" (If you don't know the names, knowledge of things is likewise lost).

I should like to thank Professor Leena Bruckner-Tuderman who kindly offered me a working place at the Department of Dermatology in Freiburg and to Professor Michael Hertl who had previously done the same in Marburg. Special thanks go to Drs. Marcel Müller and Johannes Kern from Freiburg who helped me navigate the shoals of electronic preparation of the book, to Ms. Ellen Blasig from Springer Science + Business Media who thoughtfully accompanied the progress of the work, and to Ms. Madona Samuel from SPi Global, Chennai, India, who managed the production of the book in a creative and straightforward way. On the ensuing pages, readers will find a list of persons who generously provided unpublished photographs or gave advice how to find case reports being difficult to trace. Finally, I want to thank my wife Karin who has always accompanied my clinical and scientific reflections and steadfastly helped me in coping with the ups and downs of the present literary project.

There are several different cutaneous patterns of human mosaicism, but the history of this field of knowledge began with Blaschko's pioneering work. As John Savin from Edinburgh has formulated, "Alfred Blaschko would surely have been delighted with the results achieved by reading between his lines – more are sure to follow."

Freiburg im Breisgau, Germany Rudolf Happle

Acknowledgments

The following colleagues or other persons kindly provided photographs or gave advice how to find cases and references:

David Atherton, London, UK
Jacques Ayer, Geneva, Switzerland
Eulalia Baselga, Barcelona, Spain
Herrmann Blaschko, Oxford, UK
Mario Bittar, Mendoza, Argentina
Ernesto Bonifazi, Bari, Italy
Frédéric Cambazard, St. Etienne, France
Marco Castori, Rome, Italy
WenChieh Chen, Munich, Germany
Hansjörg Cremer, Heilbronn, Germany
Franco Crovato, Genoa, Italy
Hugo Degreef, Leuven, Belgium
Elzo Folkers, Zaandam, The Netherlands
Regina Fölster-Holst, Kiel, Germany
Hansjörg Frei, Zurich, Switzerland
Alejandro García Vargas, Guadalajara, Mexico
Antonia González-Enseñat, Barcelona, Spain
Robert J. Gorlin, Minneapolis, Minnesota, USA
Henning Hamm, Würzburg, Germany
Susanne Happle, Shanghai, China
Adelaide Hebert, Houston, Texas
Helena de las Heras, Madrid, Spain
Susan M. Huson, Manchester, UK
Peter H. Itin, Basel, Switzerland
Marcel Jonkman, Groningen, The Netherlands
Hülya Kayserili, Istanbul, Turkey
Claudia Kluge, Freiburg, Germany
Arne König, Marburg, Germany
Thomas Krieg, Cologne, Germany
Gerhard Kurlemann, Münster, Germany
Michael Landthaler, Regensburg, Germany
Ulrich Langenbeck, Frankfurt, Germany
Eric Legius, Leuven, Belgium
Derek Lim, Birmingham, UK
Gérard Lorette, Tours, France

Eamonn Maher, Birmingham, UK
Wolfgang Marsch, Halle (Saale), Germany
Silvestre Martínez-García, Málaga, Spain
Beatrice Mintz, Philadelphia, Pennsylvania, USA
Georges Moulin, Lyon, France
Marcel Müller, Freiburg, Germany
Kirsti-Maria Niemi, Helsinki, Finland
Shiro Niiyama, Sagamihara, Japan
Shigeo Nishiyama, Kamakura, Japan
Arnold Oranje, Rotterdam, The Netherlands
Jean-Paul Ortonne, Nice, France
Francis Palisson, Santiago, Chile
Mauro Paradisi, Rome, Italy
Anna M. Pasmooij, Groningen, The Netherlands
Gerd Plewig, Munich, Germany
Howard Pride, Danville, Pennsylvania, USA
Hans-Dieter Rott, Erlangen, Germany
Ramón Ruiz-Maldonado, Mexico City, Mexico
Thomas Ruzicka, Munich, Germany
Aïcha Salhi, Algiers, Algeria
Walter Salzburger, Basel, Switzerland
Silvina Sartori, Santa Fe, Argentina
Maxwell B. Sauder, Ottawa, Canada
Cornelia S. Seitz, Göttingen, Germany
Theo Starink, Amsterdam, The Netherlands
Beat Steinmann, Zurich, Switzerland
Alexander Stella, Vienna, Austria
Alain Taïeb, Bordeaux, France
Iliana Tantcheva-Poor, Cologne, Germany
Mustafa Tekin, Ankara, Turkey
Sigrid Tinschert, Berlin, Germany
Uwe Töllner, Fulda, Germany
Jaime Toribio, Santiago de Compostela, Spain
Antonio Torrelo, Madrid, Spain
Adoni Urtizberea, Paris, France
Fereydoun Vakilzadeh, Hildesheim, Germany
Shyam B. Verma, Vadodara, India
Wolf I. Worret, Munich, Germany
Ching-Ying Wu, Kaohsiung, Taiwan
Hitoshi Yaguchi, Tokyo, Japan
Shehu M. Yusuf, Kano, Nigeria
Mónica Zambrano, Quito, Ecuador

Contents

1 Introduction

Since ancient times, a mosaic denotes a piece of artwork made by placing colored squares in a pattern creating a picture. In modern biology, the term mosaic is used in a metaphoric way. It means an organism that is composed of two or more genetically different cell lines originating from one homogeneous zygote. Mosaicism can occur in all pluricellular living organisms. In human genetics, a well-known example is functional mosaicism in women because one of the X chromosomes is randomly inactivated at an early developmental stage. Mosaicism can involve all organs but is most easily noted in the skin because this organ is right before our eyes.

During the past decades, the concept of mosaicism in human skin has gained increasing importance and awareness because in many cutaneous disorders a mixture of normal and aberrant cells, giving rise to alternating segments of affected and unaffected skin, has now been documented at the molecular level. Today, it is clear that every human being represents, to some degree, a mosaic. For example, evidence has been provided that all nevi reflect mosaicism and that epigenetic mosaicism involves a large number of genes in both male and female individuals.

In dermatology, the concept of mosaicism has been shown to exert explanatory power to understand the etiology and pathogenesis of both rare and common diseases. Combined clinical and molecular research on mosaic skin disorders has helped understanding the following problems:

- How to find a reasonable definition of the term nevus;
- Why different types of nevi may sometimes occur together and in close proximity to each other;
- Why some nevi that tend to occur sporadically may, by way of exception, affect several members of a family;
- Why the pattern of distribution of large nevi should neither be called "zosteriform" nor "dermatomal";
- Why the presence of a segmental form of monogenic disorders such as neurofibromatosis implies an increased risk for the next generation;
- Why we can discriminate, in autosomal dominant skin disorders, three different categories of postzygotic mosaicism;
- Why some genodermatoses occur exclusively in females;
- Why some segmentally arranged skin disorders are heritable whereas others are not;
- Why patients with a severe autosomal recessive skin disorder such as epidermolysis bullosa may sometimes develop patchy areas of completely healthy skin;
- Why common skin disorders such as psoriasis are sometimes superimposed by a pronounced linear or otherwise segmental involvement;
- In which way the giant melanocytic nevus is nosologically related to small melanocytic nevi that may be congenital or acquired;
- How skin cancer develops.

R. Happle, *Mosaicism in Human Skin*,
DOI 10.1007/978-3-642-38765-4_1, © Springer-Verlag Berlin Heidelberg 2014

If an organism originates from the fusion of two different zygotes, the result will not be a mosaic but a chimera. In order to clarify the similarities and differences between mosaicism and chimerism, the reader will find in this book also a chapter on chimerism in human skin.

Part I

The Mechanisms of Cutaneous Mosaicism

Mosaicism as a Biological Concept

2

Contents

Virtually all human skin disorders, including common diseases such as psoriasis, drug eruptions, or acne, sometimes display mosaicism. It is important to realize, however, that mosaicism also occurs as a physiological phenomenon. All human beings represent, to some degree, mosaics because today we know that mutual monoallelic expression of genes gives rise to functional mosaicism in both women and men. In other words, humans always display epigenetic mosaicism. On the other hand, all mammalian organisms will develop, during their lifetime, various forms of genomic mosaicism that may originate from diverse mechanisms such as postzygotic new mutation, mitotic recombination resulting in loss of heterozygosity, or other forms of allelic loss. Most of these mechanisms have first been studied in plants or animals. Conversely, some forms of mosaicism such as the superimposed segmental manifestation of both autosomal dominant and polygenic traits have initially been recognized in human disorders.

2.1 Historical Beginnings

During the first half of the twentieth century, mosaic patterning was first described in plants and subsequently in animals. Concepts explaining mosaic phenotypes of animals and plants were already proposed shortly after 1901, the year when the Mendelian rules of inheritance had been rediscovered. In 1904, the zoologist Valentin Häcker [20] mentioned, in an article on

R. Happle, *Mosaicism in Human Skin*,
DOI 10.1007/978-3-642-38765-4_2, © Springer-Verlag Berlin Heidelberg 2014

animal breeding, the word "Mosaikbastarde" (mosaic bastards) to denote animals showing the traits of their parents arranged in a "mosaic-like distribution" on the various parts of their body. In 1913 Collins [5] used the term "mosaic" to describe a variegated pattern of seeds in maize. It took a rather long time, however, until the concept of mosaicism was also applied to human disorders.

2.2 Mosaicism in Plants

Mosaic patches are easily seen in plants (Fig. 2.1). Since the rediscovery of the rules of Mendelian inheritance in 1901, formation of mosaicism has been reported, initially under the term "somatic segregation" [32], in various species such as maize [5, 9, 10, 38], citrus fruits [16, 52], apple [8, 32], pineapple [6], tobacco [57], and Dahlia [33].

Fig. 2.1 Mosaic flowers in (**a**) *Solanum* species (Courtesy of Dr. Ulrich Langenbeck, Friedrichsdorf, Germany) and (**b**) *Cyclamen* species

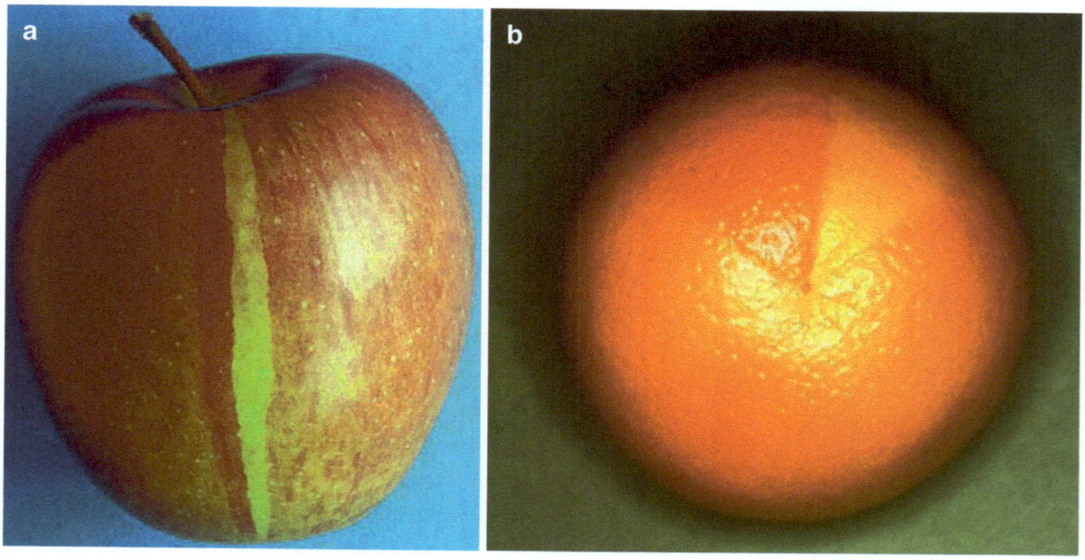

Fig. 2.2 Twin spotting (**a**) in an apple and (**b**) in an orange (**b**: Courtesy of Dr. Frédéric Cambazard, St. Etienne, France)

Fig. 2.3 Twin spot in a tomato leaf (Painted from a photograph published by Ross and Holm [48])

In 1935, Emerson et al. [11] suggested that mosaic seeds of maize "may also be interpreted as somatic mutations."

In 1937, twin spotting in maize and apple (Fig. 2.2a) was comprehensively described by Donald F. Jones from New Haven, Connecticut, and interpreted as a result of somatic crossing-over [31]. He already distinguished allelic from nonallelic twin spotting and even described "twin spots within twin spots" in seeds of maize as well as "twin stripes" in oranges, maize kernels, and apples. He concluded that "paired stripes may be expected in all fruits and flowers that are heterozygous for color in the epidermis." Jones even anticipated the concept of revertant mosaicism in the form of "reversal of recessive changes to the original dominant condition" when writing: "However, somatic crossing over offers no solution for recessive mosaics returning to the dominant condition. In a few cases this apparently does occur."

Twin spotting has been described in many other plants such as tomato (Fig. 2.3) [48], Dahlia species (Fig. 2.4) [33], tobacco [12], soybean [60], and snapdragon [27].

By investigating mosaic traits in maize, Barbara McClintock [39] developed her concept of "jumping genes" and thus opened the new field of epigenetic mosaicism, an achievement for which she later received the Nobel Prize.

Fig. 2.4 Historical picture of twin spotting in a Dahlia petal. Painted from a figure published by W.H.J. Lawrence in 1929 [33]

2.3 Mosaicism in Animals

During the first half of the past century, mosaic patterns were described in various species such as Drosophila [7, 40, 45], pigeon [28], budgerigar (Fig. 2.5) [54], domestic fowl [47, 51], rat [3], rabbit [46], and guinea pig [61]. Hollander [29] published a seminal paper on mosaic effects in domestic birds. In 1927, Curd Stern [55] reported on the experimental production of mosaicism in Drosophila.

Fig. 2.5 Mosaic plumage of a budgerigar [54] (Reproduced with permission from Revue Suisse de Zoologie, Geneva, Switzerland)

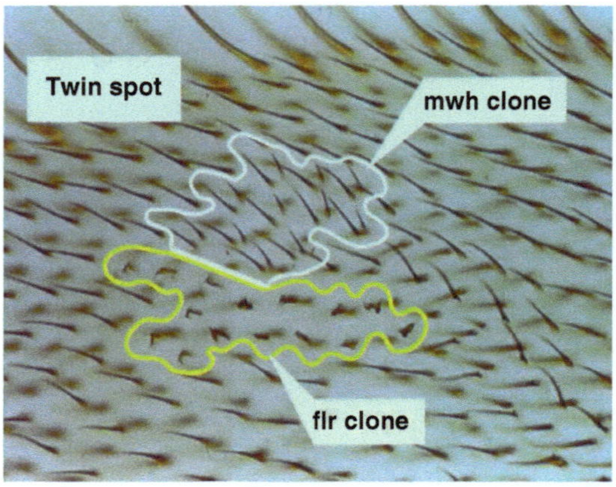

Fig. 2.6 Twin spotting in the Drosophila wing spot test. The two homozygous cell clones (*mwh* multiple wing hair, *flr* flare) originated from somatic recombination induced by a chemical mutagen (Courtesy of Dr. Hansjörg Frei, Zurich, Switzerland)

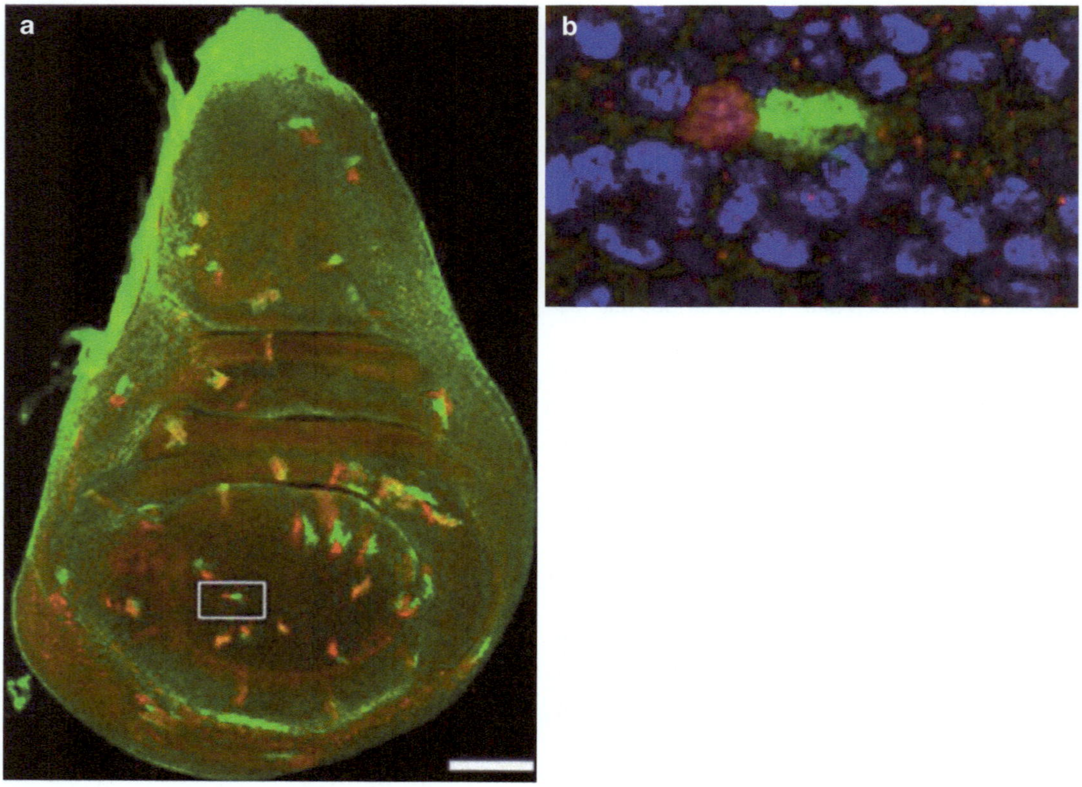

Fig. 2.7 (**a**) Imaginal wing disc of *Drosophila* showing green and red twin spots induced by the twin-spot generator technique. (**b**) *Inset* shows twin spot at the two-cell stage [19] (Reprinted with permission from Nature Publishing Group)

In 1929, John T. Patterson [44], a zoologist from Austin (Texas), first described "twin areas" produced in *Drosophila melanogaster* by x-ray treatment of eggs or young larvae. In the integument he found paired spots being "adjacent or close to each other," in the form of mutant clones of either "*yellow*" or "*singed*." Analogously, he was able to produce, in the eyes of Drosophila, "twin areas" of different mutant colors [43]. Subsequently, somatic crossing-over to explain both single mosaic spots and twin spots was extensively studied by Stern [56].

Today, the Drosophila wing spot test as developed by Graf et al. [17] is used to test various chemicals for their mutagenic or recombinogenic effects or for their action of antigenotoxicity [18]. Larvae of the animal are exposed to the test substance, and subsequently the wings of the fly are microscopically examined for the presence of single or paired spots (Fig. 2.6). Moreover, the twin-spot technique can be used to study the function of imaginal disc epithelial cells in Drosophila [2]. Twin spots have also been induced experimentally in mice [13].

More recently, an FLP recombinase that recognizes sites consisting of short DNA sequences is used in Drosophila to induce postzygotic recombination resulting in mosaicism [34, 35]. Griffin et al. [19] developed a "twin-spot generator" technique that generates paired green and red spots being detectable already as single cells (Figs. 2.7, 2.8, and 2.9). The method has been used to investigate the cellular dynamics of eye or leg regeneration in imaginal discs of Drosophila [58]. By applying the same technique, Yu et al. [62] labeled the sister clones, derived from a common neuroblast, simultaneously in different colors within the antennal lobe of Drosophila instars. In this way they were able to visualize the complete developmental sequence of different neuronal lineages and thus to get more insight into the complex embryogenesis of the brain of Drosophila.

In 1961, Mary Lyon [37] interpreted linear and patchy coat patterns of female mice carrying an X-linked mutation as a manifestation of functional X-chromosome mosaicism, a phenomenon that is today also called lyonization. In this way her name is certainly eternalized more effectively than by a Nobel Prize that she never received.

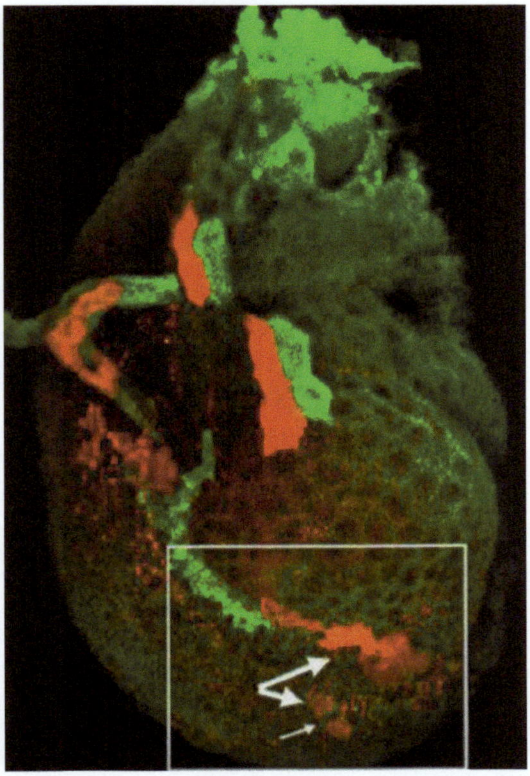

Fig. 2.8 Separation of clones in an imaginal leg disc of *Drosophila*, visualized by the twin-spot generator technique. *Large arrows* indicate separated clones. *Small arrow* indicates an almost separated clone [19] (Reprinted with permission from Nature Publishing Group)

2.4 Mosaicism in Human Skin

When compared to the advances achieved in the vegetable and animal kingdoms, the concept of mosaicism was accepted rather slowly in dermatology. During the first half of the twentieth century, the pioneering work of Alfred Blaschko on his "nevus lines" [1] (see Figs. 5.4, 5.5, and 5.6) fell into a long sleep. Research on cutaneous mosaicism of man was usually characterized by conjectures and far-fetched speculations. The linear pattern of nevi or nevoid conditions was

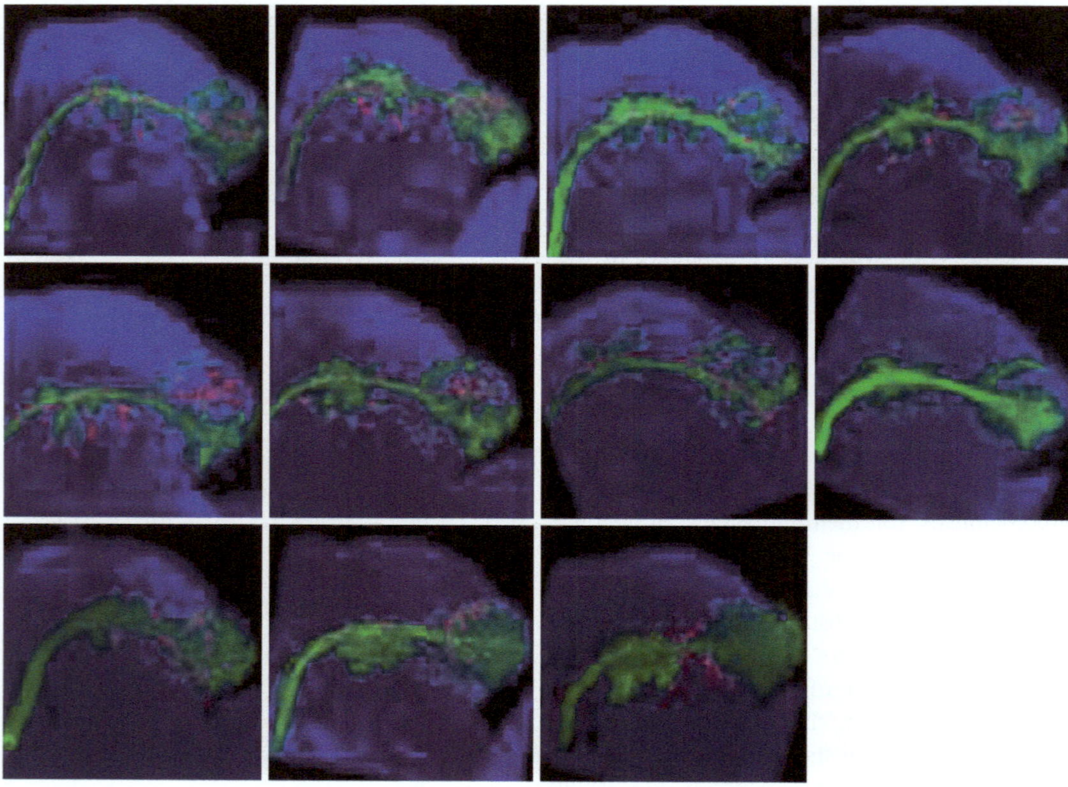

Fig. 2.9 Identification of sequentially derived neurons in the antennal lobe of a Drosophila instar by twin-spot MARCM (mosaic analysis with a repressible cell marker) [62] (Reprinted under license of Creative Commons)

generally thought to reflect the distribution of peripheral nerves [21].

In 1976, the lines of Blaschko [1] regained general awareness through an extensive review published by Robert Jackson from Toronto [30]. At the same time and independently, a comprehensive explanation of Blaschko's lines as a manifestation of genetic mosaicism was presented in Germany (Fig. 5.7) [22, 23]. A classification of several other patterns of cutaneous mosaicism was proposed in 1993 [24, 25]. At the end of the past century, molecular studies began to confirm the mosaic nature of these patterns [24, 26, 42, 49, 50, 53, 59]. Prior to the molecular era, Chemke et al. [4] had provided evidence that a case of linear hypermelanosis present in a mentally deficient patient reflected trisomy 18 mosaicism, and many reports of cytogenetic mosaicism in cases of linear pigmentary disorders had followed [41].

2.5 Mosaicism Versus Chimerism

A chimera is a composite animal or plant originating from the fusion of two or more different zygotes. Primary chimeras develop when the different components develop from fertilization or from a very early developmental stage. Secondary chimeras are formed in postnatal life by transplantation of tissues, or blood transfusion for medical purposes, or by transfer of blood cells from an embryo to the mother.

Primary chimerism may result in skin patterns resembling those observed in mosaic states [14, 15, 36]. Mosaicism usually implies the coexistence of a normal and one or more abnormal components of the skin, whereas in chimeras all of the components are completely normal. The only abnormal feature is their coexistence within one individual.

For the clinician, it is often impossible to determine whether a linear pattern of pigmentary disturbance in human skin reflects mosaicism or chimerism. As a rule of thumb, however, we can say that mosaicism occurs more frequently than chimerism.

References

1. Blaschko A (1901) Die Nervenverteilung in der Haut in ihrer Beziehung zu den Erkrankungen der Haut. Beilage zu den Verhandlungen der Deutschen Dermatologischen Gesellschaft, VII. Congress zu Breslau im Mai 1901. Wien und Leipzig, Braumüller
2. Boedigheimer MJ, Nguyen KP, Bryant PJ (1997) Expanded functions in the apical cell domain to regulate the growth rate of imaginal discs. Dev Genet 20:103–110
3. Castle WE (1922) On a transmissible tricolor variation in rats. Carnegie Inst Publ 320:51–55
4. Chemke J, Rappaport S, Etrog R (1983) Aberrant melanoblast migration associated with trisomy 18 mosaicism. J Med Genet 20:135–137
5. Collins GN (1913) Mosaic coherence of characters in seeds of maize. U S Dept Agr Cir Plant Ind Circ 132:19–21
6. Collins JL (1936) A frequently mutating gene in the pineapple *Ananas comosus* (L.). Merr Am Nat 70:467–476
7. Crew FA, Lamy R (1939) Mosaicism in Drosophila pseudo-obscura. J Genet 37:211–228
8. Dahlgren K, Ossian V (1927) Eine Sektorialchimäre vom Apfel. Hereditas 9:335–342
9. East EM, Hayes HK (1911) Inheritance in maize. Conn Agric Exp Sta Bull 167:1–142
10. Emerson RA (1917) Genetical studies of variegated pericarp in maize. Genetics 2:1–35
11. Emerson RA, Beadle GW, Fraser AC (1935) A summary of linkage studies in maize. Cornell Univ Agric Exp Sta Mem 180:1–83
12. Evans DA, Paddock EF (1976) Comparison of somatic crossing over frequency in *Nicotiana tabacum* and three other crop species. Can J Genet Cytol 18:57–65
13. Fahrig R, Steinkamp-Zucht A (1996) Co-recombinogenic and anti-mutagenic effects of diethylhexyl phthalate, inactiveness of pentachlorophenol in the spot test with mice. Mutat Res 354:59–67
14. Findlay GH, Moores PP (1980) Pigment anomalies of the skin in the human chimaera: their relation to systematized naevi. Br J Dermatol 103:489–498
15. Fitzgerald PH, Donald RA, Kirk RL (1979) A true hermaphrodite dispermic chimera with 46, XX and 46, XY karyotypes. Clin Genet 15:89–96
16. Frost HB (1926) Polyembrony, heterozygosis and chimeras in citrus. Hilgardia 1:365–402
17. Graf U, Würgler FE, Katz AJ, Frei H, Juon H, Hall CB, Kale PG (1984) Somatic mutation and recombination test in *Drosophila melanogaster*. Environ Mutagen 6:153–188
18. Graf U, Abraham SK, Guzman-Rincon J, Würgler FE (1998) Antigenotoxicity studies in *Drosophila melanogaster*. Mutat Res 402:203–209
19. Griffin R, Sustar A, Bonvin M, Binari R, del Valle Rodriguez A, Hohl AM, Bateman JR, Villalta C, Heffern E, Grunwald D, Bakal C, Desplan C, Schubiger G, Wu CT, Perrimon N (2009) The twin spot generator for differential *Drosophila lineage analysis*. Nat Methods 6:600–602
20. Haecker V (1904) Über die Ergebnisse der Bastardlehre, ihre zellengeschichtliche Begründung und ihre Bedeutung für die praktische Tierzucht. Arch Rassen Gesellschaftsbiol 1:321–338
21. Haensch R (1961) Eczema and neural factors. Observations in polyneuroradiculitis. Arch Klin Exp Dermatol 214:35–40
22. Happle R (1976) Genetic mechanisms giving rise to linear skin lesions. Joint meeting of the Vereinigung Südwestdeutscher Dermatologen and the Vereinigung Rheinisch-Westfälischer Dermatologen, Heidelberg, 8–10 Oct.
23. Happle R (1977) Genetic significance of Blaschko's lines. Z Hautkr 52:935–944
24. Happle R (1993) Mosaicism in human skin. Understanding the patterns and mechanisms. Arch Dermatol 129:1460–1470
25. Happle R (1993) Pigmentary patterns associated with human mosaicism: a proposed classification. Eur J Dermatol 3:170–174
26. Happle R (1999) Loss of heterozygosity in human skin. J Am Acad Dermatol 41:143–164
27. Harrison BJ, Carpenter R (1977) Somatic crossing-over in *Antirrhinum majus*. Heredity 38:169–189
28. Hollander WF, Cole LJ (1940) Somatic mosaics in the domestic pigeon. Genetics 25:16–40
29. Hollander WF (1944) Mosaic effects in domestic birds. Quart Rev Biol 19:285–307
30. Jackson R (1976) The lines of Blaschko: a review and reconsideration: observations of the cause of certain unusual linear conditions of the skin. Br J Dermatol 95:349–360
31. Jones DF (1937) Somatic segregation and its relation to atypical growth. Genetics 22:484–522
32. Kraus EJ (1916) Somatic segregation. J Hered 7: 2–8
33. Lawrence WJC (1929) The genetics and cytology of Dahlia species. J Genet 21:125–128
34. Lee T, Luo L (1999) Mosaic analysis with a repressible cell marker for studies of gene function in neuronal morphogenesis. Neuron 22:451–461
35. Lee T, Luo L (2001) Mosaic analysis with a repressible cell marker (MARCM) for Drosophila neural development. Trends Neurosci 24: 251–254
36. Lipsker D, Flory E, Wiesel ML, Hanau D, de la Salle H (2008) Between light and dark, the chimera comes out. Arch Dermatol 144:327–330

37. Lyon MF (1961) Gene action in the X-chromosome of the mouse (*Mus musculus* L.). Nature 190:372–373
38. McClintock B (1932) A Correlation of Ring-Shaped Chromosomes with Variegation in Zea Mays. Proc Natl Acad Sci U S A 18:677–681
39. McClintock B (1951) Chromosome organization and genic expression. Cold Spring Harb Symp Quant Biol 16:13–47
40. Morgan T, Bridges CB (1919) The origin of gynandromorphs. Carnegie Inst Wash Publ 278:1–122
41. Moss C, Savin J (1995) Dermatology and the new genetics. Blackwell Science Ltd, Osney Mead, Oxford
42. Paller AS, Syder AJ, Chan YM, Yu QC, Hutton E, Tadini G, Fuchs E (1994) Genetic and clinical mosaicism in a type of epidermal nevus. N Engl J Med 331:1408–1415
43. Patterson JT (1929) The production of mutations in somatic cells of *Drosophila melanogaster* by means of x-rays. J Exp Zool 53:327–372
44. Patterson JT (1929) Somatic segregation produced by x-rays in *Drosophila melanogaster*. Proc Natl Acad Sci U S A 16:109–111
45. Patterson JT (1930) Proof that the entire chromosome is not eliminated in the production of somatic variations by X-rays in Drosophila. Genetics 15:141–149
46. Pickard JN (1929) A brown-and-black rabbit. J Hered 20:483–484
47. Roberts E, Quisenberry JJ (1935) A Brahma-Plymouth Rock mosaic. J Hered 26:11–14
48. Ross JG, Holm G (1960) Somatic segregation in a tomato. Hereditas 46:224–230
49. Sakuntabhai A, Dhitavat J, Burge S, Hovnanian A (2000) Mosaicism for *ATP2A2* mutations causes segmental Darier's disease. J Invest Dermatol 115(6): 1144–1147
50. Savin JA (1996) Cutaneous mosaicism. QJM 89(7):489–491
51. Serebrovsky AS (1936) "Somatic segregation" in domestic fowl. J Genet 16:33–42
52. Shamel AD, Scott LB, Pomeroy CS (1918) Citrus fruit improvement: a study of bud variegation in the Valencia orange. U S Dep Agric Bull 624:1–120
53. Siegel DH (2008) Cutaneous mosaicism: a molecular and clinical review. Adv Dermatol 24:223–244
54. Steiner H (1938) Über eine halbseitige "Mutationschimäre" des Wellensittichs, *Melopsittacus undulatus* Shaw. Rev Suisse Zool 45:431–440
55. Stern C (1927) Über Chromosomenelimination bei der Taufliege. Naturwissenschaften 15:740–746
56. Stern C (1936) Somatic crossing over and segregation in Drosophila melanogaster. Genetics 21(6):625–730
57. Stino KR (1940) Inheritance in Nicotiana tabacum. XV. Carmine-white variegation. J Hered 31:19–24
58. Sustar A, Bonvin M, Schubiger M, Schubiger G (2011) Drosophila twin spot clones reveal cell division dynamics in regenerating imaginal discs. Dev Biol 356:576–587
59. Tinschert S, Naumann I, Stegmann E, Buske A, Kaufmann D, Thiel G, Jenne DE (2000) Segmental neurofibromatosis is caused by somatic mutation of the neurofibromatosis type 1 (*NF1*) gene. Eur J Hum Genet 8(6):455–459
60. Vig BK (1982) Soybean (*Glycine* max [L.] Merrill) as a short-term assay for study of environmental mutagens. A report of the U.S. Environmental Protection Agency Gene-Tox Program. Mutat Res 99:339–347
61. Wright S, Eaton ON (1926) Mutational mosaic coat patterns of the guinea pig. Genetics 11:333–351
62. Yu HH, Kao CF, He Y, Ding P, Kao JC, Lee T (2010) A complete developmental sequence of a *Drosophila* neuronal lineage as revealed by twin-spot MARCM. PLoS Biol 8(8)

3 Two Major Categories of Mosaicism

Contents

Among the mosaic skin disorders, two major categories in the form of either genomic or epigenetic mosaicism can be distinguished.

3.1 Genomic Mosaicism

Genomic mosaicism in human skin reflects the action of either autosomal or, by way of exception, X-linked genes. Most of the autosomal mosaics represent sporadic traits. By contrast, the rare cases of genomic mosaicism of an X-linked mutation occurring in a male imply that the mosaic trait can be transmitted to a daughter in the form of epigenetic mosaicism.

3.1.1 Genomic Mosaicism of Autosomes

Many human mosaics originate from a postzygotic autosomal mutation. For the purpose of genetic counseling, it is important to distinguish between lethal and nonlethal postzygotic mutations. If the underlying mutation acts as a lethal factor, the risk for the next generation is virtually nil, whereas children of a patient showing mosaicism of a nonlethal mutation run an increased risk that the same phenotype may diffusely affect their entire body.

R. Happle, *Mosaicism in Human Skin*,
DOI 10.1007/978-3-642-38765-4_3, © Springer-Verlag Berlin Heidelberg 2014

3.1.1.1 Mosaicism Caused by Loss of Heterozygosity

In an individual heterozygous for an autosomal mutation involving the skin, postzygotic loss of the corresponding wild-type allele may result in a mosaic phenotype.

Both benign and malignant skin tumors can be taken as examples of cutaneous mosaicism. Most, if not all of them, originate from LOH. In a tissue heterozygous for a mutant allele predisposing to neoplastic growth, loss of the corresponding wild-type allele may give rise to a cell being either homozygous or hemizygous for the tumor gene. Remarkably, however, the traditional two-hit mechanism [31, 106] does not appear to be valid for all skin tumors. Already prior to the molecular era, statistical data obtained by theoretical mathematics indicated that this model was too simplistic for many hereditary malignant tumors [38, 39, 175], and a four-mutation model was proposed [39, 76]. More recent molecular studies have supported this view [25, 40, 60, 133, 163]. Other forms of multihit mechanisms are likewise possible [141].

Moreover, the concept of LOH appears to be a powerful tool to explain various other forms of cutaneous mosaicism such as the type 2 segmental manifestation of autosomal dominant skin disorders, paradominant inheritance, or didymosis (twin spotting).

3.1.1.2 Genomic Mosaicism of Lethal Autosomal Mutations

Some multisystem birth defects are caused by autosomal mutations that, when present in the fertilized egg, result in early intrauterine death of the embryo. Cells affected with such mutations can only survive in an admixture with normal cells, i.e., in a mosaic state that usually originates from an early postzygotic mutation. Theoretically, such mosaics may also develop from a gametic half-chromatid mutation [119], but this hypothesis is so far unproven and its practical significance is unclear.

Mosaicism Caused by Lethal Cytogenetic Abnormalities

A numerical aberration of chromosomes or parts of a chromosome has been found in many cases of cutaneous mosaicism. In 1983, Chemke et al. [26] reported on a linear hypermelanosis reflecting trisomy 18 mosaicism. Subsequently, other authors found various types of numerical chromosome aberrations in patients with pigmentary mosaicism of the hypermelanotic type [111] or of the Ito type being characterized by hypomelanotic bands following Blaschko's lines [114]. More rarely, lethal cytogenetic aberrations have been found in the tissue of epidermal nevi [170, 172] or systematized blaschkitis [123].

Mosaic partial trisomy 13q causes a distinct neurocutaneous phenotype in the form of phylloid hypomelanosis [56]. In its nonmosaic form, trisomy 13 is a "sublethal" chromosome aberration, which means that children with this cytogenetic disorder usually die during early infancy, whereas some of them may reach adolescence or even adulthood [22].

Mosaicism Caused by Lethal Molecular Defects

Some mosaic skin disorders exclusively occur sporadically because the underlying mutation, when present in the zygote, acts as a lethal factor [65, 66]. In the developing embryo, cells carrying the mutation can only survive in a mosaic state, in close vicinity to the normal cell population. The hypothesis has now been corroborated by several molecular studies (Table 3.1).

A similar etiology has been proposed for many other mosaic traits that occur sporadically, such as Becker nevus syndrome, nevus comedonicus syndrome, encephalocraniocutaneous lipomatosis, and cutis marmorata telangiectatica congenita [73, 79] (Table 3.2). Sturge-Weber syndrome is most likely a cephalic manifestation of Sturge-Weber-Klippel-Trenaunay syndrome (see Sect. 7.4.1.1).

Delleman syndrome is a characterized by a triad of ocular, neurological, and cutaneous defects. Major features include orbital cyst, eyelid coloboma, porencephaly, cranial defects, periorbital skin tags, and multiple patchy aplastic skin lesions. The phenotype always occurs sporadically and shows a mosaic, asymmetrical arrangement of lesions, which is why the action of a lethal gene is rather likely [66].

Table 3.1 Mosaic phenotypes reflecting lethal autosomal mutations confirmed at the molecular level

Disorder	OMIM number	Reference	Section number in this book
CLOVES syndrome	612918	[113, 159]	7.3.1.1
FGFR3 epidermal nevus syndrome (García-Hafner-Happle syndrome)	No specific entry (see 162900)	[52, 143]	7.3.1.1
Megalencephaly-livedo reticularis congenita syndrome ("macrocephaly-capillary malformation syndrome")	602501	[153]	7.4.1.5
McCune-Albright syndrome	174800	[65, 161, 191]	7.2.2.3
Maffucci syndrome	614569	[2, 79]	12.2.1.2
Papular nevus spilus syndrome	No entry	[59, 78]	7.2.1.6
Proteus syndrome	176920	[66, 122]	7.3.1.5
Schimmelpenning syndrome, including phacomatosis pigmentokeratotica	163200	[58, 59, 64]	7.3.2.1
Sturge-Weber syndrome	185300	[66, 164]	7.4.1.1

Table 3.2 Other mosaic phenotypes suggesting the action of a lethal mutation surviving by mosaicism

Disorder	OMIM number	Reference	Section number in this book
Angora hair nevus syndrome (Schauder syndrome)	No entry	[160]	7.3.2.4
Becker nevus syndrome	604919	[89]	7.3.2.5
Castori syndrome	No entry	[23]	7.3.2.7
Cutis marmorata telangiectatica congenita (Van Lohuizen syndrome)	219250	[62, 67]	7.4.1.4
Delleman syndrome	164180	[66]	3.1.1.2
Encephalocraniocutaneous lipomatosis	613001	[91]	7.6.2
NEVADA syndrome	No entry	[87]	7.3.1.6
Nevus comedonicus syndrome	No entry	[68]	7.3.2.2
Nevus trichilemmocysticus syndrome	No entry	[87]	7.3.2.7
Pigmentary mosaicism of the Ito type	300337	[73]	7.2.2.1

In most cases of pigmentary mosaicism of Ito, cytogenetic analysis reveals a normal karyotype in all samples examined. Such cases can best be explained by a postzygotic point mutation or intragenic deletion that cannot be detected with the techniques so far available.

3.1.1.3 Genomic Mosaicism of Nonlethal Autosomal Mutations

According to formal genetics, we can discriminate, in autosomal dominant skin disorders, three different types of mosaicism.

Three Categories of Mosaicism of Nonlethal Mutations

In a type 1 segmental involvement, the remaining skin is healthy. By contrast, a type 2 segmental manifestation is rather pronounced and superimposed on the nonsegmental disorder, which means that the remaining skin is affected to a degree as noted in the ordinary, nonmosaic phenotype. This dichotomy of mosaic manifestations is important for genetic counseling because in the type 2 segmental involvement the next generation runs a 50 % risk of occurrence of the nonsegmental trait, whereas in the type 1 this risk is much lower.

A third category of mosaicism develops at a later stage of fetal development or during postnatal life. It is by far the most common type of mosaicism occurring in autosomal dominant traits and consists in a nonsegmental, disseminated arrangement of neoplastic or nonneoplastic skin lesions. For example, all neurofibromas and café-au-lait macules of neurofibromatosis 1 reflect the clonal outgrowth of cells having lost, at the *NF1* locus, the corresponding wild-type allele [51]. An unusual form of disseminated revertant mosaicism as noted in ichthyosis in confetti (ichthyosis variegata) [29] does likewise belong to this common category.

Type 1 Segmental Manifestation of Autosomal Dominant Traits

A segmental involvement associated with otherwise unaffected skin has been described in many autosomal dominant genodermatoses such as neurofibromatosis 1 [178], tuberous sclerosis [3, 134, 186], epidermolytic ichthyosis of Brocq [145], Hailey-Hailey disease [94], syringomatosis [200], Darier disease [157], and Ehlers-Danlos syndrome [34, 110].

It is important to realize that a type 1 segmental manifestation may involve, by way of exception, most parts of the integument, leaving only small segments of the skin unaffected (Fig. 3.1) [185].

Practical Aspect: A type 1 segmental involvement may imply gonadal mosaicism (Fig. 3.2). Therefore, patients with a type 1 segmental manifestation run an increased risk to give birth to children affected with a nonsegmental form of the same trait. Most likely, this risk will increase with the degree of segmental involvement as present in the parent.

Type 2 Segmental Manifestation of Autosomal Dominant Traits

Until the end of the twentieth century, the theory prevailed that mosaic forms of autosomal dominant skin disorders originated from postzygotic new mutations. Today, this concept is no longer valid [74]. Sometimes, the mosaic involvement is very pronounced and may be superimposed on a milder, diffuse manifestation of the same phenotype [137]. Such cases may be best explained by loss of the corresponding wild-type allele occurring at an early developmental stage, giving rise to a mosaic cell clone being either homozygous or hemizygous for the underlying mutation [70]. Based on this concept the rule of dichotomous types of segmental manifestation was developed (Fig. 3.3) [75, 77]. The major differences

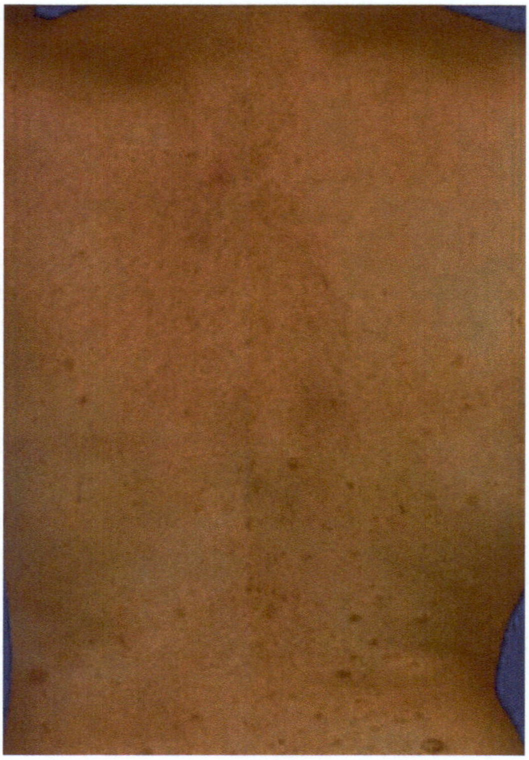

Fig. 3.1 This 38-year-old woman had lesions of neurofibromatosis 1 throughout her body but leaving a few segmental areas unaffected. Molecular proof of a postzygotic new mutation was provided by the authors [185]

between the two types of segmental involvement are summarized in Table 3.3.

The type 2 segmental manifestation should not be taken as a rarely occurring oddity. Rather, it reflects a general rule that can be observed in almost all autosomal dominant skin disorders.

Some clinical examples are summarized in Tables 3.4 and 3.5. The concept has so far been proven at the molecular level in several disorders such as Hailey-Hailey disease [149], PTEN hamartoma syndrome [82], and Darier disease [46]. Molecular corroboration in other autosomal dominant skin disorders can be expected in the near future.

A list of historical terms that were used to describe what can today be categorized as type 2 segmental manifestations is presented in Table 3.6. During the years to come, such descriptive wording should alert the readers that,

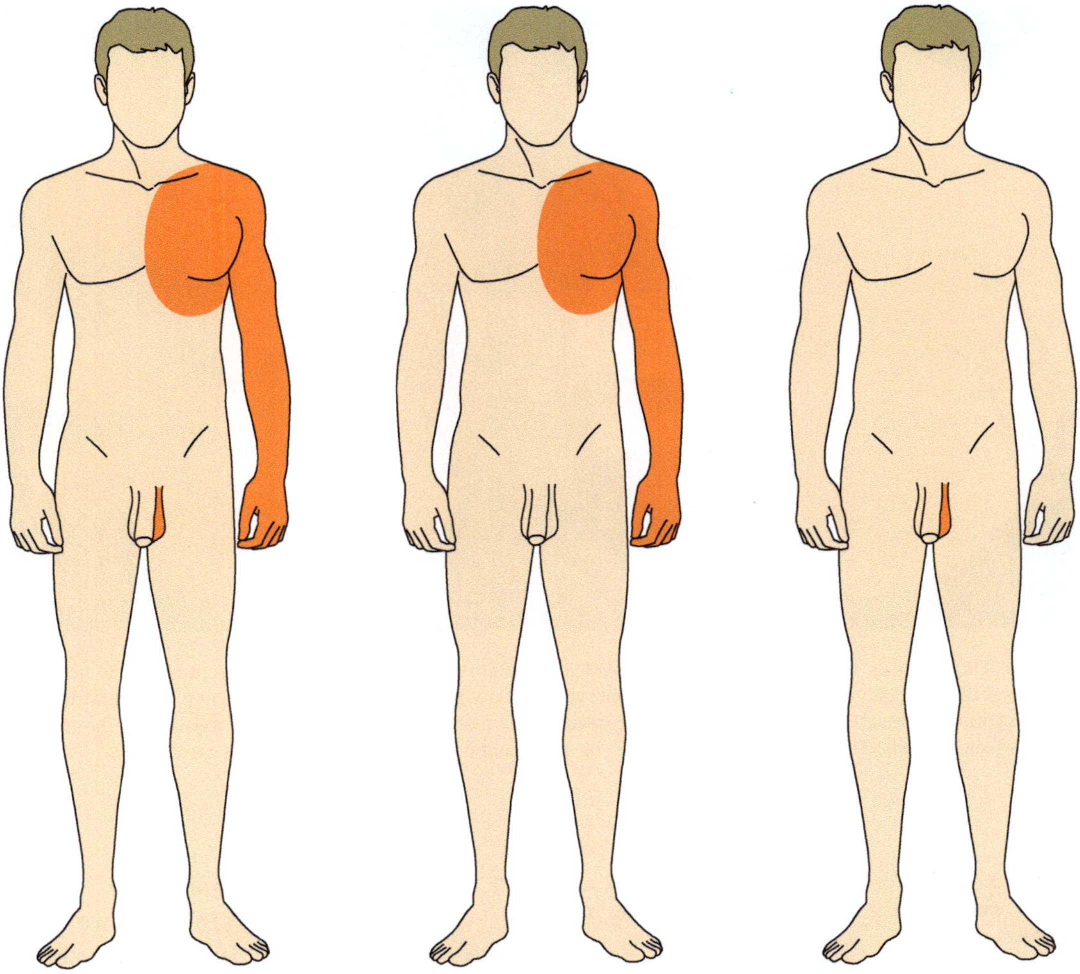

Fig. 3.2 The term "somatic mosaicism" is often used in a misleading way because it refers to only one of three possibilities. A postzygotic mutation may affect both somatic and gonadal tissues (*left*) or somatic tissue alone (*middle*) or gonadal tissue alone (*right*)

in fact, a type 2 segmental involvement may have been overlooked. One descriptive term, "progressive osseous heteroplasia," has even got its own OMIM number 166350, although it merely represents a pronounced mosaic manifestation of the autosomal dominant trait, hereditary osteoma cutis [142].

Remarkably, the proclivity to develop a type 2 segmental involvement varies to a large degree in different traits. It appears to be particularly high in leiomyomatosis, glomangiomatosis, tuberous sclerosis, neurofibromatosis 1, and disseminated superficial actinic porokeratosis (DSAP), whereas no convincing example of such superimposed segmental involvement has so far been found in another autosomal dominant trait, ichthyosis vulgaris [77].

It should be noted that the type 2 is by no means a rarely occurring phenomenon when compared with the type 1 segmental involvement. In some genodermatoses such as glomangiomatosis or DSAP, a type 2 segmental involvement is far more frequently reported in the literature. Apparently, this difference cannot

Fig. 3.3 Dichotomous types of mosaicism as noted in autosomal dominant skin disorders. In a healthy embryo, a postzygotic new mutation gives rise to a type 1 segmental involvement reflecting heterozygosity (*left*). In a heterozygous embryo, an early event of loss of heterozygosity may cause a type 2 segmental involvement being far more pronounced and superimposed on the nonsegmental phenotype (*right*)

Table 3.3 Differences between type 1 and type 2 segmental manifestations of autosomal dominant skin disorders

Type 1 segmental manifestation	Type 2 segmental manifestation
Originates from a new mutation occurring in a healthy embryo	Originates in a heterozygous embryo
Reflects heterozygosity	Reflects loss of heterozygosity
Becomes manifest at an age when the nonsegmental phenotype would appear	Becomes manifest much earlier than the nonsegmental phenotype
Degree of involvement corresponds to that of the nonsegmental phenotype	Degree of involvement is very pronounced and superimposed on the ordinary trait

simply be explained by a bias of ascertainment, although a type 1 segmental manifestation of these disorders is less conspicuous [18] and may, therefore, go either unrecognized or unreported.

3.1.1.4 A Note on the Theory of Paradominant Inheritance

Some mosaic skin disorders are assumed to reflect a lethal autosomal new mutation, which is why they usually occur sporadically. By way of exception, however, they may affect several members of a family. To explain this paradox, the concept of paradominant inheritance was proposed [71]. Heterozygous individuals would in general be clinically healthy, which is why the mutation would be transmitted unperceivably through many generations. The disorder would only become manifest when, at an early developmental stage, loss of the corresponding wild-type allele occurs in a somatic cell, giving rise to a mosaic patch of homozygous or hemizygous tissue (Fig. 3.4).

Today, however, no clinical example of paradominance can be proposed with any certitude. On the contrary, in both nevus sebaceus and nevus flammeus that were previously held to be paradigmatic of paradominant inheritance [76, 79],

Table 3.4 Autosomal dominant skin disorders with type 2 segmental manifestation confirmed at the molecular level

Disorder	OMIM number	Reference	Section number in this book
Hailey-Hailey disease	169600	[149]	10.2.3
Legius syndrome	611431	[14, 84]	10.1.16
PTEN hamartoma syndrome (Cowden disease included)	158350	[82]	10.1.20
Neurofibromatosis 1	162200	[171]	10.1.13
Darier disease	124200	[46]	10.2.2
Gorlin syndrome	109400	[181]	10.1.18

Table 3.5 Autosomal dominant skin disorders showing a type 2 segmental involvement that is so far unproven at the molecular level

Disorder	OMIM number	Section number in this book
Acanthosis nigricans	100600	10.2.5
Albright's hereditary osteodystrophy	103580 612463	10.3.4
Basal cell carcinoma, nonsyndromic hereditary multiple	605462	10.1.19
Basaloid follicular hamartoma, multiple	605827; 604451	10.1.4
Blue rubber bleb angiomatosis	112200	10.1.11
Buschke-Ollendorff syndrome	166700	10.3.2
Dyskeratosis congenita, autosomal dominant	127550; 613989;613990	10.2.7
Ehlers-Danlos syndrome, type III	130020	10.3.3
Epidermolytic ichthyosis of Brocq	113800	10.2.1
Glomangiomatosis	138000	10.1.10
Hereditary hemorrhagic telangiectasia	187300	10.4.1
Hornstein-Knickenberg syndrome (alias Birt-Hogg-Dubé syndrome)	135150	10.1.5
KID syndrome	148210	10.2.6
Leiomyomatosis, cutaneous	150800	10.1.17
Neurofibromatosis 2	101000	10.1.14
Osteomatosis cutis, hereditary	166350	10.3.5
Porokeratosis, disseminated superficial actinic	175900; 607728;612293	10.2.9
Porokeratosis of Mibelli	175800	10.2.10
Porokeratosis palmaris, plantaris et disseminata	175850	10.2.11
Trichoepithelioma, multiple	601606; 605041;612099	10.1.1
Tuberous sclerosis	191100; 631254	10.3.1
Spiradenoma, multiple	605018; 605041	10.1.7
Syringoma, multiple	186600	10.1.6

this mode of transmission has been excluded [58, 164]. Future research may show whether any paradominant trait can be found within the group of mosaic skin disorders.

The definition of paradominance includes mosaic traits originating from LOH that occurs during early embryogenesis and excludes all classical autosomal dominant traits characterized by tumors or malformations originating from LOH such as retinoblastoma or glomangiomatosis. Unfortunately, this concept has been misunderstood by several authors who believed that I wanted to create a neologism for the two-hit model of Knudson [105] and Comings [31]. They applied the word *paradominant* to disseminated lesions of classical autosomal dominant traits

Table 3.6 Historical descriptive words or phrases heralding a type 2 segmental manifestation of autosomal dominant skin disorders

Name of disorder known to show a type 2 segmental involvement	Historical descriptive terms referring to type 2 segmental lesions
Acanthosis nigricans	Acanthosis nigricans form of epidermal nevus [44]
Albright hereditary osteodystrophy	Progressive osseous heteroplasia [165] Sharply demarcated zone of subcutaneous and dermal hypoplasia with subcutaneous calcifications [104]
Blue rubber bleb angiomatosis ("blue rubber bleb nevus syndrome")	Huge angiomatous cutaneous mass [138] Unilateral angiomatous gigantism of hand or arm [49, 132] Massive pelvic hemangioma [7] Lymphangiomatosis-like growth pattern within the uterus [147]
Buschke-Ollendorff syndrome	Juvenile elastoma [6] Forme fruste of Buschke-Ollendorff syndrome [47]
Darier disease	Unilateral, linear, zosteriform epidermal nevus with acantholytic dyskeratosis [37]
Ehlers-Danlos syndrome type III	Connective tissue nevus (collagenoma) [166]
Epidermolytic ichthyosis of Brocq	Naevus verrucosus hystricoides [61] Epidermal nevus [43]
Glomangiomatosis	Congenital plaque-like glomangioma [156] Giant glomangioma [167] Giant congenital patch-like glomus tumors [197] Congenital plaque-type glomuvenous malformations [128]
Hailey-Hailey disease	Relapsing linear acantholytic dermatosis [4, 183]
Hornstein-Knickenberg syndrome (alias Birt-Hogg-Dubé syndrome)	Large connective tissue nevus of shagreen plaque type with papular fibromas of the follicular sheath [192]
Leiomyomatosis	Giant zoniform leiomyoma [21]
Marfan syndrome	Asymmetric Marfan syndrome [20, 55]
Neurofibromatosis 1	Plexiform neurofibroma [27, 109, 151] Elephantiasis neuromatosa [10, 131] Diffuse ganglioneuromatosis and plexiform neurofibroma [158]
Osteomatosis cutis, hereditary	Progressive osseous heteroplasia [142, 182] Severe congenital platelike osteoma cutis [196] Limited dermal ossification [48]
Porokeratosis, disseminated superficial actinic	Linear porokeratosis [32, 53] Congenital linear porokeratosis [173]
Porokeratosis of Mibelli	"Linear lesions were said to have been present since birth" [11] Ulcerative systematized porokeratosis (Mibelli) [152]
PTEN hamartoma syndrome (Cowden disease included)	Proteus-like syndrome [201] PTEN hamartoma of soft tissue [112] SOLAMEN syndrome (segmental overgrowth, lipomatosis, arteriovenous malformation, epidermal nevus) [24] Arteriovenous fistulas; "intramuscular vascular anomalies in our patients disrupted the muscular architecture and had excessive disorganized ectopic fat" [176] "Hemimegalencephaly as part of Jadassohn nevus sebaceus syndrome" [135]
Rhodoid nevus syndrome ("capillary malformation-arteriovenous malformation")	Arteriovenous malformation [12]
Spiradenomatosis	Congenital linear eccrine spiradenoma [155] Congenital blaschkoid eccrine spiradenoma [41]

Table 3.6 (continued)

Name of disorder known to show a type 2 segmental involvement	Historical descriptive terms referring to type 2 segmental lesions
Syringomatosis	Plaque-like syringoma [101] Linear syringomatous hamartoma [194] Localized form, clinical variant "en plaque" [129]
Tuberous sclerosis	Shagreen patch Cobblestone nevus Fibrous hamartoma of infancy [63] Folliculocystic and collagen hamartoma [180] Forme fruste of tuberous sclerosis [50]

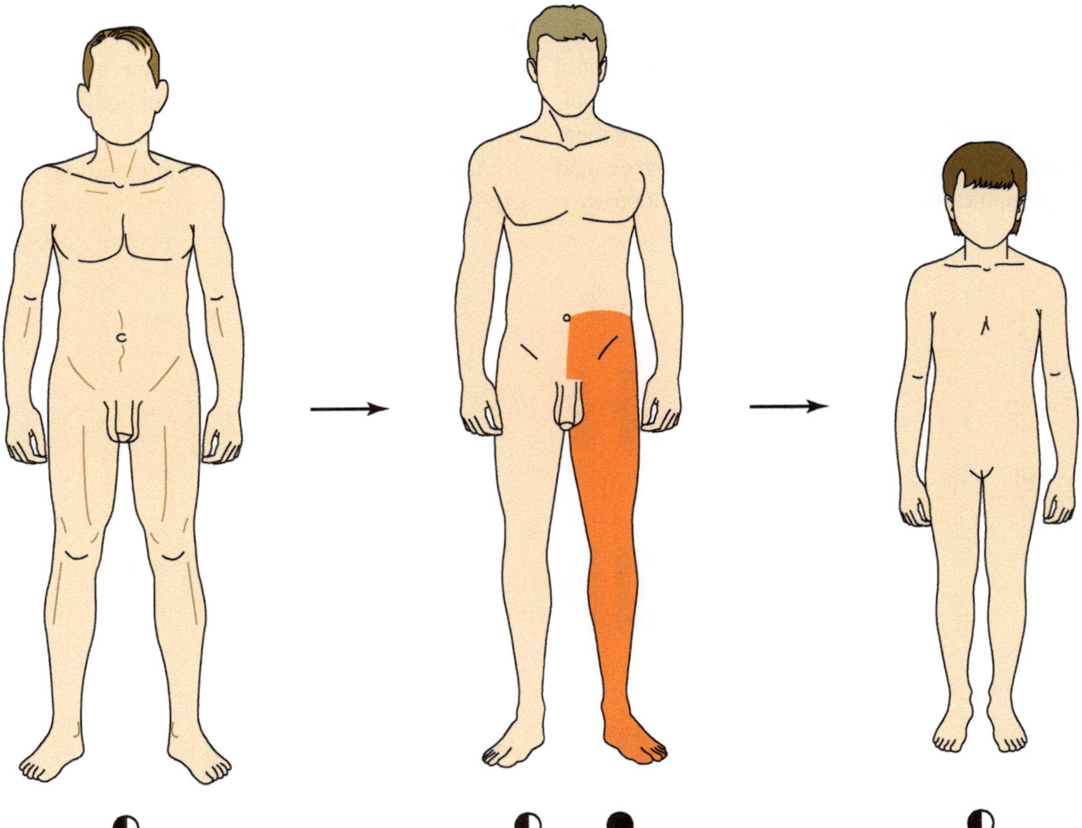

Fig. 3.4 The unproven theory of paradominant inheritance. Heterozygous individuals are phenotypically healthy. Only when loss of heterozygosity occurred in a somatic cell at an early developmental stage, the outgrowth of a homozygous or hemizygous population of cells would give rise to a segmental skin disorder

such as glomangiomatosis [16, 17, 120] or cerebral cavernous malformations [144]. In this way, the term would lose its original specific meaning and would refer to an aspect of Mendelian transmission. On the contrary, the word was created to define a particular form of non-Mendelian inheritance [85].

3.1.1.5 Didymosis (Twin Spotting)

The phenomenon of didymosis (Greek *didymos*=twin) is defined as paired patches of mutant tissue that differ genetically from each other and from the background tissue. This mechanism has extensively been studied in animals and plants [195]. An embryo may show compound heterozygosity which means that there are two different mutant alleles at the same gene locus, or transheterozygosity which means that there are two mutant alleles at different loci on either of a pair of homologous chromosomes. Then, mitotic recombination may give rise to two different homozygous daughter cells forming twin spots that are either allelic or nonallelic and usually found to be adjacent or in close proximity to each other (Fig. 3.5).

In the light of the data obtained in plants and animals (see Sects. 2.2 and 2.3), it is conceivable that didymosis should likewise occur in human skin. It took a rather long time, however, until dermatologists became interested in this phenomenon. In 1973, the geneticist James German mentioned that twin spots in the form of hyper- and hypopigmented macules may occur in Bloom syndrome [54]. His remark, however, fell into oblivion because until today no convincing photographic documentation of this phenomenon has been published. In 1990, the concept of allelic twin spotting in the form of coexistent telangiectatic nevus and nevus anemicus was proposed [90].

It should be emphasized, however, that cases of coexisting nevi often reflect an arbitrary coincidence, and that the concept of didymosis can only be applied to nevi that show a more or less regular co-occurrence. Accordingly, from a heuristic point of view, it seems reasonable to limit the concept of didymosis to disorders originating from homologous recombination, thus excluding disorders that may result from nonhomologous crossing-over [35, 184, 193].

As the underlying two different mutations are expected to be localized on the same chromosome,

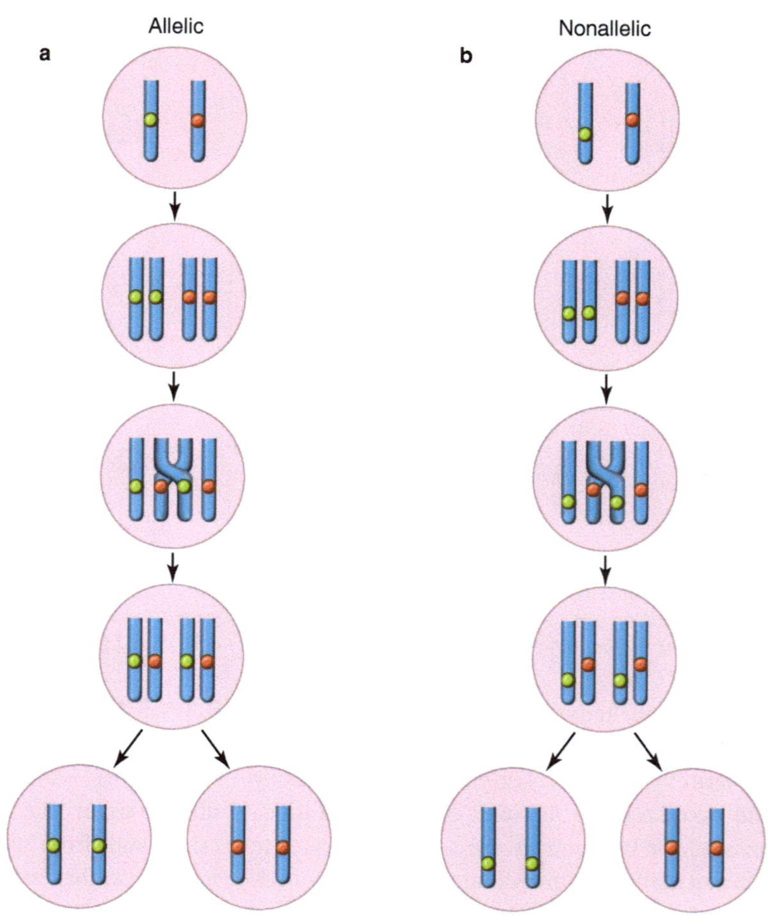

Fig. 3.5 (**a**) Allelic didymosis versus (**b**) nonallelic didymosis. In both cases, a postzygotic recombination would give rise to two homozygous daughter cells forming a pair of differently involved mosaic patches

the concept of didymosis may be important for the mapping of the involved nevi within the human genome. This forecast is supported by experiments performed in *Drosophila*. Griffin et al. [57] developed a "twin-spot generator" technique that produces paired green and red spots being detectable already as single cells [174]. This method was a further advancement of the mosaic analysis with a repressible cell marker (MARCM) as elaborated by Lee and Luo [117, 118]. By use of the twin-spot MARCM technique, it was possible to visualize, step by step, the embryonic outgrowth of different neurons within the brain of a Drosophila instar (Fig. 3.6; see also Sect. 2.3) [121, 198, 199].

Allelic Didymosis

Proposed examples of allelic didymosis are summarized in Table 3.7.

The Theory of Nonallelic Didymosis

Until recently, the concept of nonallelic twin spotting was considered to explain phacomatosis pigmentokeratotica [88], different types of phacomatosis pigmentovascularis [81], and several other cases of binary genodermatoses [1, 124, 154]. Meanwhile, however, molecular research has shown that phacomatosis pigmentokeratotica is caused by one single postzygotic mutation being present in a heterozygous state [59], which is incompatible with the hypothesis of twin spotting. Similarly, the lesions of phacomatosis cesioflammea will almost certainly turn out to be heterozygous for a *GNAQ* mutation [164].

Future studies may show whether a mosaic skin disorder exemplifying nonallelic didymosis exists. Because several examples of nonallelic twin spotting have been described in *Drosophila melanogaster* (see Sect. 8.2), it seems possible that a similar phenomenon will also be found in human skin.

3.1.1.6 Revertant Mosaicism

In patients with an autosomal recessive disorder, a revertant mutation may give rise to a clone of heterozygous cells that have regained, either partly or completely, their normal physiological function. In a woman with generalized atrophic benign epidermolysis bullosa caused by compound heterozygosity at the *COL17A1* locus, Jonkman et al. [98] found several patchy areas of healthy skin and provided molecular proof that

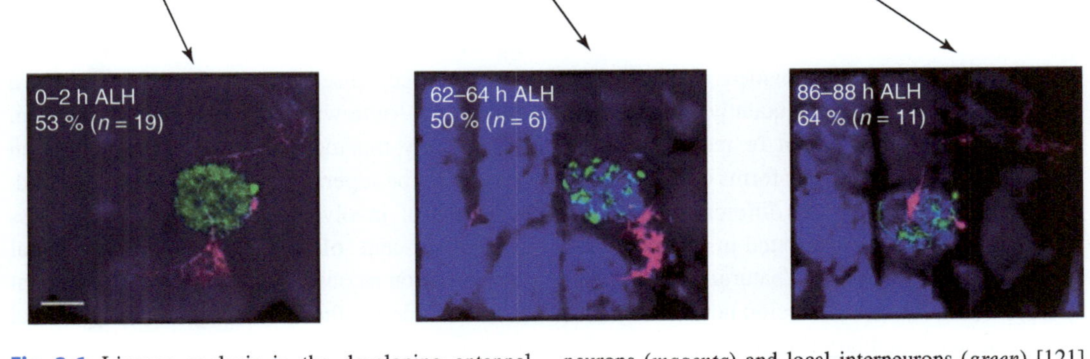

Fig. 3.6 Lineage analysis in the developing antennal lobe of Drosophila. Binary cell fate decisions produce, in a stereotyped sequence, different pairs of projection neurons (*magenta*) and local interneurons (*green*) [121] (Reprinted under Creative Commons License)

Table 3.7 Possible examples of allelic didymosis in human skin

Name of didymosis	Section number in this book
Nevus flammeus twinned with nevus anemicus	8.1.1.1
Nevus vascularis mixtus	8.1.1.2
Cutis tricolor, Ruggieri-Happle type	8.1.2.1
Cutis tricolor parvimaculata	8.1.2.2
Cutis tricolor, Blaschko-linear type	8.1.2.3
Didymotic excessive and absent involvement in epidermolytic ichthyosis of Brocq	8.1.3
Didymotic excessive and absent involvement in Darier disease	8.1.4

the clinically unaffected skin contained keratinocytes carrying a back mutation and, therefore, producing small amounts of normal collagen type XVII.

Similar examples of "natural gene therapy" by revertant mosaicism have been described in Bloom syndrome [42], leukocyte adherence deficiency type 1 [179], and Wiskott-Aldrich syndrome [36, 187]. Moreover, RAG1-deficient severe combined immunodeficiency has been found to be mitigated, due to revertant T-cell mosaicism, to Omenn syndrome [188].

Either the back mutations represent true revertants that may result from a reverse point mutation, postzygotic crossing-over, or gene conversion, thus restoring the amino sequence [98], or a revertant clone carries, inside or outside the mutant gene, a second-site mutation that may originate from addition or deletion of a base pair, a suppressor mutation, or chromosomal loss or gain [96].

Evidence has been provided that revertant mosaicism occurs rather frequently in some skin disorders of either dominant or recessive inheritance. In autosomal recessive forms of epidermolysis bullosa, even two or three different spontaneous back mutations have been found in the same individual [146]. Such cases of "natural gene therapy" may serve as a model to develop new strategies of treating severe hereditary skin disorders [97, 115].

In some Mendelian disorders such as ichthyosis in confetti (ichthyosis variegata) or Kindler syndrome, multiple events of mitotic recombination or slipped mispairing may give rise to numerous patches of healthy skin [29, 103, 116]. In these disorders, manifold reversion mechanisms appear to be rather the rule than an exceptional event (see Chap. 11).

3.1.2 Genomic X-Chromosome Mosaicism in Male Patients

In men with a normal 46,XY karyotype, a postzygotic mutation involving the skin may be localized on the X chromosome and give rise to cutaneous mosaicism. For example, in three out of four male patients affected with incontinentia pigmenti, Kenwrick et al. [99] described a classical *NEMO* deletion and a normal 46,XY karyotype. Such male patients can transmit the trait to their daughters. Similarly, a postzygotic *NSDHL* mutation was documented in a boy with a normal XY gonosome constitution and typical features of CHILD syndrome, a disorder that is known to be inherited as an X-linked dominant male-lethal trait [108]. Focal dermal hypoplasia has likewise been reported in 46,XY males [15, 45], and postzygotic mosaicism was confirmed in such cases by molecular analysis [13, 189].

3.1.3 Superimposed Segmental Manifestation of Polygenic Skin Disorders

Common skin disorders such as psoriasis or atopic dermatitis have a polygenic basis, rendering the individual more prone to develop the disorder. Such diseases may sometimes show a linear or otherwise segmental distribution. Remarkably, this mosaic manifestation may often be found to be superimposed on a less pronounced, symmetrical involvement. Although such cases are reminiscent of a superimposed segmental manifestation as observed in autosomal dominant skin disorders, the term "type 2 segmental involvement" is not appropriate because in polygenic diseases it appears to be elusive to recognize a "type 1 segmental involvement" (Fig. 3.7).

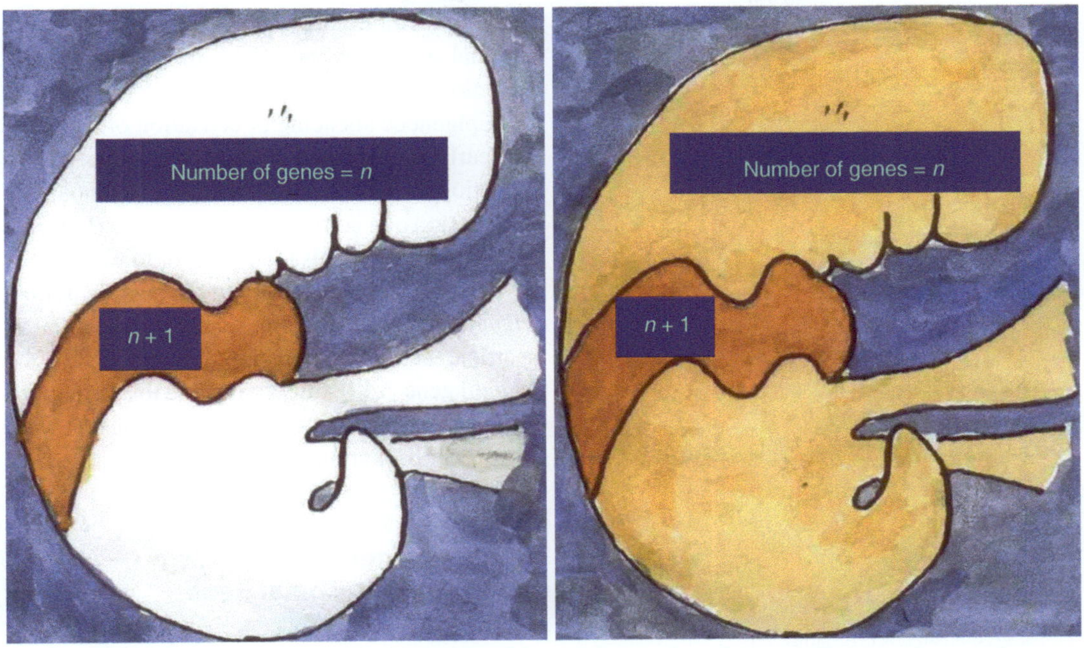

Fig. 3.7 In polygenic skin disorders, we can only make a clinical discrimination between an (**a**) isolated or (**b**) superimposed mosaic manifestation. In both cases, the number of predisposing genes is unknown. Hence, we can only say that in both cases of segmental involvement, the number of predisposing genes is $n+1$. In other words, a distinction between a "type 1" and "type 2" segmental manifestation is impossible

In fact, we can never be sure that a patient showing an isolated segmental manifestation of a common disease like psoriasis will not develop during life, in addition, some ordinary nonsegmental lesions of the same disease. Therefore, in polygenic disorders it is appropriate to use the less specific terms "isolated" versus "superimposed" segmental involvement [83].

In polygenic skin diseases, two different mutational events may cause a superimposed segmental manifestation. Either LOH for one of the predisposing heterozygous alleles may occur at an early developmental stage, or a postzygotic new mutation may occur at an additional gene locus and increase the number of heterozygous predisposing alleles.

The concept of superimposed segmental involvement was first proposed for psoriasis in 1991 [69] and later applied to many other common skin diseases (Table 3.8) [83] (see Sect. 12.3.7). Most of these mosaic manifestations are arranged along Blaschko's lines. In vitiligo, however, a flag-like distribution is noted.

The theory of superimposed segmental manifestation of polygenic skin disorders has so far not been proven at the molecular level. It offers, however, a reasonable explanation as to why:

- "Mixed" cases of segmental and nonsegmental lesions may occur.
- The mosaic involvement tends to become manifest at a rather young age and may even be present at birth.
- Segmentally arranged lesions are notoriously difficult to treat.
- Additional nonsegmental lesions may develop later in life.
- Family members may be affected with nonsegmental lesions.

For the purpose of molecular studies, the concept opens a new way of research because a superimposed segmental manifestation tells us something about the primary role of skin epitopes in psoriasis and numerous other acquired disorders such as atopic dermatitis, lichen planus, granuloma annulare, systemic lupus erythematosus, pemphigus vulgaris, erythema multiforme,

Table 3.8 Polygenic skin disorders reported to be superimposed by pronounced segmental lesions

Disorder	Section number in this book
Acne vulgaris	12.3.7.1.8
Atopic dermatitis	12.3.7.1.3
Bullous pemphigoid	12.3.7.1.15
Cherry angiomas	12.3.7.3
Dermatomyositis	12.3.7.1.13
Drug eruption, common	12.3.7.1.20
Drug eruption, fixed	12.3.7.1.21
Erythema multiforme	12.3.7.1.19
Exanthema of childhood, lateralized	12.3.7.1.22
Graft-versus-host reaction	12.3.7.1.16
Granuloma annulare	12.3.7.1.18
Lichen nitidus	12.3.7.1.7
Lichen planus	12.3.7.1.5
Lichen planopilaris	12.3.7.1.6
Lupus erythematosus, discoid	12.3.7.1.9
Lupus erythematosus profundus	12.3.7.1.10
Lupus erythematosus, subacute cutaneous	12.3.7.1.11
Lupus erythematosus, systemic	12.3.7.1.12
Morphea, disseminated	12.3.7.1.17
Pemphigus vulgaris	12.3.7.1.14
Prurigo, chronic	12.3.7.1.4
Psoriasis vulgaris	12.3.7.1.1
Pustular psoriasis	12.3.7.1.2
Vitiligo	12.3.7.2

or drug eruptions. A comparison of tissue samples from the involved segment and the remaining skin by application of presently available chip techniques may help elucidate the predisposing genes of such genetically complex disorders.

3.2 Epigenetic Mosaicism

In contrast to genomic mosaicism that results from a change within the DNA sequence, epigenetic mosaicism originates from the action of epimutations that are silencing or activating the expression of a neighboring gene by methylation or demethylation [95]. In principle, all genes may be susceptible to epimutation. Unlike a mutation, an epimutation may rather easily revert to a normal state, thus reshuffling the components of functional mosaicism [130].

Epigenetic silencing or activation of a gene at an early developmental stage may give rise to functional mosaicism reflecting monoallelic expression of a color mutation in plants (Fig. 3.8) [150, 169] or animals [140, 190] or to variation of symmetry in plants [33]. In the *viable yellow agouti* mutant of the mouse, an intracisternal A particle acts as an epimutation modifying the *agouti* gene [148], thus giving rise to a variegated coat pattern of linear or patchy areas of *yellow* on an *agouti* background (Fig. 3.9).

3.2.1 Epigenetic Mosaicism of Autosomal Genes

More than 40 % of human DNA is derived from retrotransposons [168]. Most of this retroviral material remains silent, but some transposons are highly active during embryogenesis or even later in life.

The hereditary *brindle* trait of dogs is a perfect morphological counterpart of Blaschko-linear pigmentary mosaicism of humans (Fig. 3.10). *Brindle* is under the control of a retrotransposon [100]. Another example of epigenetic mosaicism is the *merle* patterning of dogs (Fig. 3.11) that was found to reflect a retrotransposon insertion in the SILV gene [30]. Heterozygous dogs have an auditory-pigmentation disorder that is very similar to Waardenburg syndrome. Perhaps this human dominant trait does likewise reflect the action of an epimutation.

In man, the principle of epigenetic mosaicism has been proposed to explain some cases of familial occurrence of ILVEN or pigmentary mosaicism (Fig. 3.12) [80, 93]. An unusual case of systematized linear hypermelanosis involving four members of a family [28] can be taken as an additional example of autosomal monoallelic expression [86]. A report on "hypomelanosis of Ito" affecting two siblings and their mother [139] may likewise be explained by autosomal epigenetic transmission. On the other hand, in a case of "linear and whorled nevoid

Fig. 3.8 Epigenetic mosaicism in *Petunia hybrida*. (**a**) Flower of a plant containing the *Psl* transposable element but showing virtually no expression; (**b**) flower of a plant showing low expression of *Psl*; (**c, d**) flowers of plants showing high expression of *Psl* [169] (Reprinted with permission from John Wiley & Sons)

hypermelanosis" in three successive generations [136], the alternative assumption of heterozygosity for ectodermal dysplasia of Zonana (see Chap. 4) cannot be excluded because all of the affected individuals were females.

3.2.2 Epigenetic Mosaicism of X Chromosomes

In female somatic cells one of the two X chromosomes becomes inactivated at an early stage of embryonic development. Without this mechanism the amount of X-chromosomal gene products would be twice as high as in male cells. Apparently, during mammalian phylogenesis it was necessary to develop a compensation for this considerable difference between the genders. Such dosage compensation is provided by random inactivation of either the maternal or the paternal X chromosome [126]. In a given clone, the determination which of the two X chromosomes is inactivated will be maintained in all daughter cells throughout life (Fig. 3.13). This mechanism results in a mosaic of two functionally different populations of cells.

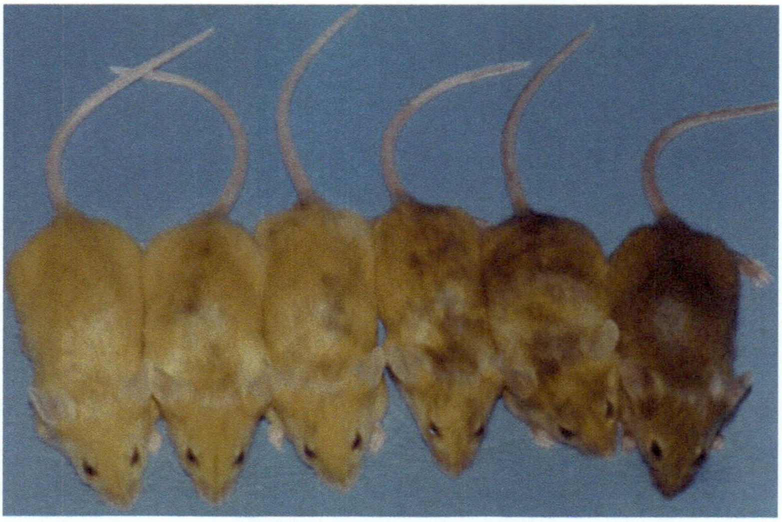

Fig. 3.9 A^{vy} mice are epigenetic mosaics reflecting activity of a retrotransposon. Their coats vary in a continuous spectrum from full *yellow*, through variegated/agouti, to full agouti ("pseudoagouti") [140] (Reprinted with permission from Nature Publishing Group)

Fig. 3.10 *Brindle* trait in a dog (Reproduced from Harlis.jpg, Creative Commons)

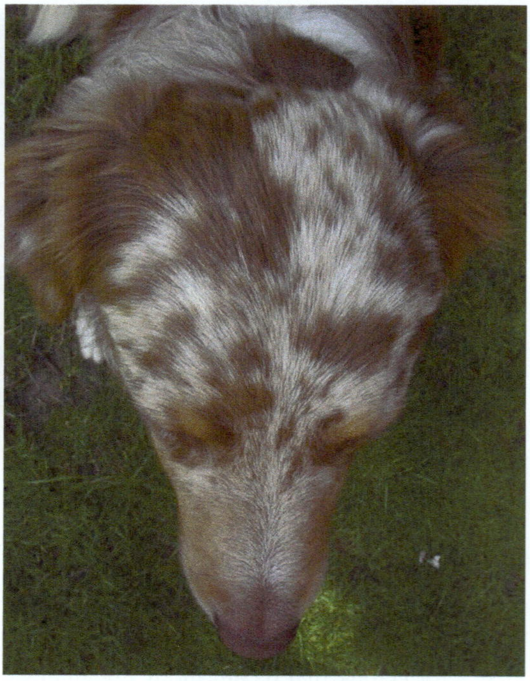

Fig. 3.11 *Merle* trait in a dog (Courtesy of Mrs. Claudia Kluge, Freiburg, Germany)

Today, it is likely that X inactivation is under the control of LINE-1 retrotransposons ("long interspersed nuclear elements") that have been found to be overabundant in the region of the X inactivation center at Xq13 [8, 127]. Hence, the Lyon effect can be taken as a particular form of epigenetic mosaicism.

3.2.2.1 Functional X-Chromosome Mosaicism in Female Patients

For dermatology the Lyon effect is of particular significance because in women, functional X-chromosome mosaicism may give rise to a patchy or linear or otherwise segmental pattern of skin lesions [79, 102]. A classical example is incontinentia pigmenti (Fig. 3.14). This X-linked trait occurs almost exclusively in girls because the underlying *NEMO* mutation represents a lethal factor for hemizygous male embryos. Most of the affected girls show mosaic skin lesions arranged along Blaschko's lines (see Sect. 12.1).

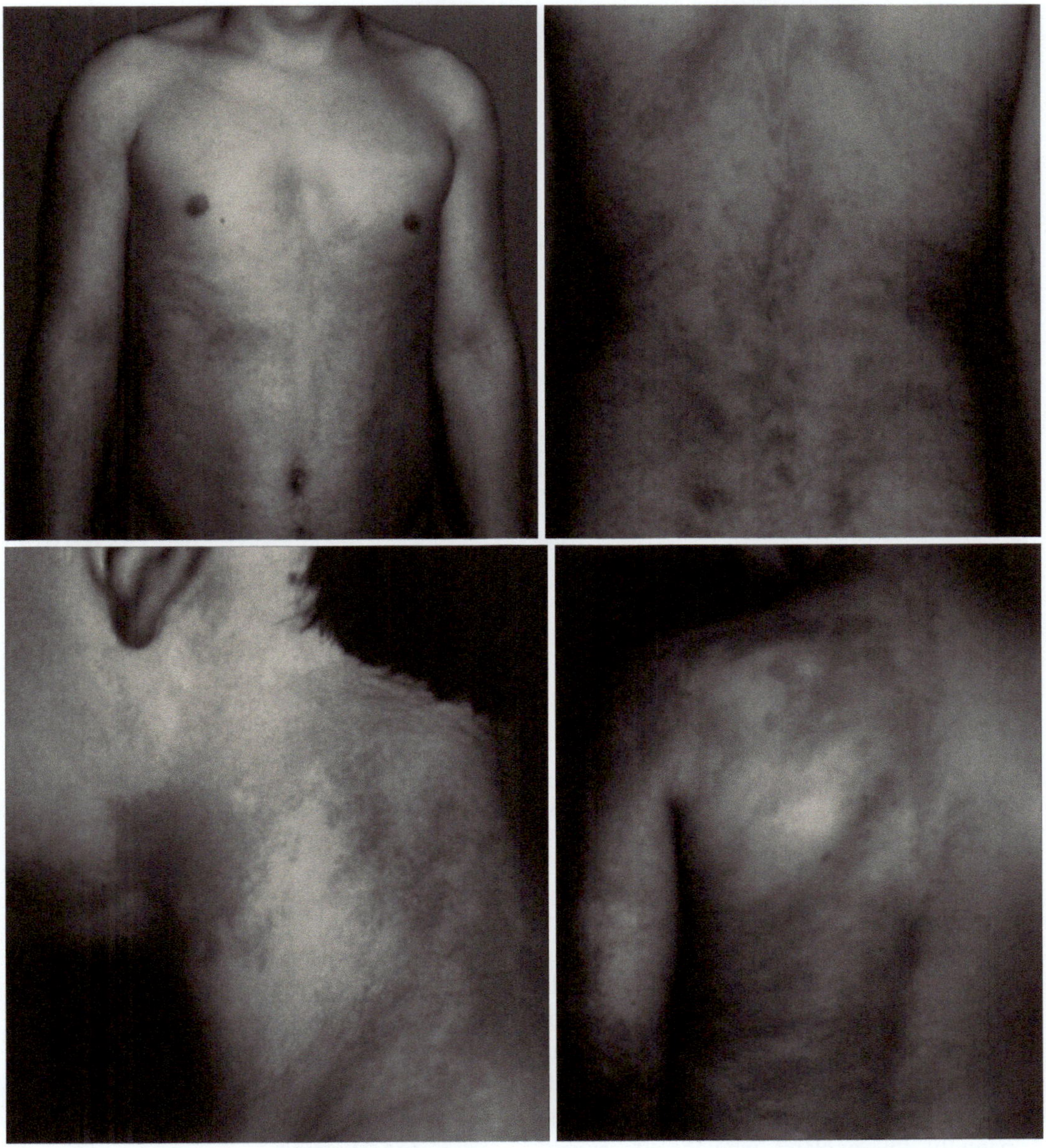

Fig. 3.12 Blaschko-linear hypermelanosis in a young man (*above*) and his half brother (*below*) [92]. Such familial occurrence can best be explained by autosomal epigenetic mosaicism (Reprinted with permission from John Wiley and Sons)

Interestingly, Mary Lyon had based her hypothesis of X inactivation on the observation of similar linear or patchy coat patterns in mice heterozygous for X-linked genes [125].

Why do women live longer?

X-linked male-lethal traits such as incontinentia pigmenti can be taken as an extreme example illustrating the general rule that women live longer than men. The X chromosome contains about 5 % of our genetic information, whereas the Y chromosome of men contains, by

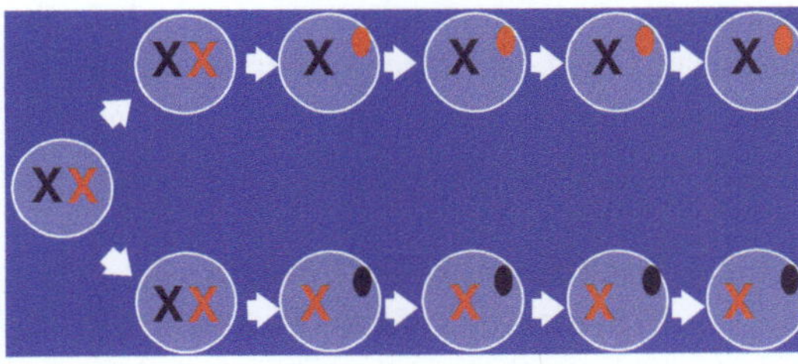

Fig. 3.13 In a female embryo, either the paternal or the maternal X chromosome is inactivated at an early developmental stage. This decision occurs at random and is maintained in all daughter cells, which gives rise to two functionally different cell clones ("lyonization")

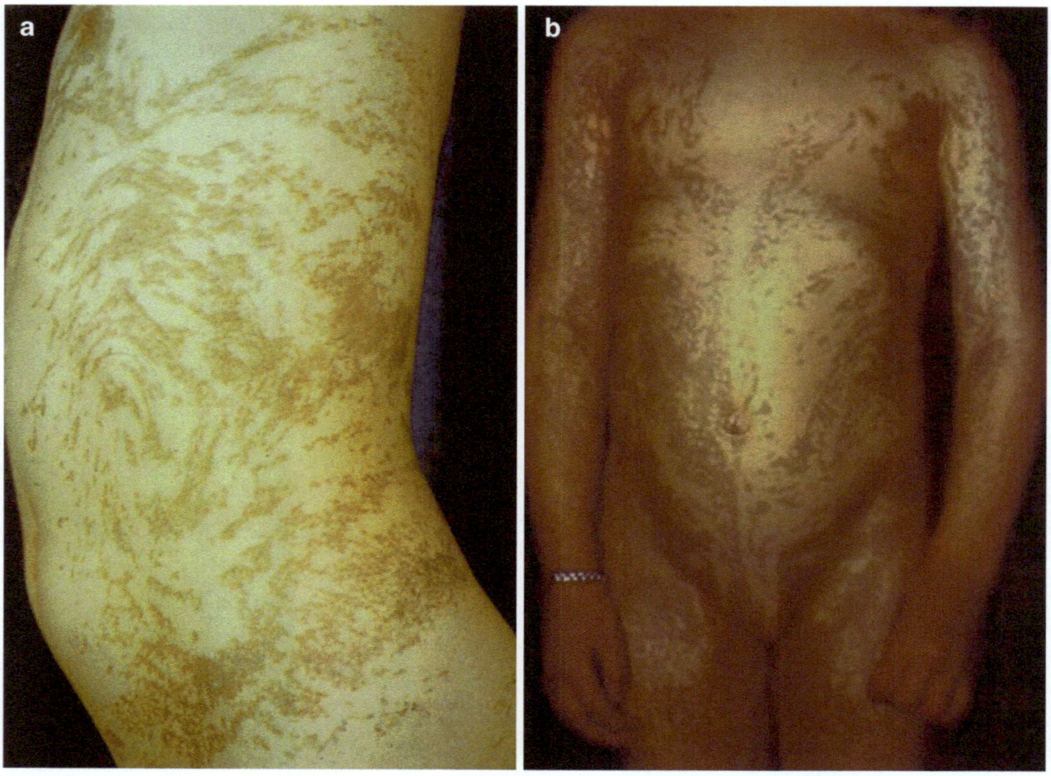

Fig. 3.14 Linear hyperpigmentation reflecting lyonization in girls with incontinentia pigmenti. (**a**) Lateral view; (**b**) frontal view (**a**: Courtesy of Dr. Shigeo Nishiyama, Kamakura, Japan; **b**: courtesy of Dr. Ramón Ruiz-Maldonado, Mexico City, Mexico)

way of comparison, pitifully little information. It seems to be the "meaning" of X inactivation to counterbalance this disadvantage of the male gender. Notwithstanding, the mechanism of lyonization accounts for the fact that women can respond more flexibly to heat, cold, hunger, emotional stress, and viral or bacterial attacks. Apparently, this explains why the life expectancy of women is at least 6 years higher than that of men. Hence, the "unisex" tariffs of life insurances, as presently imposed by the European Court of Justice, reflect modern irrationalism that ignores unchangeable genetic inequalities.

3.2.2.2 Functional X-Chromosome Mosaicism in Male Patients

By way of exception, X-linked male-lethal phenotypes such as incontinentia pigmenti may be observed in male patients. Some of them have a 46,XY karyotype and their disease reflects genomic mosaicism (see Sect. 3.1.2), whereas others have a gonosome constitution XXY that gives rise to functional X-chromosome mosaicism as noted in females [19].

3.2.2.3 X-Linked Genes Escaping Inactivation

Although most X-linked human genes are randomly inactivated in the form of the Lyon effect, it is today clear that many other gene loci being interspersed along the entire X chromosome escape from X inactivation. As a consequence, women carrying a mutation at such loci tend to show a completely normal phenotype, as noted in X-linked recessive ichthyosis or in macular dystrophy of Mendes da Costa. Or they may show a diffuse but rather mild and symmetrical involvement as observed in Bazex-Dupré-Christol syndrome [72].

Paradoxically, such nonsegmental involvement is also noted in women with keratosis follicularis spinulosa decalvans [9, 72, 162] although the underlying mutation involves the MBTPS2 gene [5] that is known to show lyonization in women with IFAP syndrome [107, 177].

References

1. Afsar FS, Aktas S, Ortac R (2007) Becker's naevus and segmental naevus depigmentosus: an example of twin spotting? Australas J Dermatol 48:224–226
2. Amary MF, Damato S, Halai D, Eskandarpour M, Berisha F, Bonar F, McCarthy S, Fantin VR, Straley KS, Lobo S, Aston W, Green CL, Gale RE, Tirabosco R, Futreal A, Campbell P, Presneau N, Flanagan AM (2011) Ollier disease and Maffucci syndrome are caused by somatic mosaic mutations of *IDH1* and *IDH2*. Nat Genet 43:1262–1265
3. Anliker MD, Dummer R, Burg G (1997) Unilateral agminated angiofibromas: a segmental expression of tuberous sclerosis? Dermatology 195:176–178
4. Arora S, Arora G, Ranjan P (2005) Relapsing linear acantholytic dermatosis in a four-year-old boy. Indian J Dermatol Venereol Leprol 71:351–353
5. Aten E, Brasz LC, Bornholdt D, Hooijkaas IB, Porteous ME, Sybert VP, Vermeer MH, Vossen RH, van der Wielen MJ, Bakker E, Breuning MH, Grzeschik KH, Oosterwijk JC, den Dunnen JT (2010) Keratosis follicularis spinulosa decalvans is caused by mutations in *MBTPS2*. Hum Mutat 31:1125–1133
6. Atherton DJ, Wells RS (1982) Juvenile elastoma and osteopoikilosis (the Buschke-Ollendorf syndrome). Clin Exp Dermatol 7:109–113
7. Atten MJ, Ahmed S, Attar BM, Richter H 3rd, Mehta B (2000) Massive pelvic hemangioma in a patient with blue rubber bleb nevus syndrome. South Med J 93:1122–1125
8. Bailey JA, Carrel L, Chakravarti A, Eichler EE (2000) Molecular evidence for a relationship between LINE-1 elements and X chromosome inactivation: the Lyon repeat hypothesis. Proc Natl Acad Sci U S A 97:6634–6639
9. Bellet JS, Kaplan AL, Selim MA, Olsen EA (2008) Keratosis follicularis spinulosa decalvans in a family. J Am Acad Dermatol 58:499–502
10. Bingham EA, Burrows D (1984) Neurofibromatosis (elephantiasis neuromatosa). Br J Dermatol 111:65–66
11. Bloom D, Abramowitz EW (1943) Porokeratosis Mibelli. Arch Derm Syphilol 47:1–15
12. Boon LM, Mulliken JB, Vikkula M (2005) RASA1: variable phenotype with capillary and arteriovenous malformations. Curr Opin Genet Dev 15:265–269
13. Bornholdt D, Oeffner F, König A, Happle R, Alanay Y, Ascherman J, Benke PJ, Boente Mdel C, van der Burgt I, Chassaing N, Ellis I, Francisco CR, Della Giovanna P, Hamel B, Has C, Heinelt K, Janecke A, Kastrup W, Loeys B, Lohrisch I, Marcelis C, Mehraein Y, Nicolas ME, Pagliarini D, Paradisi M, Patrizi A, Piccione M, Piza-Katzer H, Prager B, Prescott K, Strien J, Utine GE, Zeller MS, Grzeschik KH (2009) *PORCN* mutations in focal dermal hypoplasia: coping with lethality. Hum Mutat 30:E618–E628
14. Brems H, Chmara M, Sahbatou M, Denayer E, Taniguchi K, Kato R, Somers R, Messiaen L, De Schepper S, Fryns JP, Cools J, Marynen P, Thomas G, Yoshimura A, Legius E (2007) Germline loss-of-function mutations in *SPRED1* cause a neurofibromatosis 1-like phenotype. Nat Genet 39:1120–1126
15. Broughton WL, Weaver JE, Bibro MC, White BJ (1982) Focal dermal hypoplasia: ocular manifestations in a male. J Pediatr Ophthalmol Strabismus 19:314–317
16. Brouillard P, Ghassibe M, Penington A, Boon LM, Dompmartin A, Temple IK, Cordisco M, Adams D, Piette F, Harper JI, Syed S, Boralevi F, Taieb A, Danda S, Baselga E, Enjolras O, Mulliken JB, Vikkula M (2005) Four common glomulin mutations cause two thirds of glomuvenous malformations ("familial glomangiomas"): evidence for a founder effect. J Med Genet 42:e13
17. Brouillard P, Vikkula M (2007) Genetic causes of vascular malformations. Hum Mol Genet 16:R140–R149
18. Buhl T, Wienrich BG, Sieblist C, Schön MP, Seitz CS (2010) Development of segmental superficial actinic porokeratosis during immunosuppressive therapy for pemphigus vulgaris. Acta Derm Venereol 90:212–213
19. Buinauskaite E, Buinauskiene J, Kucinskiene V, Strazdiene D, Valiukeviciene S (2010) Incontinentia pigmenti in a male infant with Klinefelter syndrome: a case report and review of the literature. Pediatr Dermatol 27:492–495
20. Burgio RG, Martini A, Cetta G, Zanaboni G, Vitellaro L, Danesino C (1988) Asymmetric Marfan syndrome. Am J Med Genet 30:905–909
21. Camacho F, Jorquera E, Vasquez FJ, Hevia A (1994) Giant zoniform leiomyoma: light and electron microscopy study. Eur J Dermatol 4:384–386
22. Carey J (2010) Trisomy 18 and trisomy 13 syndromes. In: Cassidy SB, Allanson JE (eds) Management of genetic syndromes. Wiley-Blackwell, Hoboken
23. Castori M, Annessi G, Castiglia D, Buffa V, Paradisi A, Cascone P, Zambruno G, Grammatico P, Paradisi M (2010) Systematized organoid epidermal nevus with eccrine differentiation, multiple facial and oral congenital scars, gingival synechiae, and blepharophimosis: a novel epidermal nevus syndrome. Am J Med Genet A 152A:25–31
24. Caux F, Plauchu H, Chibon F, Faivre L, Fain O, Vabres P, Bonnet F, Selma ZB, Laroche L, Gérard

M, Longy M (2007) Segmental overgrowth, lipomatosis, arteriovenous malformation and epidermal nevus (SOLAMEN) syndrome is related to mosaic *PTEN* nullizygosity. Eur J Hum Genet 15:767–773
25. Chao LY, Huff V, Tomlinson G, Riccardi VM, Strong LC, Saunders GF (1993) Genetic mosaicism in normal tissues of Wilms' tumour patients. Nat Genet 3:127–131
26. Chemke J, Rappaport S, Etrog R (1983) Aberrant melanoblast migration associated with trisomy 18 mosaicism. J Med Genet 20:135–137
27. Chen TC, Kuo WR, Chai CY, Juan KH (1991) Huge plexiform neurofibroma of the head and liver–case report. Gaoxiong Yi Xue Ke Xue Za Zhi 7:650–656
28. Chernosky ME, Anderson DE, Chang JP, Shaw MW, Romsdahl MM (1971) Familial progressive hyperpigmentation. Arch Dermatol 103:581–591 passim
29. Choate KA, Lu Y, Zhou J, Choi M, Elias PM, Farhi A, Nelson-Williams C, Crumrine D, Williams ML, Nopper AJ, Bree A, Milstone LM, Lifton RP (2010) Mitotic recombination in patients with ichthyosis causes reversion of dominant mutations in *KRT10*. Science 330:94–97
30. Clark LA, Wahl JM, Rees CA, Murphy KE (2006) Retrotransposon insertion in *SILV* is responsible for merle patterning of the domestic dog. Proc Natl Acad Sci U S A 103:1376–1381
31. Comings DE (1973) A general theory of carcinogenesis. Proc Natl Acad Sci U S A 70:3324–3328
32. Commens CA, Shumack SP (1987) Linear porokeratosis in two families with disseminated superficial actinic porokeratosis. Pediatr Dermatol 4:209–214
33. Cubas P, Vincent C, Coen E (1999) An epigenetic mutation responsible for natural variation in floral symmetry. Nature 401:157–161
34. Cullen SI (1979) Localized Ehlers-Danlos syndrome. Arch Dermatol 115:332–333
35. D'Angelo CS, Gajecka M, Kim CA, Gentles AJ, Glotzbach CD, Shaffer LG, Koiffmann CP (2009) Further delineation of nonhomologous-based recombination and evidence for subtelomeric segmental duplications in 1p36 rearrangements. Hum Genet 125:551–563
36. Davis BR, Yan Q, Bui JH, Felix K, Moratto D, Muul LM, Prokopishyn NL, Blaese RM, Candotti F (2010) Somatic mosaicism in the Wiskott-Aldrich syndrome: molecular and functional characterization of genotypic revertants. Clin Immunol 135:72–83
37. Demetree JW, Lang PG, St Clair JT (1979) Unilateral, linear, zosteriform epidermal nevus with acantholytic dyskeratosis. Arch Dermatol 115:875–877
38. Den Otter W, Koten JW, Van der Vegt BJ, Beemer FA, Boxma OJ, De Graaf PW, Derkinderen DJ, Hill FW, Huber J, Klein WR et al (1990) Hereditary cancer and its clinical implications: a view. Anticancer Res 10:489–495
39. Derkinderen DJ, Boxma OJ, Koten JW, Den Otter W (1990) Stochastic theory of oncogenesis. Anticancer Res 10:497–504
40. Dolan K, Garde J, Gosney J, Sissons M, Wright T, Kingsnorth AN, Walker SJ, Sutton R, Meltzer SJ, Field JK (1998) Allelotype analysis of oesophageal adenocarcinoma: loss of heterozygosity occurs at multiple sites. Br J Cancer 78:950–957
41. Ekmekci TR, Koslu A, Sakiz D (2005) Congenital blaschkoid eccrine spiradenoma on the face. Eur J Dermatol 15:73–74
42. Ellis NA, Ciocci S, German J (2001) Back mutation can produce phenotype reversion in Bloom syndrome somatic cells. Hum Genet 108:167–173
43. Eng AM, Brody P, Rhee HL, Bronson DM (1991) Congenital ichthyosiform erythroderma and epidermal nevus. Int J Dermatol 30:284–287
44. Ersoy-Evans S, Sahin S, Mancini AJ, Paller AS, Guitart J (2006) The acanthosis nigricans form of epidermal nevus. J Am Acad Dermatol 55:696–698
45. Fjellner B (1979) Focal dermal hypoplasia in a 46, XY male. Int J Dermatol 18:812–815
46. Fölster-Holst R, Nellen RG, Jensen JM, Poblete-Gutiérrez P, Steijlen PM, Schwarz T, Happle R, Van Geel M, Frank J (2012) Molecular genetic support for the rule of dichotomy in type 2 segmental Darier disease. Br J Dermatol 166:464–466
47. Foo CC, Kumarasinghe SP (2005) Juvenile elastoma: a forme fruste of the Buschke-Ollendorff syndrome? Australas J Dermatol 46:250–252
48. Foster CM, Levin S, Levine M, Mukherjee A, Costa JL, Eanes ED, Triche T, Zasloff M (1986) Limited dermal ossification: clinical features and natural history. J Pediatr 109:71–76
49. Fretzin DF, Potter B (1965) Blue rubber bleb nevus. Arch Intern Med 116:924–929
50. Garcia Muret MP, Pujol RM, de Moragas JM (1998) Multiple and unilateral angiofibromas of the face: forme fruste of Bourneville tuberous sclerosis. Ann Dermatol Venereol 125:325–327
51. Garcia-Linares C, Fernandez-Rodriguez J, Terribas E, Mercade J, Pros E, Benito L, Benavente Y, Capella G, Ravella A, Blanco I, Kehrer-Sawatzki H, Lazaro C, Serra E (2011) Dissecting loss of heterozygosity (LOH) in neurofibromatosis type 1-associated neurofibromas: Importance of copy neutral LOH. Hum Mutat 32:78–90
52. García-Vargas A, Hafner C, Pérez-Rodríguez AG, Rodríguez-Rojas LX, González-Esqueda P, Stoehr R, Hernández-Torres M, Happle R (2008) An epidermal nevus syndrome with cerebral involvement caused by a mosaic *FGFR3* mutation. Am J Med Genet A 146A:2275–2279
53. Garg T, Ramchander, Varghese B, Barara M, Nangia A (2011) Generalized linear porokeratosis: a rare entity with excellent response to acitretin. Dermatol Online J 17:3
54. German J (1973) Genetic disorders associated with chromosomal instability and cancer. J Invest Dermatol 60:427–434
55. Godfrey M, Olson S, Burgio RG, Martini A, Valli M, Cetta G, Hori H, Hollister DW (1990) Unilateral microfibrillar abnormalities in a case of asymmetric Marfan syndrome. Am J Hum Genet 46:661–671
56. González-Enseñat MA, Vicente A, Poo P, Catalá V, Pérez-Iribarne MM, Fuster C, Geán E, Happle R

(2009) Phylloid hypomelanosis and mosaic partial trisomy 13: two cases that provide further evidence of a distinct clinicogenetic entity. Arch Dermatol 145:576–578
57. Griffin R, Sustar A, Bonvin M, Binari R, del Valle Rodriguez A, Hohl AM, Bateman JR, Villalta C, Heffern E, Grunwald D, Bakal C, Desplan C, Schubiger G, Wu CT, Perrimon N (2009) The twin spot generator for differential Drosophila lineage analysis. Nat Methods 6:600–602
58. Groesser L, Herschberger E, Ruetten A, Ruivenkamp C, Lopriore E, Zutt M, Langmann T, Singer S, Klingseisen L, Schneider-Brachert W, Toll A, Real FX, Landthaler M, Hafner C (2012) Postzygotic HRAS and KRAS mutations cause nevus sebaceous and Schimmelpenning syndrome. Nat Genet 44:783–787
59. Groesser L, Herschberger E, Sagrera A, Shwayder T, Flux K, Ehmann L, Wollenberg A, Torrelo A, Bagazgoitia L, Diaz-Ley B, Tinschert S, Oschlies I, Siner S, Mickler M, Toll A, Landthaler M, Real FX, Hafner C (2013) Phacomatosis pigmentokeratotica is caused by a postzygotic *HRAS* mutation in a multipotent progenitor cell. J Invest Dermatol 133:1998–2003
60. Hadfield KD, Newman WG, Bowers NL, Wallace A, Bolger C, Colley A, McCann E, Trump D, Prescott T, Evans DG (2008) Molecular characterisation of *SMARCB1* and *NF2* in familial and sporadic schwannomatosis. J Med Genet 45:332–339
61. Hadlich J, Linse R (1989) Keratosen mit granulärer Degeneration und ihre Beziehungen zueinander. 2. Mitteilung: Heterophänie von epidermolytic Hyperkeratosis (Erythrodermia ichthyosiformis bullosa). Naevus verrucosus hystricoides und Keratosis palmoplantaris cum degeneratione granulosa Vörner. Dermatol Monatsschr 175:418–424
62. Hamm H (1999) Cutaneous mosaicism of lethal mutations. Am J Med Genet 85:342–345
63. Han HJ, Lim GY, You CY (2009) A large infiltrating fibrous hamartoma of infancy in the abdominal wall with rare associated tuberous sclerosis. Pediatr Radiol 39:743–746
64. Happle R (1986) Cutaneous manifestation of lethal genes. Hum Genet 72:280
65. Happle R (1986) The McCune-Albright syndrome: a lethal gene surviving by mosaicism. Clin Genet 29:321–324
66. Happle R (1987) Lethal genes surviving by mosaicism: a possible explanation for sporadic birth defects involving the skin. J Am Acad Dermatol 16:899–906
67. Happle R (1990) Nichterbliche Genodermatosen. Hautarzt 41:104–109
68. Happle R (1991) How many epidermal nevus syndromes exist? A clinicogenetic classification. J Am Acad Dermatol 25:550–556
69. Happle R (1991) Somatic recombination may explain linear psoriasis. J Med Genet 28:337
70. Happle R (1991) Somatic recombination may explain linear porokeratosis associated with disseminated superficial actinic porokeratosis. Am J Med Genet 39:237
71. Happle R (1992) Paradominant inheritance: a possible explanation for Becker's pigmented hairy nevus. Eur J Dermatol 2:39–40
72. Happle R (1992) Cutaneous manifestation of X-linked genes escaping inactivation. Clin Exp Dermatol 17:69
73. Happle R (1993) Mosaicism in human skin: understanding the patterns and mechanisms. Arch Dermatol 129:1460–1470
74. Happle R (1996) Segmental forms of autosomal dominant skin disorders: different types of severity reflect different states of zygosity. Am J Med Genet 66:241–242
75. Happle R (1997) A rule concerning the segmental manifestation of autosomal dominant skin disorders. Review of clinical examples providing evidence for dichotomous types of severity. Arch Dermatol 133:1505–1509
76. Happle R (1999) Loss of heterozygosity in human skin. J Am Acad Dermatol 41:143–164
77. Happle R (2001) Type 2 segmental manifestation of autosomal dominant skin diseases: development of a new formal genetic concept. Hautarzt 52:283–287
78. Happle R (2002) Speckled lentiginous nevus syndrome: delineation of a new distinct neurocutaneous phenotype. Eur J Dermatol 12:133–135
79. Happle R (2002) Dohi Memorial Lecture. New aspects of cutaneous mosaicism. J Dermatol 29:681–692
80. Happle R (2002) Transposable elements and the lines of Blaschko: a new perspective. Dermatology 204:4–7
81. Happle R (2005) Phacomatosis pigmentovascularis revisited and reclassified. Arch Dermatol 141:385–388
82. Happle R (2007) Type 2 segmental Cowden disease vs Proteus syndrome. Br J Dermatol 156:1089–1090
83. Happle R (2007) Superimposed segmental manifestation of polygenic skin disorders. J Am Acad Dermatol 57:690–699
84. Happle R (2009) Superimposed segmental manifestation of both rare and common cutaneous disorders: a new paradigm. Actas Dermosifiliogr 100:77–85
85. Happle R (2009) What is paradominant inheritance? J Med Genet 46:648
86. Happle R (2009) Monoallelic expression on autosomes may explain an unusual heritable form of pigmentary mosaicism: a historical case revisited. Clin Exp Dermatol 34:834–837
87. Happle R (2010) The group of epidermal nevus syndromes Part II. Less well defined phenotypes. J Am Acad Dermatol 63:25–30; quiz 31–22
88. Happle R, Hoffmann R, Restano L, Caputo R, Tadini G (1996) Phacomatosis pigmentokeratotica: a melanocytic-epidermal twin nevus syndrome. Am J Med Genet 65:363–365
89. Happle R, Koopman RJ (1997) Becker nevus syndrome. Am J Med Genet 68:357–361

90. Happle R, Koopman R, Mier PD (1990) Hypothesis: vascular twin naevi and somatic recombination in man. Lancet 335:376–378
91. Happle R, Steijlen PM (1993) Encephalocraniocutaneous lipomatosis: a non-hereditary mosaic phenotype. Hautarzt 44:19–22
92. Horn D, Happle R, Neitzel H, Kunze J (2002) Pigmentary mosaicism of the hyperpigmented type in two half-brothers. Am J Med Genet 112:65–69
93. Horn D, Rommeck M, Sommer D, Körner H (1997) Phylloid pigmentary pattern with mosaic trisomy 13. Pediatr Dermatol 14:278–280
94. Hwang LY, Lee JB, Richard G, Uitto JJ, Hsu S (2003) Type 1 segmental manifestation of Hailey-Hailey disease. J Am Acad Dermatol 49:712–714
95. Jiang YH, Bressler J, Beaudet AL (2004) Epigenetics and human disease. Annu Rev Genomics Hum Genet 5:479–510
96. Jonkman MF (1999) Revertant mosaicism in human genetic disorders. Am J Med Genet 85:361–364
97. Jonkman MF, Pasmooij AM (2009) Revertant mosaicism–patchwork in the skin. N Engl J Med 360:1680–1682
98. Jonkman MF, Scheffer H, Stulp R, Pas HH, Nijenhuis M, Heeres K, Owaribe K, Pulkkinen L, Uitto J (1997) Revertant mosaicism in epidermolysis bullosa caused by mitotic gene conversion. Cell 88:543–551
99. Kenwrick S, Woffendin H, Jakins T, Shuttleworth SG, Mayer E, Greenhalgh L, Whittaker J, Rugolotto S, Bardaro T, Esposito T, D'Urso M, Soli F, Turco A, Smahi A, Hamel-Teillac D, Lyonnet S, Bonnefont JP, Munnich A, Aradhya S, Kashork CD, Shaffer LG, Nelson DL, Levy M, Lewis RA (2001) Survival of male patients with incontinentia pigmenti carrying a lethal mutation can be explained by somatic mosaicism or Klinefelter syndrome. Am J Hum Genet 69:1210–1217
100. Kerns JA, Cargill EJ, Clark LA, Candille SI, Berryere TG, Olivier M, Lust G, Todhunter RJ, Schmutz SM, Murphy KE, Barsh GS (2007) Linkage and segregation analysis of black and brindle coat color in domestic dogs. Genetics 176:1679–1689
101. Kikuchi I, Idemori M, Okazaki M (1979) Plaque type syringoma. J Dermatol 6:329–331
102. Kim CA, Konig A, Bertola DR, Albano LM, Gattas GJ, Bornholdt D, Leveleki L, Happle R, Grzeschik KH (2005) CHILD syndrome caused by a deletion of exons 6–8 of the NSDHL gene. Dermatology 211:155–158
103. Kiritsi D, He Y, Pasmooij AM, Onder M, Happle R, Jonkman MF, Bruckner-Tuderman L, Has C (2012) Revertant mosaicism in a human skin fragility disorder results from slipped mispairing and mitotic recombination. J Clin Invest 122:1742–1746
104. Klaassens M, Blom EW, Schrander JJ, Ris-Stalpers C, Nieuwenhuijzen Kruseman AC, van Steensel MA, Schrander-Stumpel CT (2010) Unique skin changes in a case of Albright hereditary osteodystrophy caused by a rare *GNAS1* mutation. Br J Dermatol 162:690–694
105. Knudson AG Jr (1971) Mutation and cancer: statistical study of retinoblastoma. Proc Natl Acad Sci U S A 68:820–823
106. Knudson AG Jr, Hethcote HW, Brown BW (1975) Mutation and childhood cancer: a probabilistic model for the incidence of retinoblastoma. Proc Natl Acad Sci U S A 72:5116–5120
107. König A, Happle R (1999) Linear lesions reflecting lyonization in women heterozygous for IFAP syndrome (ichthyosis follicularis with atrichia and photophobia). Am J Med Genet 85:365–368
108. König A, Happle R, Bornholdt D, Engel H, Grzeschik KH (2000) Mutations in the NSDHL gene, encoding a 3beta-hydroxysteroid dehydrogenase, cause CHILD syndrome. Am J Med Genet 90:339–346
109. Korf BR (1999) Plexiform neurofibromas. Am J Med Genet 89:31–37
110. Krieg T, Landthaler M, Ihme A, Wolff HH, Braun-Falco O (1984) Cutis hyperelastica circumscripta. A case report with a morphologic study. Hautarzt 35:415–417
111. Kubota Y, Shimura Y, Shimada S, Tamaki K, Amamiya S (1992) Linear and whorled nevoid hypermelanosis in a child with chromosomal mosaicism. Int J Dermatol 31:345–347
112. Kurek KC, Howard E, Tennant LB, Upton J, Alomari AI, Burrows PE, Chalache K, Harris DJ, Trenor CC 3rd, Eng C, Fishman SJ, Mulliken JB, Perez-Atayde AR, Kozakewich HP (2012) PTEN hamartoma of soft tissue: a distinctive lesion in PTEN syndromes. Am J Surg Pathol 36:671–687
113. Kurek KC, Luks VL, Ayturk UM, Alomari AI, Fishman SJ, Spencer SA, Mulliken JB, Bowen ME, Yamamoto GL, Kozakewich HP, Warman ML (2012) Somatic mosaic activating mutations in *PIK3CA* cause CLOVES syndrome. Am J Hum Genet 90:1108–1115
114. Küster W, König A (1999) Hypomelanosis of Ito: no entity, but a cutaneous sign of mosaicism. Am J Med Genet 85:346–350
115. Lai-Cheong JE, McGrath JA, Uitto J (2011) Revertant mosaicism in skin: natural gene therapy. Trends Mol Med 17:140–148
116. Lai-Cheong JE, Moss C, Parsons M, Almaani N, McGrath JA (2012) Revertant mosaicism in Kindler syndrome. J Invest Dermatol 132: 730–732
117. Lee T, Luo L (1999) Mosaic analysis with a repressible cell marker for studies of gene function in neuronal morphogenesis. Neuron 22:451–461
118. Lee T, Luo L (2001) Mosaic analysis with a repressible cell marker (MARCM) for Drosophila neural development. Trends Neurosci 24:251–254
119. Lenz W (1975) Half chromatid mutations may explain incontinentia pigmenti in males. Am J Hum Genet 27:690–691
120. Limaye N, Boon LM, Vikkula M (2009) From germline towards somatic mutations in the pathophysiology of vascular anomalies. Hum Mol Genet 18:R65–R74

121. Lin S, Kao CF, Yu HH, Huang Y, Lee T (2012) Lineage analysis of Drosophila lateral antennal lobe neurons reveals notch-dependent binary temporal fate decisions. PLoS Biol 10:e1001425
122. Lindhurst MJ, Sapp JC, Teer JK, Johnston JJ, Finn EM, Peters K, Turner J, Cannons JL, Bick D, Blakemore L, Blumhorst C, Brockmann K, Calder P, Cherman N, Deardorff MA, Everman DB, Golas G, Greenstein RM, Kato BM, Keppler-Noreuil KM, Kuznetsov SA, Miyamoto RT, Newman K, Ng D, O'Brien K, Rothenberg S, Schwartzentruber DJ, Singhal V, Tirabosco R, Upton J, Wientroub S, Zackai EH, Hoag K, Whitewood-Neal T, Robey PG, Schwartzberg PL, Darling TN, Tosi LL, Mullikin JC, Biesecker LG (2011) A mosaic activating mutation in *AKT1* associated with the Proteus syndrome. N Engl J Med 365:611–619
123. Lipsker D, Cribier B, Girard-Lemaire F, Flori E, Grosshans E (2000) Genetic mosaicism in an acquired inflammatory dermatosis following the lines of Blaschko. Arch Dermatol 136:805–807
124. Lopez-Barrantes O, Torrelo A, Mediero IG, Zambrano A, Happle R (2002) Nevoid hypertrichosis and hypomelanosis. Eur J Dermatol 12:583–585
125. Lyon MF (1961) Gene action in the X-chromosome of the mouse (Mus musculus L.). Nature 190:372–373
126. Lyon MF (1972) X-chromosome inactivation and developmental patterns in mammals. Biol Rev Camb Philos Soc 47:1–35
127. Lyon MF (1998) X-chromosome inactivation: a repeat hypothesis. Cytogenet Cell Genet 80:133–137
128. Mallory SB, Enjolras O, Boon LM, Rogers E, Berk DR, Blei F, Baselga E, Ros AM, Vikkula M (2006) Congenital plaque-type glomuvenous malformations presenting in childhood. Arch Dermatol 142:892–896
129. Marini MA, Saponaro AE, Parra LSR (2004) Siringomas múltiples: forma localizada, variedad clínica "en placa, axilar unilateral". Arch Argent Dermatol 54:7–75
130. Martin DI, Ward R, Suter CM (2005) Germline epimutation: a basis for epigenetic disease in humans. Ann N Y Acad Sci 1054:68–77
131. Martínez-García S, Vera-Casaño A, Eloy-García Carrasco C, del Boz-González J, Martínez-Pilar L, Crespo-Erchiga V (2008) Elephantiasis neuromatosa in a patient with neurofibromatosis type 1. J Eur Acad Dermatol Venereol 22:103–105
132. McCarthy JC, Goldberg MJ, Zimbler S (1982) Orthopaedic dysfunction in the blue rubber-bleb nevus syndrome. J Bone Joint Surg Am 64:280–283
133. McDonald JM, Douglass EC, Fisher R, Geiser CF, Krill CE, Strong LC, Virshup D, Huff V (1998) Linkage of familial Wilms' tumor predisposition to chromosome 19 and a two-locus model for the etiology of familial tumors. Cancer Res 58:1387–1390
134. McGrae JD Jr, Hashimoto K (1996) Unilateral facial angiofibromas–a segmental form of tuberous sclerosis. Br J Dermatol 134:727–730
135. Merks JH, de Vries LS, Zhou XP, Nikkels P, Barth PG, Eng C, Hennekam RC (2003) PTEN hamartoma tumour syndrome: variability of an entity. J Med Genet 40:e111
136. Metta AK, Ramachandra S, Sadath N, Manupati S (2011) Linear and whorled nevoid hypermelanosis in three successive generations. Indian J Dermatol Venereol Leprol 77:403
137. Mezzadra G (1965) Leiomioma cutaneo multiplo ereditario: studio di un caso sistematizzato in soggetto maschile appartenente a famiglia portatrice di leiomiomatosi cutanea e fibromiomatosi uterina. Minerva Dermatol 40:388–393
138. Monrigal E, Gallot D, James I, Hameury F, Vanlieferinghen P, Guibaud L (2009) Venous malformation of the soft tissue associated with blue rubber bleb nevus syndrome: prenatal imaging and impact on postnatal management. Ultrasound Obstet Gynecol 34:730–732
139. Montagna P, Procaccianti G, Galli G, Ripamonti L, Patrizi A, Baruzzi A (1991) Familial hypomelanosis of Ito. Eur Neurol 31:345–347
140. Morgan HD, Sutherland HG, Martin DI, Whitelaw E (1999) Epigenetic inheritance at the agouti locus in the mouse. Nat Genet 23:314–318
141. Oberg AL, French AJ, Sarver AL, Subramanian S, Morlan BW, Riska SM, Borralho PM, Cunningham JM, Boardman LA, Wang L, Smyrk TC, Asmann Y, Steer CJ, Thibodeau SN (2011) MiRNA expression in colon polyps provides evidence for a multihit model of colon cancer. PLoS One 6
142. Online Mendelian Inheritance in Man (OMIM). Available at: http://www.ncbi.nlm.nih.gov/omim/. Accessed 16 Aug 2013
143. Ousager LB, Bygum A, Hafner C (2012) Identification of a novel S249C *FGFR3* mutation in a keratinocytic epidermal nevus syndrome. Br J Dermatol 167:202–204
144. Pagenstecher A, Stahl S, Sure U, Felbor U (2009) A two-hit mechanism causes cerebral cavernous malformations: complete inactivation of CCM1, CCM2 or CCM3 in affected endothelial cells. Hum Mol Genet 18:911–918
145. Paller AS, Syder AJ, Chan YM, Yu QC, Hutton E, Tadini G, Fuchs E (1994) Genetic and clinical mosaicism in a type of epidermal nevus. N Engl J Med 331:1408–1415
146. Pasmooij AM, Pas HH, Deviaene FC, Nijenhuis M, Jonkman MF (2005) Multiple correcting *COL17A1* mutations in patients with revertant mosaicism of epidermolysis bullosa. Am J Hum Genet 77:727–740
147. Patel RC, Zynger DL, Laskin WB (2009) Blue rubber bleb nevus syndrome: novel lymphangiomatosis-like growth pattern within the uterus and immunohistochemical analysis. Hum Pathol 40:413–417
148. Perry WL, Copeland NG, Jenkins NA (1994) The molecular basis for dominant yellow agouti coat color mutations. Bioessays 16:705–707
149. Poblete-Gutiérrez P, Wiederholt T, König A, Jugert FK, Marquardt Y, Rübben A, Merk HF, Happle R, Frank J (2004) Allelic loss underlies type 2 segmental Hailey-Hailey disease, providing molecular

confirmation of a novel genetic concept. J Clin Invest 114:1467–1474
150. Quattrocchio F, Wing J, van der Woude K, Souer E, de Vetten N, Mol J, Koes R (1999) Molecular analysis of the anthocyanin2 gene of petunia and its role in the evolution of flower color. Plant Cell 11:1433–1444
151. Rallis E, Ragiadakou D (2009) Giant plexiform neurofibroma in a patient with neurofibromatosis type I. Dermatol Online J 15:7
152. Razack EM, Natarajan M (1977) Ulcerative systematized porokeratosis (Mibelli). Arch Dermatol 113:1583–1584
153. Rivière JB, Mirzaa GM, O'Roak BJ, Beddaoui M, Alcantara D, Conway RL, St-Onge J, Schwartzentruber JA, Gripp KW, Nikkel SM, Worthylake T, Sullivan CT, Ward TR, Butler HE, Kramer NA, Albrecht B, Armour CM, Armstrong L, Caluseriu O, Cytrynbaum C, Drolet BA, Innes AM, Lauzon JL, Lin AE, Mancini GM, Meschino WS, Reggin JD, Saggar AK, Lerman-Sagie T, Uyanik G, Weksberg R, Zirn B, Beaulieu CL, Majewski J, Bulman DE, O'Driscoll M, Shendure J, Graham JM Jr, Boycott KM, Dobyns WB (2012) De novo germline and postzygotic mutations in *AKT3*, *PIK3R2* and *PIK3CA* cause a spectrum of related megalencephaly syndromes. Nat Genet 44:934–940
154. Rodriguez-Diaz E, Alvarez-Cuesta CC, Blanco S, Galache C, Hidalgo Y (2006) Becker's nevus associated with epidermal nevus: another example of twin spotting? Actas Dermosifiliogr 97:200–202
155. Rodríguez-Martín M, Sánchez González R, Sáez-Rodríguez M, García-Bustinduy M, Martín-Herrera A, Noda-Cabrera A (2009) An unusual case of congenital linear eccrine spiradenoma. Pediatr Dermatol 26:180–183
156. Rodríguez-Martín M, Sánchez R, Sáez-Rodríguez M, García-Bustínduy M, Sidro M, Pérez N, González S, Alvarez H, Noda A (2008) Congenital plaque-like glomangioma associated with superficial hypertrichosis. J Am Acad Dermatol 58:S92–S93
157. Sakuntabhai A, Dhitavat J, Burge S, Hovnanian A (2000) Mosaicism for *ATP2A2* mutations causes segmental Darier's disease. J Invest Dermatol 115:1144–1147
158. Salvitti M, Celestino F, Gerocarni Nappo S, Caione P (2013) Diffuse ganglioneuromatosis and plexiform neurofibroma of the urinary bladder: an uncommon cause of severe urological disease in an infant. J Pediatr Urol 9(3):e131–e133
159. Sapp JC, Turner JT, van de Kamp JM, van Dijk FS, Lowry RB, Biesecker LG (2007) Newly delineated syndrome of congenital lipomatous overgrowth, vascular malformations, and epidermal nevi (CLOVE syndrome) in seven patients. Am J Med Genet A 143A:2944–2958
160. Schauder S, Hanefeld F, Noske UM, Zoll B (2000) Depigmented hypertrichosis following Blaschko's lines associated with cerebral and ocular malformations: a new neurocutaneous, autosomal lethal gene syndrome from the group of epidermal naevus syndromes? Br J Dermatol 142:1204–1207
161. Schwindinger WF, Francomano CA, Levine MA (1992) Identification of a mutation in the gene encoding the alpha subunit of the stimulatory G protein of adenylyl cyclase in McCune-Albright syndrome. Proc Natl Acad Sci U S A 89:5152–5156
162. Sequeira FF, Jayaseelan E (2011) Keratosis follicularis spinulosa decalvans in a female. Indian J Dermatol Venereol Leprol 77:325–327
163. Sestini R, Bacci C, Provenzano A, Genuardi M, Papi L (2008) Evidence of a four-hit mechanism involving *SMARCB1* and *NF2* in schwannomatosis-associated schwannomas. Hum Mutat 29:227–231
164. Shirley MD, Tang H, Gallione CJ, Baugher JD, Frelin LP, Cohen B, North PE, Marchuk DA, Comi AM, Pevsner J (2013) Sturge-Weber syndrome and port-wine stains caused by somatic mutation in *GNAQ*. N Engl J Med 368:1971–1979
165. Shore EM, Ahn J, Jan de Beur S, Li M, Xu M, Gardner RJ, Zasloff MA, Whyte MP, Levine MA, Kaplan FS (2002) Paternally inherited inactivating mutations of the GNAS1 gene in progressive osseous heteroplasia. N Engl J Med 346:99–106
166. Sidwell RU, Francis N, Grahame R, Pope FM, Bunker CB (2003) Connective tissue naevus (collagenoma) in a patient with benign joint hypermobility syndrome (Ehlers-Danlos syndrome type III). Clin Exp Dermatol 28:323–325
167. Simpson JR (1974) Giant glomangioma. Br J Dermatol 90:229–231
168. Smit AF (1999) Interspersed repeats and other mementos of transposable elements in mammalian genomes. Curr Opin Genet Dev 9:657–663
169. Snowden KC, Napoli CA (1998) Psl: a novel Spm-like transposable element from Petunia hybrida. Plant J 14:43–54
170. Sobey GJ, Quarrell OW, Williams S, McGrath HM (2007) Mosaic chromosome 6 trisomy in an epidermal nevus. Pediatr Dermatol 24:144–146
171. Steinmann K, Kluwe L, Friedrich RE, Mautner VF, Cooper DN, Kehrer-Sawatzki H (2009) Mechanisms of loss of heterozygosity in neurofibromatosis type 1-associated plexiform neurofibromas. J Invest Dermatol 129:615–621
172. Stosiek N, Ulmer R, von den Driesch P, Claussen U, Hornstein OP, Rott HD (1994) Chromosomal mosaicism in two patients with epidermal verrucous nevus. Demonstration of chromosomal breakpoint. J Am Acad Dermatol 30:622–625
173. Suh DH, Lee HS, Kim SD, Cho KH, Kim KH, Park KC (2000) Coexistence of disseminated superficial porokeratosis in childhood with congenital linear porokeratosis. Pediatr Dermatol 17:466–468
174. Sustar A, Bonvin M, Schubiger M, Schubiger G (2011) Drosophila twin spot clones reveal cell division dynamics in regenerating imaginal discs. Dev Biol 356:576–587
175. Sutherland JV, Bailar JC 3rd (1984) The multihit model of carcinogenesis: etiologic implications for colon cancer. J Chronic Dis 37:465–480

176. Tan WH, Baris HN, Burrows PE, Robson CD, Alomari AI, Mulliken JB, Fishman SJ, Irons MB (2007) The spectrum of vascular anomalies in patients with *PTEN* mutations: implications for diagnosis and management. J Med Genet 44:594–602
177. Tang L, Liang J, Wang W, Yu L, Yao Z (2011) A novel mutation in *MBTPS2* causes a broad phenotypic spectrum of ichthyosis follicularis, atrichia, and photophobia syndrome in a large Chinese family. J Am Acad Dermatol 64:716–722
178. Tinschert S, Naumann I, Stegmann E, Buske A, Kaufmann D, Thiel G, Jenne DE (2000) Segmental neurofibromatosis is caused by somatic mutation of the neurofibromatosis type 1 (NF1) gene. Eur J Hum Genet 8:455–459
179. Tone Y, Wada T, Shibata F, Toma T, Hashida Y, Kasahara Y, Koizumi S, Yachie A (2007) Somatic revertant mosaicism in a patient with leukocyte adhesion deficiency type 1. Blood 109:1182–1184
180. Torrelo A, Hadj-Rabia S, Colmenero I, Piston R, Sybert VP, Hilari-Carbonell H, Hernandez-Martin A, Ferreres JC, Vano-Galvan S, Azorin D, de Salamanca JE, Requena L, Bodemer C, Happle R, Garcia-Patos V, Fraitag S (2012) Folliculocystic and collagen hamartoma of tuberous sclerosis complex. J Am Acad Dermatol 66:617–621
181. Torrelo A, Hernández-Martín A, Bueno E, Colmenero I, Rivera I, Requena L, Happle R, González-Sarmiento R (2013) Molecular proof of type 2 mosaicism in Gorlin syndrome. Br J Dematol. doi:10.1111/bjd.12458
182. Urtizberea JA, Testart H, Cartault F, Boccon-Gibod L, Le Merrer M, Kaplan FS (1998) Progressive osseous heteroplasia. Report of a family. J Bone Joint Surg Br 80:768–771
183. Vakilzadeh F, Kolde G (1985) Relapsing linear acantholytic dermatosis. Br J Dermatol 112: 349–355
184. van Rijk A, Bloemendal H (2003) Molecular mechanisms of exon shuffling: illegitimate recombination. Genetica 118:245–249
185. Vandenbroucke I, van Doorn R, Callens T, Cobben JM, Starink TM, Messiaen L (2004) Genetic and clinical mosaicism in a patient with neurofibromatosis type 1. Hum Genet 114:284–290
186. Verhoef S, Vrtel R, van Essen T, Bakker L, Sikkens E, Halley D, Lindhout D, van den Ouweland A (1995) Somatic mosaicism and clinical variation in tuberous sclerosis complex. Lancet 345:202
187. Wada T, Schurman SH, Jagadeesh GJ, Garabedian EK, Nelson DL, Candotti F (2004) Multiple patients with revertant mosaicism in a single Wiskott-Aldrich syndrome family. Blood 104:1270–1272
188. Wada T, Yasui M, Toma T, Nakayama Y, Nishida M, Shimizu M, Okajima M, Kasahara Y, Koizumi S, Inoue M, Kawa K, Yachie A (2008) Detection of T lymphocytes with a second-site mutation in skin lesions of atypical X-linked severe combined immunodeficiency mimicking Omenn syndrome. Blood 112:1872–1875
189. Wang X, Reid Sutton V, Omar Peraza-Llanes J, Yu Z, Rosetta R, Kou YC, Eble TN, Patel A, Thaller C, Fang P, Van den Veyver IB (2007) Mutations in X-linked *PORCN*, a putative regulator of Wnt signaling, cause focal dermal hypoplasia. Nat Genet 39:836–838
190. Waterland RA, Jirtle RL (2003) Transposable elements: targets for early nutritional effects on epigenetic gene regulation. Mol Cell Biol 23:5293–5300
191. Weinstein LS (2006) G(s)alpha mutations in fibrous dysplasia and McCune-Albright syndrome. J Bone Miner Res 21:P120–P124
192. Weintraub R, Pinkus H (1977) Multiple fibrofolliculomas (Birt-Hogg-Dubé) associated with a large connective tissue nevus. J Cutan Pathol 4:289–299
193. Weterings E, Chen DJ (2008) The endless tale of non-homologous end-joining. Cell Res 18:114–124
194. White JM, Short K, Salisbury JR, Fuller LC (2006) A novel case of linear syringomatous hamartoma. Clin Exp Dermatol 31:222–224
195. Whitehouse HLK (1882) Genetic recombination: understanding the mechanisms. Wiley, Chichester
196. Yeh GL, Mathur S, Wivel A, Li M, Gannon FH, Ulied A, Audi L, Olmstead EA, Kaplan FS, Shore EM (2000) GNAS1 mutation and Cbfa1 misexpression in a child with severe congenital platelike osteoma cutis. J Bone Miner Res 15:2063–2073
197. Yoon TY, Lee HT, Chang SH (1999) Giant congenital multiple patch-like glomus tumors. J Am Acad Dermatol 40:826–828
198. Yu HH, Chen CH, Shi L, Huang Y, Lee T (2009) Twin-spot MARCM to reveal the developmental origin and identity of neurons. Nat Neurosci 12:947–953
199. Yu HH, Kao CF, He Y, Ding P, Kao JC, Lee T (2010) A complete developmental sequence of a Drosophila neuronal lineage as revealed by twin-spot MARCM. PLoS Biol 8:e1000461
200. Yung CW, Soltani K, Bernstein JE, Lorincz AL (1981) Unilateral linear nevoidal syringoma. J Am Acad Dermatol 4:412–416
201. Zhou XP, Marsh DJ, Hampel H, Mulliken JB, Gimm O, Eng C (2000) Germline and germline mosaic *PTEN* mutations associated with a Proteus-like syndrome of hemihypertrophy, lower limb asymmetry, arteriovenous malformations and lipomatosis. Hum Mol Genet 9:765–768

4 Relationship Between Hypomorphic Alleles and Mosaicism of Lethal Mutations

Contents

Some genes may harbor mutations that are deleterious to such degree that affected embryos can only survive in a mosaic state. For example, male embryos hemizygous for incontinentia pigmenti are dying in utero. Remarkably, however, other mutations within the same gene give rise to ectodermal dysplasia of Zonana, a disorder that is so mild that hemizygous males can survive (Table 4.1). They have the full-blown phenotype including dysgammaglobulinemia and recurrent infections [1, 13]. Female carriers only show a systematized linear pattern of pigmentary disturbance that should not be confused with incontinentia pigmenti which represents a male-lethal trait, whereas ectodermal dysplasia of Zonana is caused by a hypomorphic *NEMO* mutation [12].

Similarly, Milunsky et al. [11] described a boy with "a severe atypical phenotype for X-linked dominant Conradi-Hünermann-Happle syndrome." The child had a multisystem birth defect but no epiphyseal stippling nor any other clinical sign of Conradi-Hünermann-Happle syndrome, although an *EBP* mutation was found. His mother who carried the same mutation did not show any clinical abnormality. Therefore, one reader argued that this was not a severe form of that X-linked male-lethal disorder but a quite different X-linked recessive phenotype being caused by a hypomorphic *EBP* allele resulting in survival of the boy [7, 8]. Subsequently similar cases were described [5], and the term MEND syndrome (*m*ale *E*BP disorder with *n*eurological *d*efects) was proposed [2].

CHILD syndrome (Fig. 4.1) is an X-linked dominant, male-lethal trait caused by *NSDHL*

R. Happle, *Mosaicism in Human Skin*,
DOI 10.1007/978-3-642-38765-4_4, © Springer-Verlag Berlin Heidelberg 2014

Table 4.1 Genes that may harbor either lethal or hypomorphic alleles

Locus	Gene and its protein	Mosaic phenotype caused by lethal alleles	Nonmosaic phenotype caused by hypomorphic alleles	Is heterozygosity for a hypomorphic allele clinically recognizable?
Xq28	*NEMO* IKK-gamma	Incontinentia pigmenti	Ectodermal dysplasia of Zonana [13]	Yes
Xp11.22–p11.23	*EBP* Sterol-Δ^8-isomerase	Conradi-Hünermann-Happle syndrome	MEND syndrome [2, 7]	No
Xq28	*NSDHL* 3ß-hydroxysteroid dehydrogenase	CHILD syndrome [9]	CK syndrome [4, 10]	No
4p16.3	*FGFR3* Fibroblast growth factor receptor 3	FGFR3 epidermal nevus syndrome [6]	Autosomal dominant acanthosis nigricans [3]	Yes

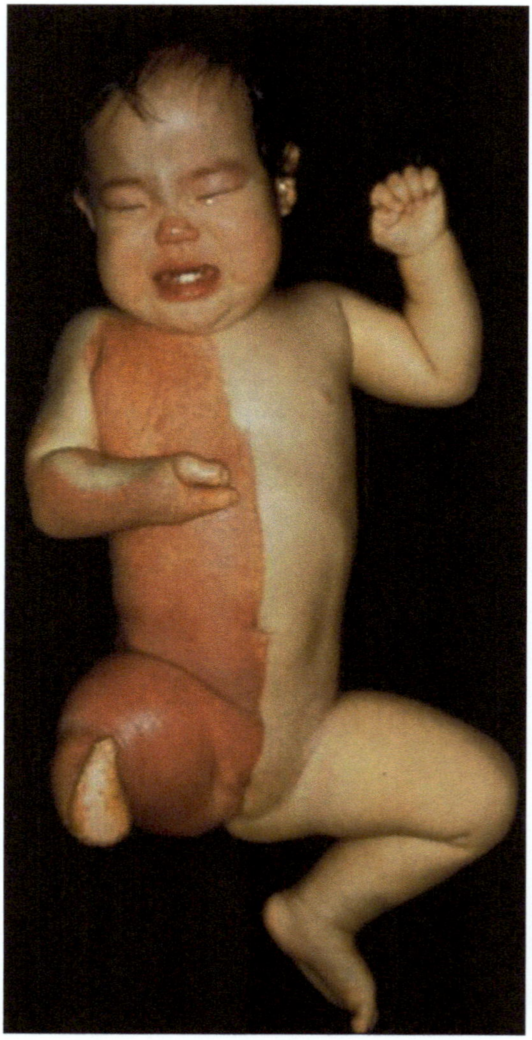

Fig. 4.1 CHILD syndrome being caused by an X-linked male-lethal mutation in *NSDHL* (Courtesy of Dr. Hitoshi Yaguchi, Tokyo, Japan)

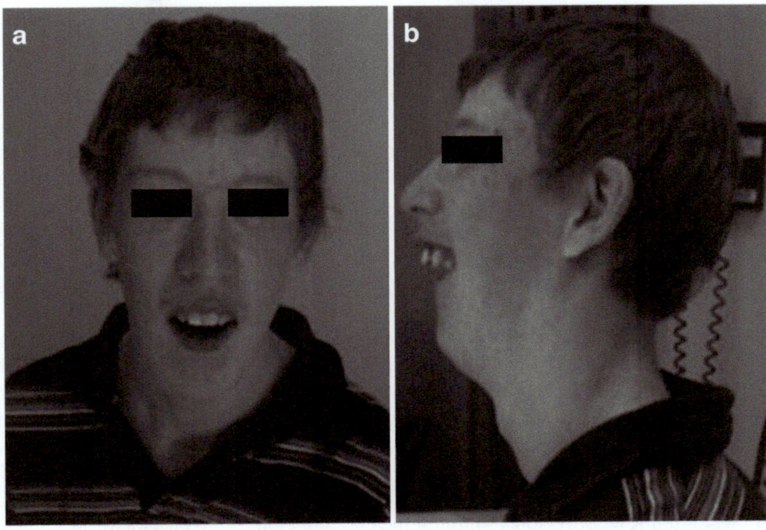

Fig. 4.2 CK syndrome being caused by a nonlethal, hypomorphic mutation in *NSDHL*. Physical features of this 17-year-old male patient are notable for (**a**) a long thin face, epicanthic folds, and almond-shaped eyes and (**b**) small jaw, high nasal bridge, and posteriorly rotated ears [4] (Reprinted with permission from John Wiley and Sons, USA)

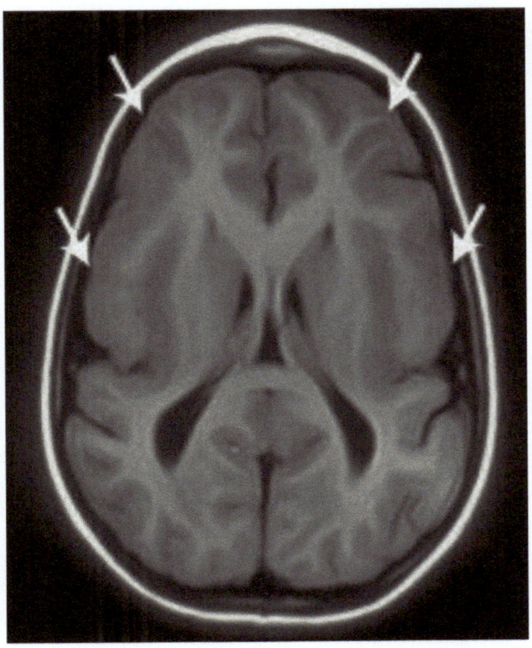

Fig. 4.3 MRI scan of a 6-year-old patient with CK syndrome showing a simplified gyral pattern in the frontal and parietal cortex (*arrows*) [10] (Reprinted with permission from Elsevier Limited, UK)

mutations [9], whereas CK syndrome, an X-linked recessive trait characterized by dysmorphic facial appearance, microcephaly, and cortical defects with mental deficiency and seizures (Figs. 4.2 and 4.3), was found to be caused by hypomorphic *NSDHL* alleles [4, 10]. Female carriers of such mutations appear to be phenotypically healthy.

Another example is autosomal dominant acanthosis nigricans. It is caused by *FGFR3* mutations that can be categorized as "hypomorphic" because a more severe R248C mutation of the FGFR3 gene gives rise either to nonmosaic thanatophoric dysplasia or to a mosaic phenotype in the form of FGFR3 epidermal nevus syndrome [6].

References

1. Aradhya S, Courtois G, Rajkovic A, Lewis RA, Levy M, Israel A, Nelson DL (2001) Atypical forms of incontinentia pigmenti in male individuals result from mutations of a cytosine tract in exon 10 of NEMO (IKK-gamma). Am J Hum Genet 68:765–771
2. Arnold AW, Bruckner-Tuderman L, Has C, Happle R (2012) Conradi-Hünermann-Happle syndrome in males versus MEND syndrome (male EBP disorder with neurological defects). Br J Dermatol 166:1309–1313
3. Berk DR, Spector EB, Bayliss SJ (2007) Familial acanthosis nigricans due to K650T *FGFR3* mutation. Arch Dermatol 143:1153–1156
4. du Souich C, Chou A, Yin J, Oh T, Nelson TN, Hurlburt J, Arbour L, Friedlander R, McGillivray BC, Tyshchenko N, Rump A, Poskitt KJ, Demos MK, Van Allen MI, Boerkoel CF (2009) Characterization of a new X-linked mental retardation syndrome with microcephaly, cortical malformation, and thin habitus. Am J Med Genet A 149A:2469–2478
5. Furtado LV, Bayrak-Toydemir P, Hulinsky B, Damjanovich K, Carey JC, Rope AF (2010) A novel X-linked multiple congenital anomaly syndrome

associated with an *EBP* mutation. Am J Med Genet A 152A:2838–2844
6. Garcia-Vargas A, Hafner C, Pérez-Rodríguez AG, Rodríguez-Rojas LX, González-Esqueda P, Stoehr R, Hernández-Torres M, Happle R (2008) An epidermal nevus syndrome with cerebral involvement caused by a mosaic *FGFR3* mutation. Am J Med Genet A 146A:2275–2279
7. Happle R (2003) Hypomorphic alleles within the EBP gene cause a phenotype quite different from Conradi-Hünermann-Happle syndrome. Am J Med Genet A 122A:279; author reply 280
8. Happle R (2011) A novel X-linked phenotype caused by hypomorphic *EBP* mutations. Am J Med Genet A 155A:1770–1771; author reply 1772
9. König A, Happle R, Bornholdt D, Engel H, Grzeschik KH (2000) Mutations in the NSDHL gene, encoding a 3beta-hydroxysteroid dehydrogenase, cause CHILD syndrome. Am J Med Genet 90:339–346
10. McLarren KW, Severson TM, du Souich C, Stockton DW, Kratz LE, Cunningham D, Hendson G, Morin RD, Wu D, Paul JE, An J, Nelson TN, Chou A, DeBarber AE, Merkens LS, Michaud JL, Waters PJ, Yin J, McGillivray B, Demos M, Rouleau GA, Grzeschik KH, Smith R, Tarpey PS, Shears D, Schwartz CE, Gecz J, Stratton MR, Arbour L, Hurlburt J, Van Allen MI, Herman GE, Zhao Y, Moore R, Kelley RI, Jones SJ, Steiner RD, Raymond FL, Marra MA, Boerkoel CF (2010) Hypomorphic temperature-sensitive alleles of *NSDHL* cause CK syndrome. Am J Hum Genet 87:905–914
11. Milunsky JM, Maher TA, Metzenberg AB (2003) Molecular, biochemical, and phenotypic analysis of a hemizygous male with a severe atypical phenotype for X-linked dominant Conradi-Hunermann-Happle syndrome and a mutation in *EBP*. Am J Med Genet A 116A:249–254
12. Smahi A, Courtois G, Rabia SH, Döffinger R, Bodemer C, Munnich A, Casanova JL, Israël A (2002) The NF-kappaB signalling pathway in human diseases: from incontinentia pigmenti to ectodermal dysplasias and immune-deficiency syndromes. Hum Mol Genet 11:2371–2375
13. Zonana J, Elder ME, Schneider LC, Orlow SJ, Moss C, Golabi M, Shapira SK, Farndon PA, Wara DW, Emmal SA, Ferguson BM (2000) A novel X-linked disorder of immune deficiency and hypohidrotic ectodermal dysplasia is allelic to incontinentia pigmenti and due to mutations in IKK-gamma (*NEMO*). Am J Hum Genet 67:1555–1562

Part II

The Patterns of Cutaneous Mosaicism

5 Six Archetypical Patterns

Contents

Six archetypical patterns of cutaneous mosaicism can so far be distinguished: (1) lines of Blaschko, (2) checkerboard pattern, (3) phylloid pattern, (4) patchy pattern without midline separation, (5) lateralization pattern, and (6) the sash-like pattern (Figs. 5.1 and 5.2). These distinct types of archetypical arrangement should not be conflated with the various shapes of the individual smaller mosaic skin lesions that may be round, oval, oblong, or triangular and may or may not have indented borders [69].

5.1 Lines of Blaschko

In 1901 Alfred Blaschko (Fig. 5.3) published his atlas on linear skin diseases, containing the well-known diagram of what he called the "nevus lines" [6]. From more than 170 case reports sent to him from colleagues of many European countries, he depicted an archetypical pattern of lines by applying woollen threads to a plaster of Paris classical statue that he had bought in Berlin from an Italian street vendor. He wrote: "I have tried … to delineate, from the total collection of available cases, a system of lines on the surface of the body representing the typical pattern that the linear nevi follow." These lines form a characteristic V figure reminiscent of a fountain on the back (Fig. 5.4) and an S figure on the lateral aspects of the trunk (Fig. 5.5). Sometimes they produce whorls on the trunk or limbs (Fig. 5.6). On the arms and legs, they run in a less distinctive, perpendicular pattern. Blaschko clearly stated that

R. Happle, *Mosaicism in Human Skin*,
DOI 10.1007/978-3-642-38765-4_5, © Springer-Verlag Berlin Heidelberg 2014

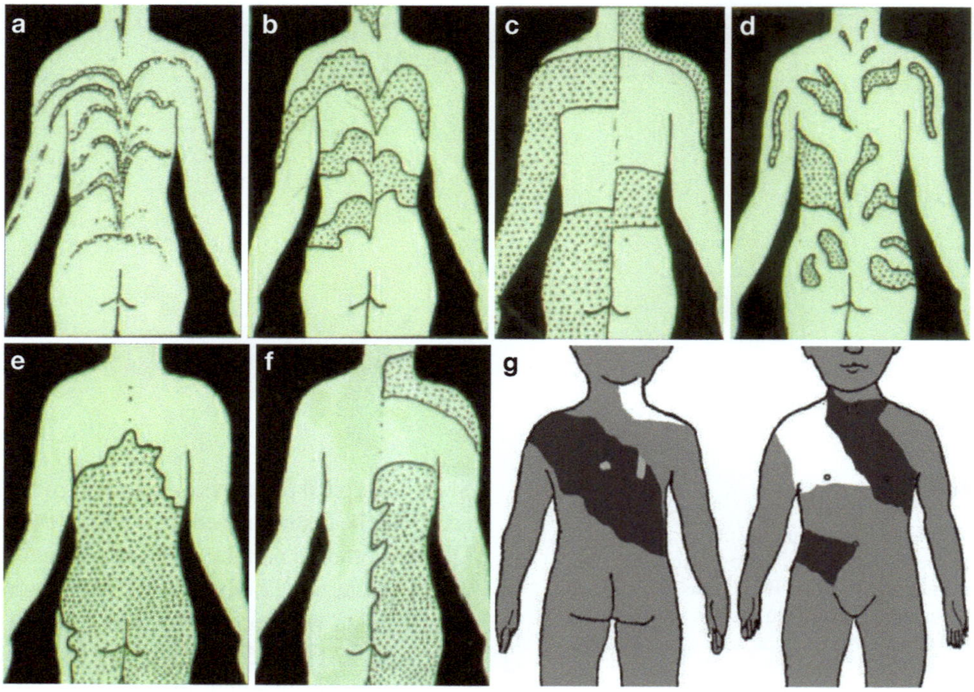

Fig. 5.1 Archetypical patterns of cutaneous mosaicism. (**a**) Type 1a, lines of Blaschko arranged in narrow bands; (**b**) type 1b, lines of Blaschko arranged in broad bands; (**c**) type 2, checkerboard pattern; (**d**) type 3, phylloid pattern; (**e**) type 4, large non-slanting patches without midline separation; (**f**) type 5, lateralization pattern; (**g**) type 6, sash-like pattern (Ruggieri-Happle type)

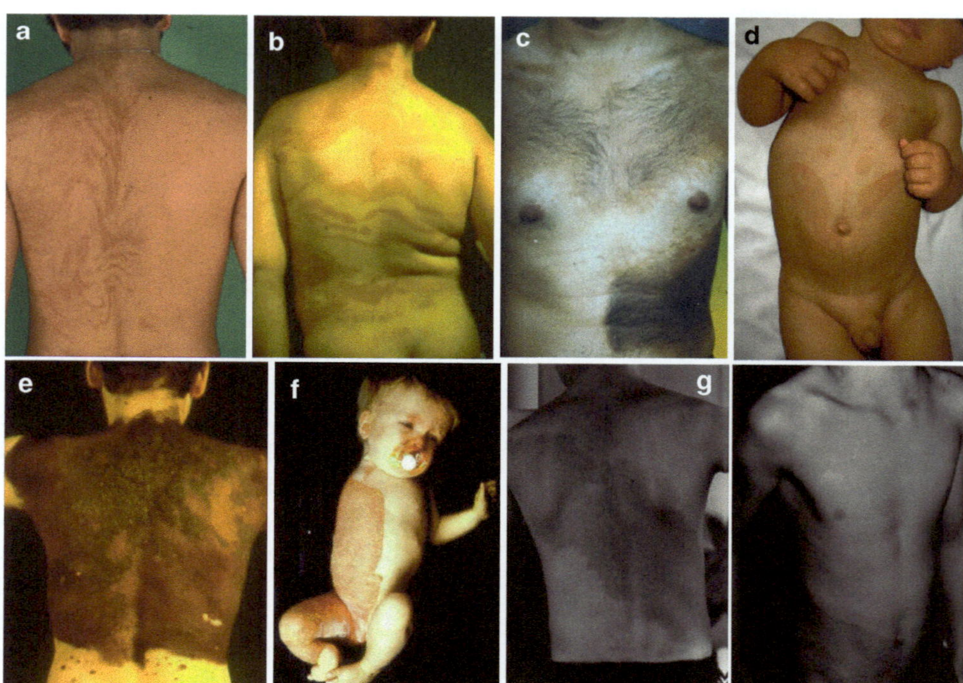

Fig. 5.2 Clinical examples of the archetypical patterns as shown in Fig. 5.1. (**a**) Blaschko-linear hypermelanosis in narrow bands; (**b**) McCune-Albright syndrome (Courtesy of Dr. Jean-Paul Ortonne, Nice, France); (**c**) systematized Becker nevus (Courtesy of Dr. Aïcha Salhi, Algiers, Algeria); (**d**) phylloid hypermelanosis [37] (Reprinted with permission from John Wiley & Sons, USA); (**e**) giant melanocytic nevus; (**f**) CHILD syndrome; (**g**) cutis tricolor of the Ruggieri-Happle type [60]

Fig. 5.3 Alfred Blaschko (Courtesy of the late Dr. Hermann Blaschko, Oxford, UK)

this system of lines was in no way related to the zones of radicular innervation. "The linear nevi are a sequela of developmental disturbances for which it is not necessary to assume a preceding disorder of the nervous system." He assumed that "during the initial stage of embryogenesis, rather intense movements and shifts of the individual cutaneous territories occur along these lines."

Blaschko had published his essential message already in 1895 [1, 5]: "Why should we resort to a pathological event – that is, in top of that, hypothetical, – in another far distant organ system and take the skin disorder as a secondary phenomenon, if our need of causality is satisfied by the simple and certainly plausible assumption of a growth disturbance within the epidermis itself… Not to mention that the influence of the nervous system on processes of growth is not at all proven as yet and more than doubtful for the sequential stages of embryonic life wherein the nerves will almost certainly play a minor role only… But I believe to be able to postulate with certainty that the epidermis, especially of the trunk, is divided into a series of more or less parallel territories… If we assume that, during the months when differentiation takes place, a disturbing factor would

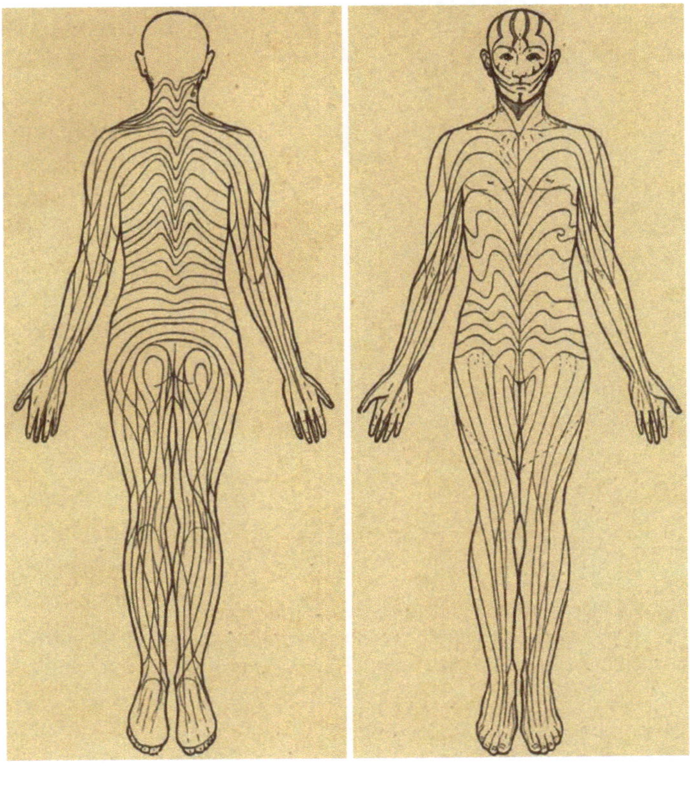

Fig. 5.4 Blaschko's original drawing of his "nevus lines" [6]. The head was mapped in a rather cursory way and left in part completely free because of lack of information

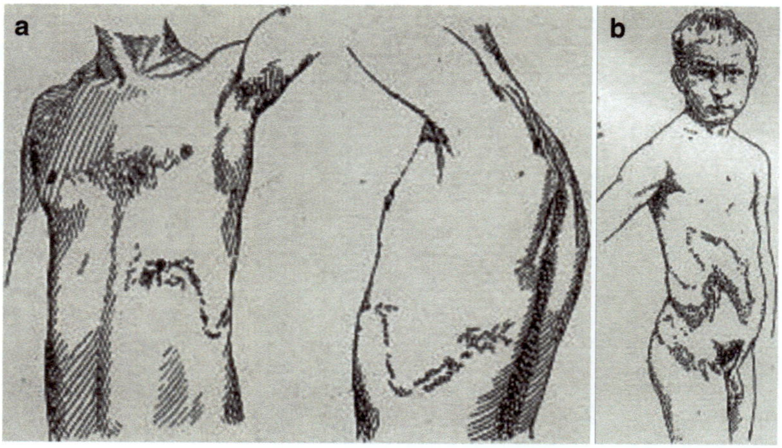

Fig. 5.5 Blaschko's lines forming characteristic S figures on the lateral aspect of the trunk (From Blaschko's atlas. (**a**) Plate XII, Fig. 4 "Nerve nevus. University Hospital Breslau". (**b**) Plate XII, Fig. 2 "Naevus verrucosus linearis, from Alexander-Blaschko" [see 1,5] [6])

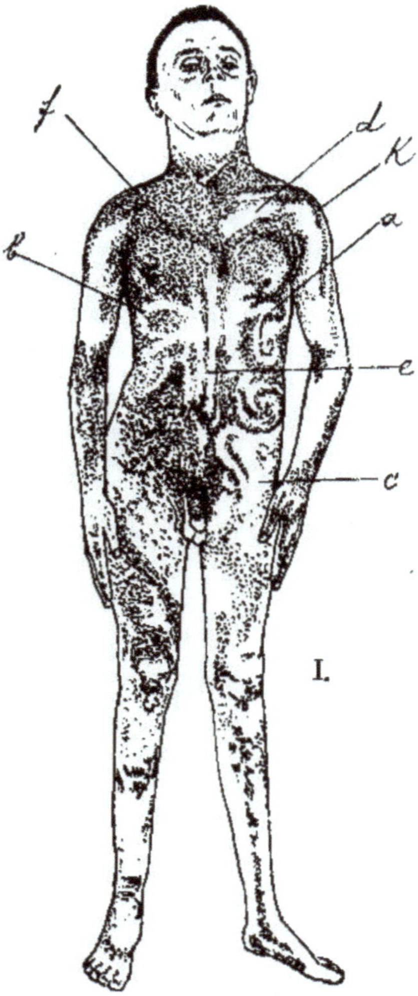

Fig. 5.6 Systematized epidermal nevus ("ichthyosis hystrix Hebrae after Gassmann") producing multiple whorls on the trunk (From Blaschko's atlas, Plate XV, Fig. 4 [6])

interfere, then it would easily be possible that only particular territories are involved, whereas intervening regions are healthy" [5]. Blaschko didn't know anything about genes and mutations, but by choosing the words "*a disturbing factor*" he anticipated the explanatory genetic mechanisms as later brought to light in the second half of the past century.

In 1901, Montgomery [53] presented similar ideas but by no means he did this "at the same time" [62]. On the contrary, he explicitly referred to Blaschko's thoughts [1, 5] when arguing that linear nevi do apparently not follow the course of nerves or blood vessels, Voigt's lines, the lines of cleavage of the skin, or the metameres of the body. He favored Blaschko's idea that "the streaks may be due to the streams or trend of growth of the tissues and to the adaptation of the embryonic sutures…" Accordingly, "a nævus linearis originates at an early stage in the development of the fœtus, when the embryonic layers are still a plastic mass….Imagining the affected cells or groups of cells to lie in the plastic mass like currants in dough, one can see that such a group lying in the region which will afterwards become the back of the neck might be pulled toward the median line when the skin closes over the neural canal, and its individual constituents become scattered along this line as the fœtus elongates… Another group situated over the place where a limb will afterwards bud out, would be stretched along in a line with the budding limb, and the line would tend to follow all the twists and turns of the limb as it grew out, exactly as Kaposi has so graphically described."

During the first half of the past century, however, Blaschko's idea was not understood by others and fell into oblivion. This may in part be explained by the fact that his pioneering atlas carried the misleading title "The nerve distribution in the skin in its relationship to the diseases of the skin" [6]. This wording simply reflected an erroneous belief of those professors who had invited Blaschko to speak at the 7th Congress of the German Dermatological Society held in Breslau in May 1901. As a consequence, only those who opened the book had a chance to become aware of the author's message that there was *no* relationship between the nervous system and the "nevus lines."

For example, Haensch [23] reported in 1961 a case of what is today called "blaschkitis" and argued: "Because of the segmental arrangement, a disorder of the corresponding spinal root must have been present as a factor determining the dermatosis… The configuration of the eczema shows in an unusually clear and pure way the topographically determining role of neural factors in the origin of eczema." And in 1964, the Swiss anatomist Töndury [68] wrote: "Hence we can state that the skin, apart from a short period of segmentation of the corium anlage within the dorsal area of the embryo, is not segmented at any time of its development." If this statement would be true, the pattern of Blaschko's lines would remain unexplained. As a remote echo of such erroneous assumptions, we find even today reports of linear nevi said to be arranged in a "zosteriform" or "dermatomal" pattern [2, 14–16, 72].

In 1969 MacDonald and Sims [48], being unaware of Blaschko's work, studied photographs and descriptions as presented in case reports, as well as 16 cases in their own records and stated that "no association could be found between the lesions and anatomical features of the skin." They concluded that "the epidermis is not…inherently segmental."

Remarkably, however, we find Blaschko's diagram of the "nevus lines" in two Dutch textbooks on skin diseases that appeared during the 1940s [9, 63] and in a German edition of one of these monographs [64].

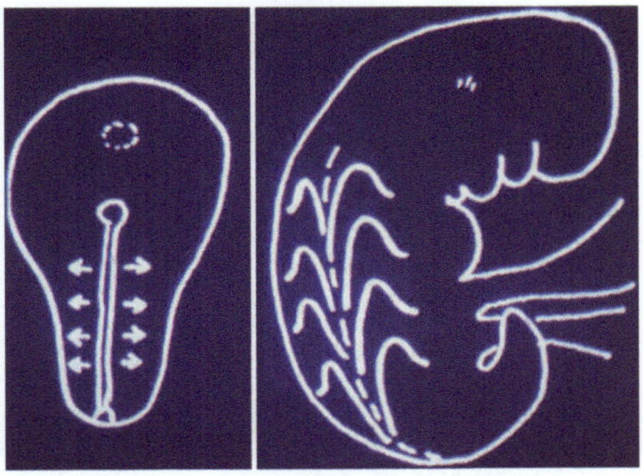

Fig. 5.7 Explanation of the pattern of Blaschko's lines. At an early developmental stage, an admixture of mutant and normal progenitor cells is arranged along the primitive streak. Their transversal proliferation interferes with the longitudinal growth and flexion of the embryo, giving rise to a fountain-like pattern on the back [30] (Reprinted with permission from S. Karger AG, Basel, Switzerland)

In 1965, Curth and Warburton [12] discussed the relationship between the Lyon hypothesis and incontinentia pigmenti. They were unaware of Blaschko's work and stated that "application of the Lyon theory does not explain (1) the usual limitation of the pigment to certain regions, (2) its patterned rather than patchy distribution, and (3) the presence of the same pattern in the few males who manifest the disease."

During the late 1970s Blaschko's lines were rediscovered independently in Canada and Germany. In 1976, Jackson [38] wrote that "the embryological explanation of Blaschko's lines is not at all clear. Other markers in addition to the skin findings are needed to determine the time and the nature of the change responsible for these lines…I have been unable even to make a guess at what stage of development the changes occur which could provide a mechanism by which the localization of Blaschko's lines is determined. It would be helpful to tie in Blaschko's lines with some other dateable embryological event…" At the same time, such explanatory mechanisms and dateable events were postulated in the form of X-chromosome inactivation, early postzygotic mutations, and gametic half-chromatid mutations [25, 26, 28]. This concept is illustrated in Fig. 5.7 [30]. When doing this, I was put on the right track

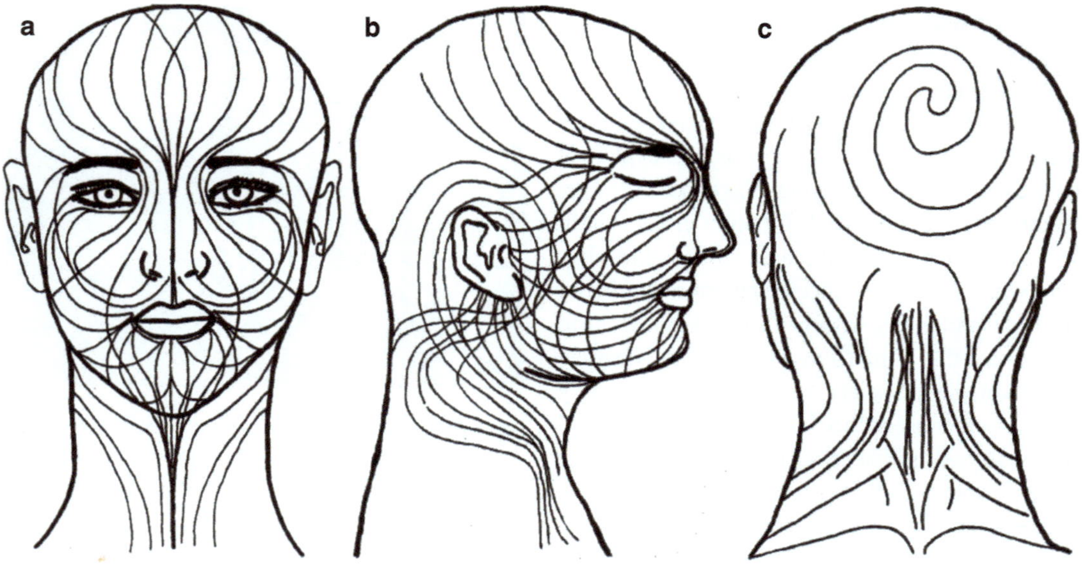

Fig. 5.8 Lines of Blaschko on the head and neck. (**a**) Frontal view; (**b**) lateral view; (**c**) dorsal view [33] (Reprinted with permission from Elsevier Limited, UK)

by the thoughts of the great geneticist Widukind Lenz who, without mentioning Blaschko's lines, had already proposed in 1970 to explain the linear lesions of incontinentia pigmenti by the mechanism of X inactivation, and systematized epidermal nevi by somatic mutations [46].

Because of lack of informative cases, Blaschko had left in his diagram the scalp as a "terra incognita" [6]. On the face and the anterior aspect of the neck, the "nevus lines" were, for similar reasons, drawn in a rather cursory way (Fig. 5.4). One hundred years later, these blank areas were filled in by use of 187 relevant figures collected from the literature (Fig. 5.8a–c) [33]. On the face the lines of Blaschko show a sandglass-like arrangement converging on the nasal root. Remarkably, however, a definite crossing of lines, sometimes even at an angle of 90°, was noted. Apparently, the direction of embryonic movements varies on the head to a large degree. So far it is unclear whether different skin disorders give rise to such different linear patterns.

It has been argued that the lines of Blaschko are of epidermal origin and thus may exclusively reflect genes expressed in keratinocytes or melanocytes [54, 55]. If so, it would be difficult or even impossible to explain the linear patterns of primarily mesodermal disorders such as focal dermal hypoplasia (see Sect. 12.1.2), linear atrophoderma of Moulin [13], or linear progressive fibromatosis [49, 65]. One may resort to the auxiliary hypothesis that these dermal defects occur under the control of the keratinocytes, but it is far more likely that the primordial skin fibroblasts proliferate and migrate in a way similar to that of the epidermal precursor cells. Future research will show which view is correct.

The lines of Blaschko often reflect epigenetic mosaicism. In this form the linear pattern can be inherited from one generation to the next one. A classical example is the Lyon effect of X inactivation giving rise to linear skin lesions in incontinentia pigmenti, focal dermal hypoplasia, Conradi-Hünermann-Happle syndrome, and many other X-linked phenotypes (see Sect. 12.1). On the other hand, familial cases of linear hyperpigmentation suggesting autosomal epigenetic mosaicism have likewise been found. Such cases can be best explained by monoallelic expression of an autosomal pigmentary gene (see Sect. 3.2.1).

The concept that Blaschko's lines reflect the clonal outgrowth of embryonic cells (Fig. 5.7) is today still refused by some authors [20]. By suggesting an analogy between Blaschko's lines and the stripes of zebras or zebra fish, Gilmore [21] assumed that all nevi are secondary to "chemical prepatterns" produced by a "reaction–diffusion process": "The distribution of the nevus over the skin will be independent of the presence of the abnormal clone, but not of the presence of the activated gene of the abnormal clone… Finally, the chemical prepattern may be the trigger for a spatially dependent selective proliferation of one cell type over another. An interesting hypothesis is that the pattern-forming process itself may be pathological, so that in people unaffected by naevoid skin disease the lines of Blaschko *do not exist*" [20]. Conversely, we concur with Alfred Blaschko that his "nevus lines" do likewise exist in normal individuals. The occurrence of acquired linear skin disorders such as linear psoriasis or linear lichen planus provides convincing evidence that Gilmore's hypothesis cannot be true. The lines of Blaschko are invisibly present throughout life in the form of a "spatial prepattern." This view is also supported by the *brindle* patches as noted in American bulldogs (Fig. 5.9).

Coat patterns that are, to some degree, reminiscent of Blaschko's lines have been described in X-linked murine traits [10] and in chimeric mice (Fig. 5.10) [52] or in transgenic animals [44]. On the other hand, the brindle trait of dogs is an exact counterpart of the human lines of Blaschko (see Sect. 3.2.1) [11, 34, 41].

Fig. 5.9 American bulldog with brindle patches, indicating that the lines of Blaschko are invisibly present in the skin of all mammals (Photo by *sannse*, Creative Commons)

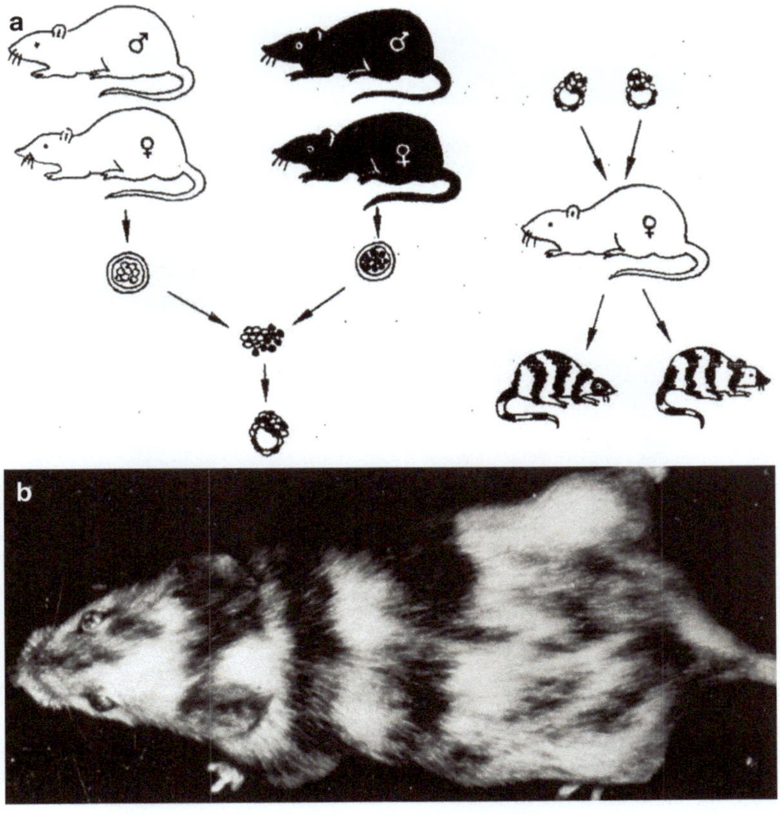

Fig. 5.10 Analogous patterns in chimeric mice. (**a**) Experimental procedure for producing "allophenic"mice according to Mintz [28, 52]; (**b**) chimeric mouse showing systematized light and dark bands [52] (Reprinted with permission obtained from the author)

Theoretically, a systematized pattern of Blaschko's lines may also be explained by a gametic half-chromatid mutation [47], but this mechanism is less likely and has so far not been proven.

For heuristic purposes, two subtypes of Blaschko's lines characterized by either narrow bands ("type 1a") or rather broad bands ("type 1b") have been separated. This somewhat artificial distinction appears to be useful, but at this point in time, we should still be open for other approaches of subclassification.

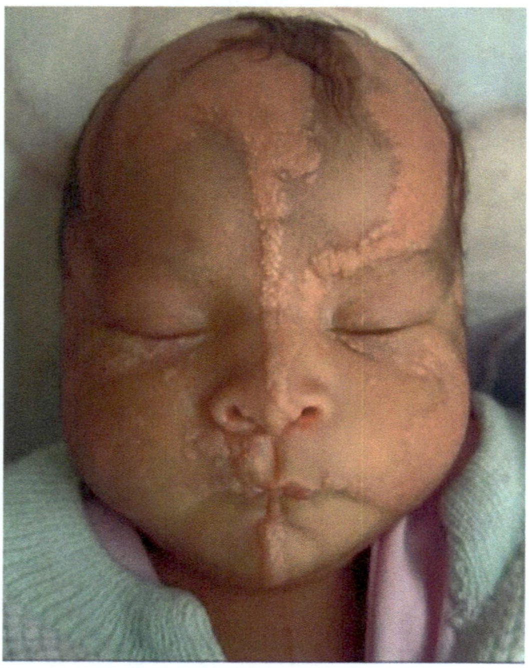

Fig. 5.11 Systematized sebaceous nevus following Blaschko's lines (Courtesy of Dr. Mónica Zambrano, Quito, Ecuador)

5.1.1 Lines of Blaschko, Narrow Bands

This archetypical pattern is followed by many hereditary or nonhereditary skin disorders that are reviewed in detail in Part III. Usually, the pattern of Blaschko's lines is easily discernible (Fig. 5.11) [7]. Sometimes, epidermal nevi being systematized along these lines may also involve the oral mucosa (Fig. 5.12). In a sebaceous nevus, the narrow bands may by chance be contiguous in some regions, giving rise to a large plaque of nonlinear appearance, especially on the scalp (Fig. 5.13).

5.1.2 Lines of Blaschko, Broad Bands

A classical example of this subtype is the pigmentary disorder associated with McCune-Albright syndrome [31]. The segmental light-brown macules are broad to such degree that the linear pattern is sometimes difficult to identify. It seems not justified, however, to say that the pattern of

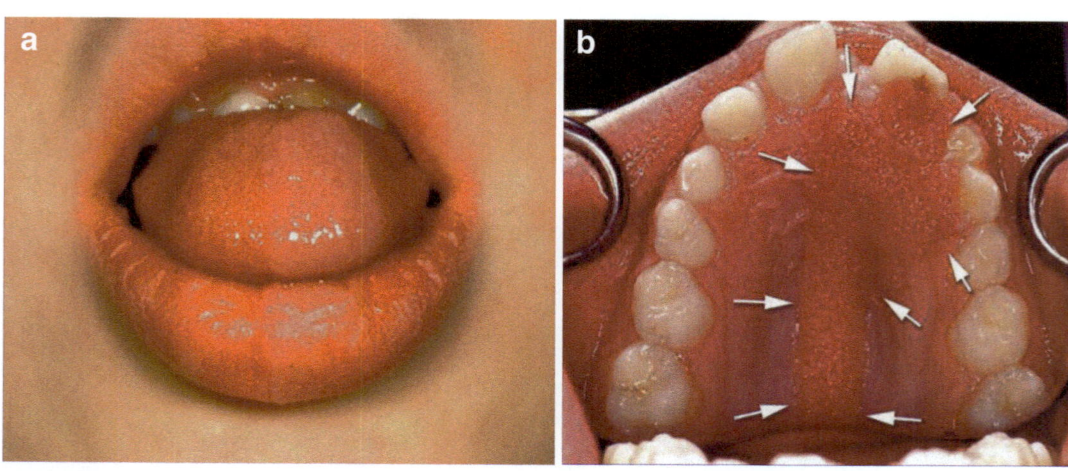

Fig. 5.12 Intraoral lesions following Blaschko's lines in two boys with (**a**) systematized keratinocytic nevus and (**b**) facial nevus sebaceus. *Arrows* indicate the borders of the nevus [70] (**a**: Courtesy of the late Dr. Robert J. Gorlin, Minneapolis, Minnesota, USA; **b**: Reprinted with permission from Elsevier Limited, UK)

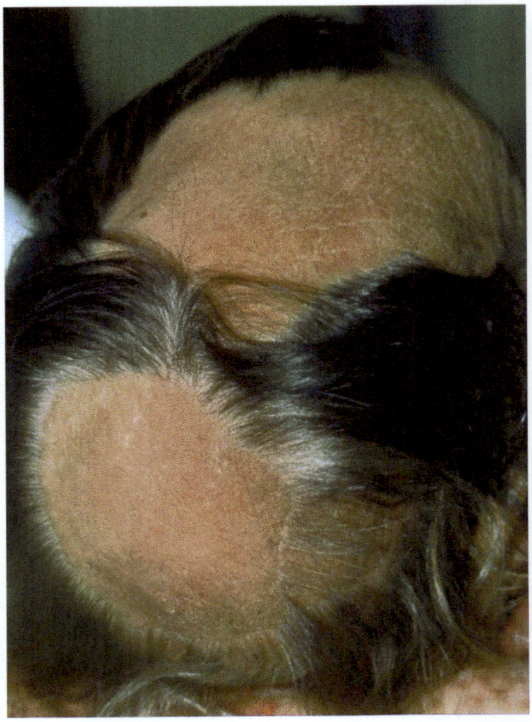

Fig. 5.13 Scalp involvement of a child with nevus sebaceous. In this particular area, the segmental lesions may be broad to such degree that a linear pattern is no longer discernible

Blaschko's lines is only "occasionally observed in the McCune-Albright syndrome" [20]. On the contrary, it should be borne in mind that this archetypical pattern is virtually always present in that disorder.

5.1.3 Analogy of Blaschko's Lines in Other Organs

In focal dermal hypoplasia, the long bones show a peculiar longitudinal striation that can be taken as an extracutaneous manifestation of functional X-chromosome mosaicism (Fig. 5.14) [27]. Similar bone lesions are noted in another X-linked dominant trait, osteopathia striata with cranial sclerosis [40, 57].

A lyonization pattern in the form of longitudinal enamel defects is noted in women heterozygous for amelogenesis imperfecta [71] and in girls affected with focal dermal hypoplasia (Fig. 5.15) [4].

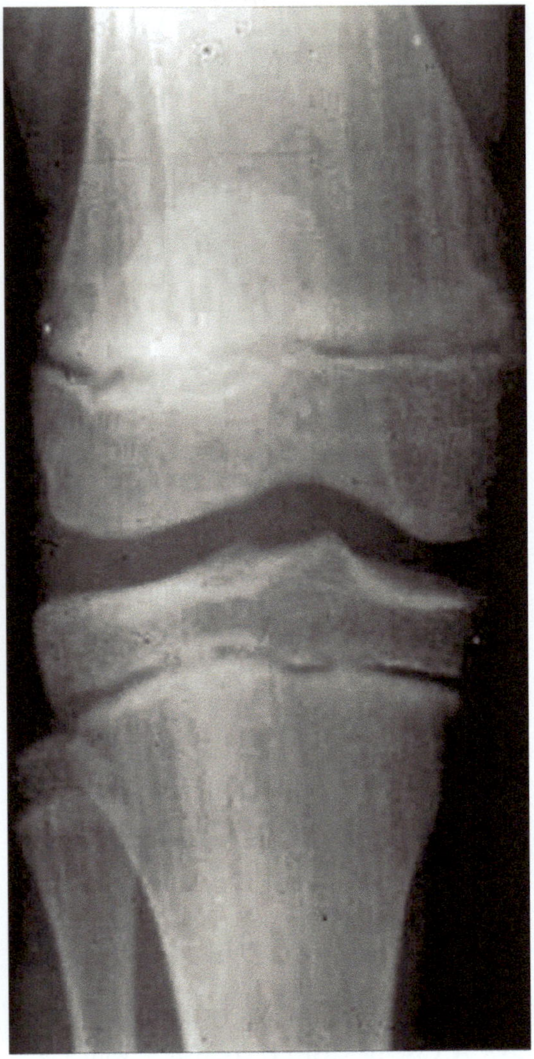

Fig. 5.14 Striation of the long bones in a patient with focal dermal hypoplasia

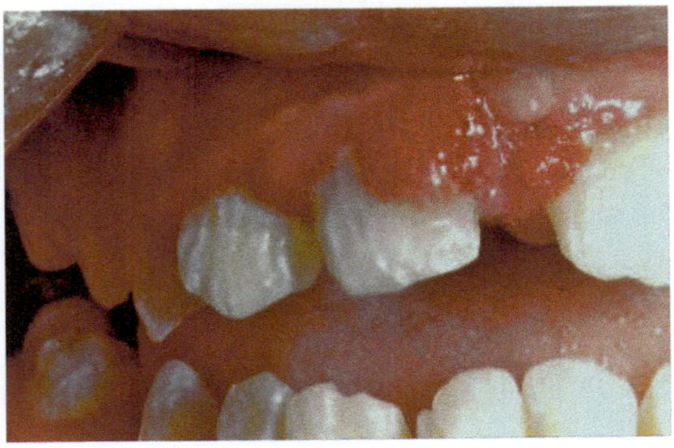

Fig. 5.15 Linear enamel defects reflecting lyonization in focal dermal hypoplasia [4] (Reprinted with permission from Elsevier Limited, UK)

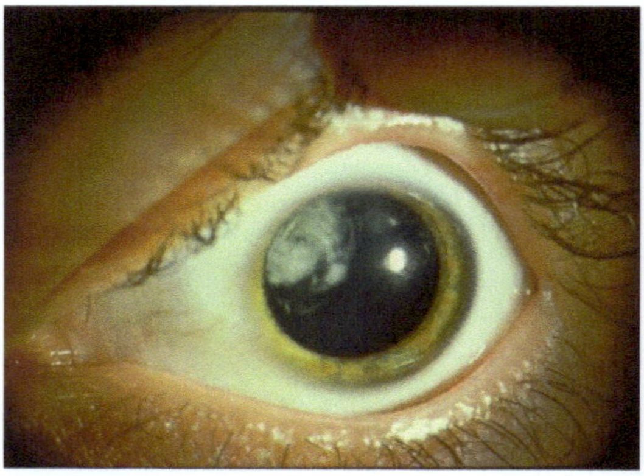

Fig. 5.16 Sectorial cataract reflecting lyonization in Conradi-Hünermann-Happle syndrome [29] (Reprinted with permission from Elsevier Limited, UK)

Fig. 5.17 Radial pattern of retinal hypopigmentation in a female carrier of oculocutaneous albinism (Courtesy of Dr. Hans Dieter Rott, Erlangen, Germany)

Sectorial cataracts reflecting lyonization are noted in Conradi-Hünermann-Happle syndrome (Fig. 5.16) [29] and in women heterozygous for another X-linked trait, Lowe syndrome [43, 58].

In female carriers of X-linked oculocutaneous albinism, the retina shows a radial pattern of hypopigmentation, which has likewise been taken as an ocular analogy of Blaschko's lines (Fig. 5.17). The retina of female carriers for dichromasia shows localized color blindness with a radial arrangement [8]. A similar pattern of the retina has been described in pigmentary mosaicism of the Ito type [59] and, more recently, in cases of grouped congenital hypertrophy of the retinal pigment epithelium (CHRPE) (Fig. 5.18) [51]. A case of CHRPE coexisting with linear hypermelanosis of the ipsilateral arm supports the proposed analogy [50].

5.1.4 Blaschko's Lines in Animals

The lines of Blaschko are an archetypical pattern of cutaneous development noted in many different mammalian species. It has been documented as a hereditary trait called "brindle" in dogs, horses, and cattle including the dwarf zebu (Fig. 5.19a). In horses, the brindle pattern has also been shown to reflect chimerism [24]. It should be borne in mind that the arrangement of these lines is not exactly the same in various mammals. For example, the human Blaschko lines are arranged almost horizontally in the lumbar region, whereas in the dwarf zebu they form an inverted V figure in the analogous region of the body (Fig. 5.19b). In mice, the linear segments may be rather broad (Fig. 5.10b), but animal experiments strongly suggest that this pattern can likewise be taken as an analogy of Blaschko's lines as noted in humans (Fig. 5.20) [39, 44].

5.1.5 Analogy of Blaschko's Lines in the Murine Brain

Tan et al. [66, 67] studied the development of the brain of mice by application of an X-inactivated transgenic marker and demonstrated that the murine neocortex has a radial columnar structure

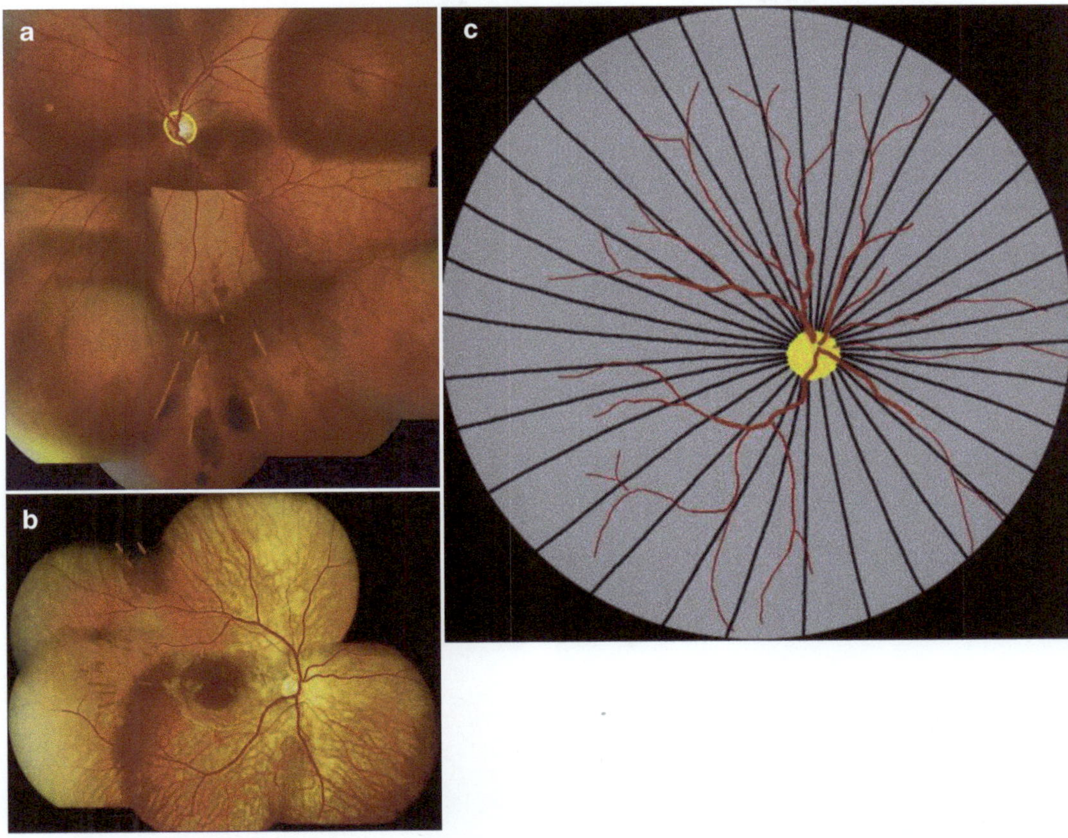

Fig. 5.18 (**a**, **b**) Fundus images of grouped congenital hypertrophy of the retinal pigment epithelium. Note the radial growth pattern; (**c**) final map of developmental lines of the retinal pigmentary epithelium, in analogy to the diagram of Blaschko's lines [51] (Reprinted with permission from Elsevier Limited, UK)

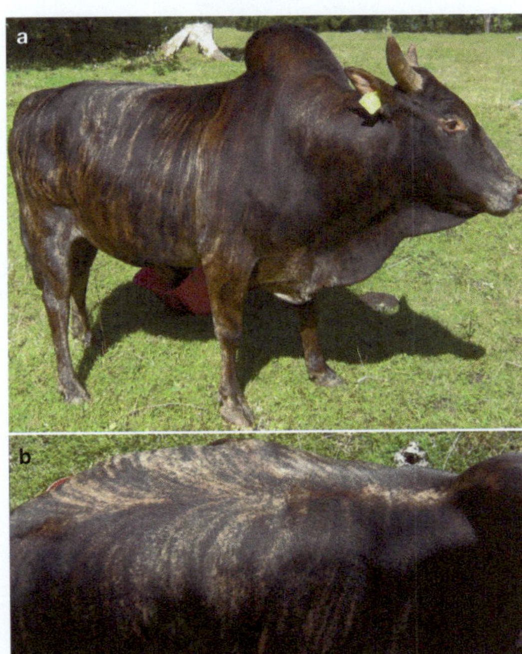

Fig. 5.19 Brindle trait in a dwarf zebu. (**a**) Note analogy to the human Blaschko lines; (**b**) on the rump, the lines form an inverted V figure

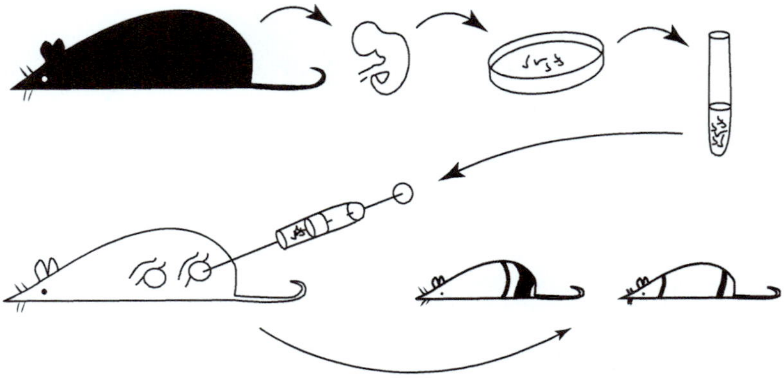

Fig. 5.20 In an experiment reported by Jaenisch [39], cultured neural tube cells obtained from embryos of agouti mice are injected into postimplantation embryos of an albino strain. The injected cells find their way into the host's neural tube to participate in cutaneous embryogenesis, resulting in striated litter

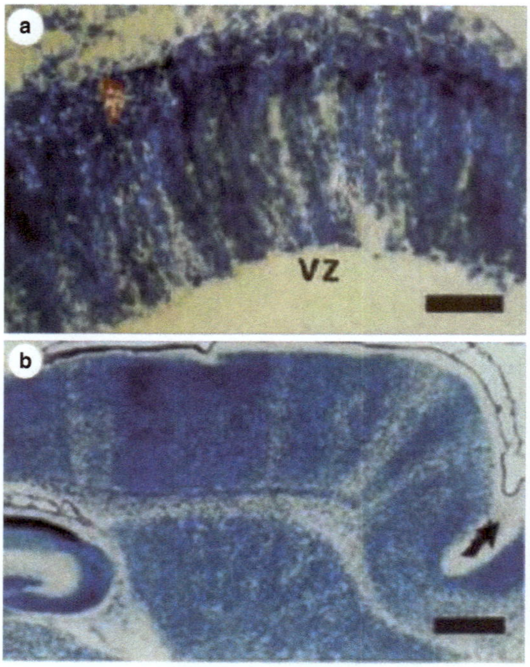

Fig. 5.21 Functional columnar mosaicism in the brain of female mice, visualized by β-galactosidase expression after insertion of an X-inactivated transgenic marker. (**a**) Radial alignment of blue and white columns in the cerebral wall of an embryo; (**b**) similar pattern in the neocortex of an adult female mouse. Many tangentially dispersed cells are likewise noted but do not obscure the radial growth pattern [66] (Reprinted with permission from Nature Publishing Group)

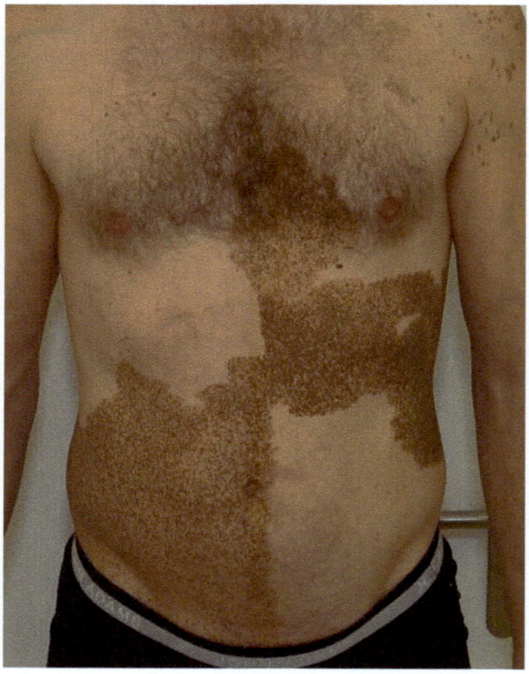

Fig. 5.22 Macular nevus spilus arranged in a checkerboard pattern (Courtesy of Dr. Sabine Wever, Basel, Switzerland)

of randomly alternating color. Although there was a significant contribution of tangentially dispersed cells that did not respect clonal borders, the columnar structure remained clearly discernible (Fig. 5.21).

5.2 Checkerboard Pattern

The word checkerboard is metaphoric and should not be taken literally. This type of mosaic arrangement can also be called "flag-like" or "block-like." It is characterized by alternating squares of aberrant tissue. Examples include mosaic hypomelanosis [3] or hypermelanosis, macular nevus spilus, papular nevus spilus (Fig. 5.22),

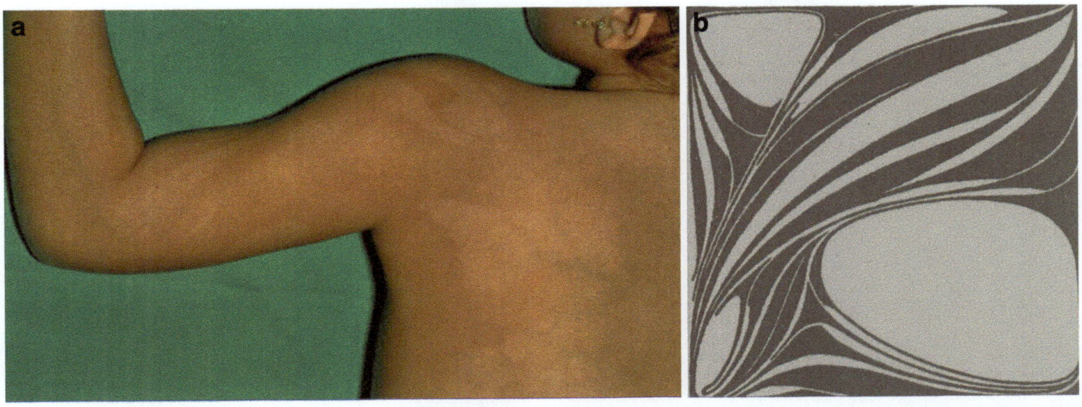

Fig. 5.23 (**a**) Phylloid hypomelanosis in a young girl (Courtesy of Dr. Antonia González-Enseñat, Barcelona, Spain); (**b**) a floral pattern in the art nouveau style

and the lateralized types of capillary nevi such as nevus flammeus and nevus roseus. Such lesions are often reported as showing a "dermatomal" or "zosteriform" arrangement, but so far no example of a nevus following such pattern can be found in the literature. The checkerboard pattern has also been described in cases of primary chimerism [18, 19].

5.3 Phylloid Pattern

The word "phylloid" means leaf-like. This pattern is characterized by leaf-shaped or oblong macules reminiscent of the floral ornaments of a Jugendstil painting (Fig. 5.23). Large pear-shaped macules and lesions resembling the asymmetrical leaves of a begonia may likewise be present.

In some cases a definite midline separation is noted, whereas in other patients the dorsal or ventral midline does not appear to be respected. So far it is not clear whether such differences justify the delineation of different subtypes of the phylloid pattern.

Phylloid hypomelanosis appears to represent a well-defined entity reflecting mosaic trisomy 13q [17, 22, 32] (see Sect. 7.2.2.6). By contrast, phylloid hypermelanosis should, according to present knowledge, not be taken as an entity, but as a cutaneous sign of various states of cytogenetic or molecular mosaicism [35, 37, 56] (see Sect. 7.2.2.7). Future research may show whether some distinct entities can be delineated within this group of disorders.

5.4 Large Patches Without Midline Separation

In contrast to most other mosaic skin disorders, large or giant melanocytic nevi do not respect the dorsal and ventral midline (Fig. 5.24). The term "giant congenital melanocytic nevus" is redundant because giant nevi are always congenital. Mosaicism was quite certain from a clinical point of view [36] and has now been proven at the molecular level [42].

5.5 Lateralization Pattern

The CHILD nevus displays two different patterns of distribution. A unilateral diffuse involvement with a sharp midline demarcation, being particularly strict on the ventral aspect of the trunk, is an almost pathognomonic arrangement (see Sect. 7.3.1.9). The face tends to be spared. On the other hand, the CHILD nevus can be arranged along Blaschko's lines. Such lesions tend to show a marked preponderance on one side of the body.

The two patterns are often intermingled. For example, the unilateral diffuse involvement may be

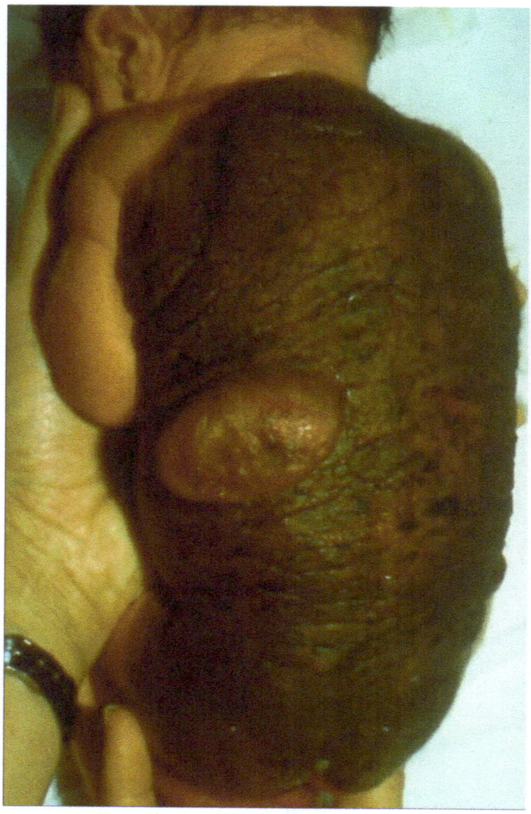

Fig. 5.24 Giant melanocytic nevus in a newborn. The large hypopigmented node was removed and found to be benign

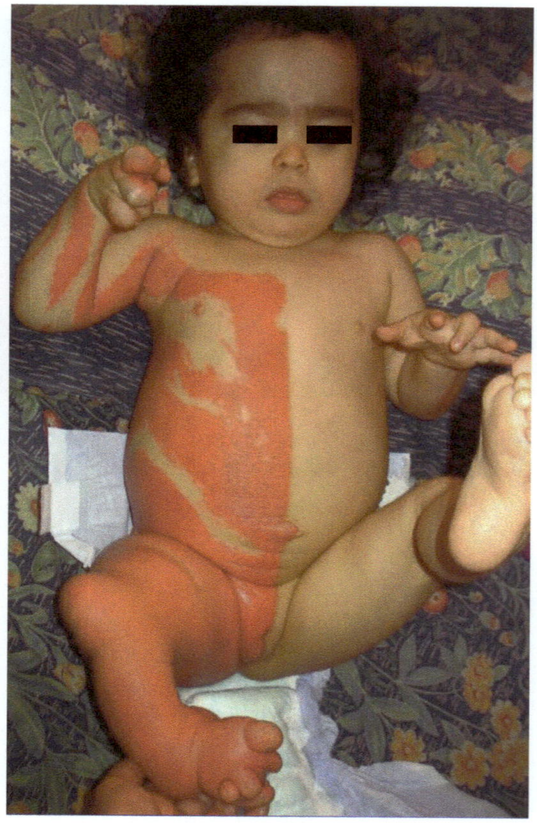

Fig. 5.25 CHILD syndrome. In this girl, the unilateral involvement is interrupted by Blaschko-linear areas of unaffected skin. Note the small linear lesion on the left forearm (Courtesy of Dr. Francis Palisson, Santiago, Chile)

interrupted by several narrow lines of unaffected skin (Fig. 5.25), or additional linear lesions may involve ipsilateral or contralateral areas of the skin.

5.6 Sash-Like Pattern

In two boys affected with cutis tricolor, Ruggieri [60] described a peculiar pigmentary disorder characterized by large oblique hyper- or hypopigmented macules reminiscent of a sash (Fig. 5.26) and by large round or flag-like areas of hyper- or hypopigmention. This pattern that does not respect the dorsal and ventral midline appears to be a characteristic feature of Ruggieri-Happle syndrome [61]. The arrangement differs from all other mosaic patterns so far known.

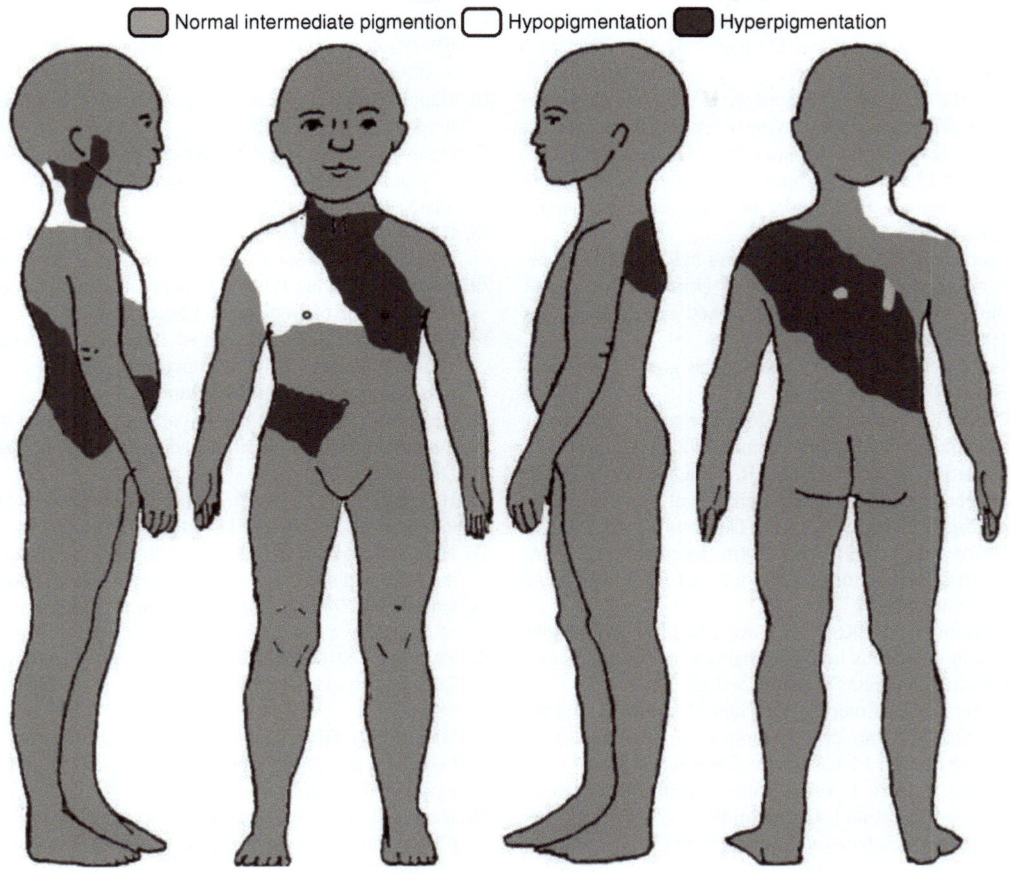

Fig. 5.26 Sash-like pattern as observed in Ruggieri-Happle syndrome [60]

References

1. Alexander A, Blaschko A (1895) Ein Fall von Naevus linearis (Ichthyosis linearis) unius lateris. Dermatol Zeitschr 2:343–361
2. Alfonso-Trujillo I, Arteaga-Hernandez E, Perez-Suarez JC (2009) Eccrine spiradenoma in a zosteriform distribution: presentation of a case. Actas Dermosifiliogr 100:619–620
3. Babilas P, Schreml S, Landthaler M, Vogt T (2009) A 12-month-old boy with impaired pigmentation. Diagnosis: nevus depigmentosus. Pediatr Ann 38:617–621
4. Balmer R, Cameron AC, Ades L, Aldred MJ (2004) Enamel defects and Lyonization in focal dermal hypoplasia. Oral Surg Oral Med Oral Pathol Oral Radiol Endod 98:686–691
5. Blaschko A (1895) Bemerkungen zu vorstehendem Aufsatz [Alexander A (1985) Ein Fall von Naevus linearis (Ichthyosis linearis) unius lateris. Dermatol Zeitschr 2:343–361]. Dermatol Zeitschr 2:361–372
6. Blaschko A (1901) Die Nervenverteilung in der Haut in ihrer Beziehung zu den Erkrankungen der Haut. Beilage zu den Verhandlungen der Deutschen Dermatologischen Gesellschaft, VII. Congress zu Breslau im Mai 1901. Wien und Leipzig, Braumüller

7. Bolognia JL, Orlow SJ, Glick SA (1994) Lines of Blaschko. J Am Acad Dermatol 31:157–190; quiz 190–192
8. Born G, Grützner P, Hemminger H (1976) Evidence for reduced colour vision in carriers of congenital colour vision deficiencies (author's transl). Hum Genet 32:189–196
9. Carol WLL (1944) Leerboek der Huidziekten. Scheltema & Holkema's Boekhandel en Uitgeversmaatschappij, Amsterdam
10. Cattanach BM, Wolfe HG, Lyon MF (1972) A comparative study of the coats of chimaeric mice and those of heterozygotes for X-linked genes. Genet Res 19:213–228
11. Cattanach BM (2005) Finding the gene for brindle. Boxer Ring 2:28–32
12. Curth HO, Warburton D (1965) The genetics of incontinentia pigmenti. Arch Dermatol 92:229–235
13. Danarti R, Bittar M, Happle R, König A (2003) Linear atrophoderma of Moulin: postulation of mosaicism for a predisposing gene. J Am Acad Dermatol 49:492–498
14. Elston DM (1999) Zosteriform distribution of acantholytic dyskeratotic epidermal nevus? J Am Acad Dermatol 40:647
15. Engelman DE, Kotz EA 3rd, Maize JC Sr (2007) Linear cutaneous lupus erythematosus in the lines of Blaschko. Pediatr Dermatol 24:125–129
16. Englander L, Emer JJ, McClain D, Amin B, Turner RB (2011) A rare case of multiple segmental eccrine spiradenomas. J Clin Aesthet Dermatol 4:38–44
17. Faletra F, Berti I, Tommasini A, Pecile V, Cleva L, Alberini E, Bruno I, Gasparini P (2012) Phylloid pattern of hypomelanosis closely related to chromosomal abnormalities in the 13q detected by SNP array analysis. Dermatology 225:294–297
18. Findlay GH, Moores PP (1980) Pigment anomalies of the skin in the human chimaera: their relation to systematized naevi. Br J Dermatol 103:489–498
19. Fitzgerald PH, Donald RA, Kirk RL (1979) A true hermaphrodite dispermic chimera with 46, XX and 46, XY karyotypes. Clin Genet 15:89–96
20. Gilmore S, Maini P (2006) Viewpoint 3 in: what is the biological basis of pattern formation in skin lesions? Exp Dermatol 15:557–559
21. Gilmore SJ (2010) Patterns in naevoid skin disease: development, disease and modelling. Exp Dermatol 19:240–245
22. González-Enseñat MA, Vicente A, Poo P, Catalá V, Pérez-Iribarne MM, Fuster C, Geán E, Happle R (2009) Phylloid hypomelanosis and mosaic partial trisomy 13: two cases that provide further evidence of a distinct clinicogenetic entity. Arch Dermatol 145:576–578
23. Haensch R (1961) Eczema and neural factors. Observations in polyneuroradiculitis. Arch Klin Exp Dermatol 214:35–40
24. Hamilton C (2006) One in a million. Am Quarter Horse J 52–55
25. Happle R (1976) Genetic mechanisms giving rise to linear skin lesions. Joint meeting of the Vereinigung Südwestdeutscher Dermatologen and the Vereinigung Rheinisch-Westfälischer Dermatologen, Heidelberg, 8–10 Oct
26. Happle R (1977) Genetic significance of Blaschko's lines. Z Hautkr 52:935–944
27. Happle R, Lenz W (1977) Striation of bones in focal dermal hypoplasia: manifestation of functional mosaicism? Br J Dermatol 96:133–135
28. Happle R (1978) Genetic interpretation of linear skin abnormalities. Hautarzt 29:357–363
29. Happle R, Küchle HJ (1983) Sectorial cataract: a possible example of lyonisation. Lancet 2:919–920
30. Happle R (1987) The lines of Blaschko: a developmental pattern visualizing functional X-chromosome mosaicism. Curr Probl Dermatol 17:5–18
31. Happle R (1993) Pigmentary patterns associated with human mosaicism: a proposed classification. Eur J Dermatol 3:170–174
32. Happle R (2000) Phylloid hypomelanosis is closely related to mosaic trisomy 13. Eur J Dermatol 10:511–512
33. Happle R, Assim A (2001) The lines of Blaschko on the head and neck. J Am Acad Dermatol 44:612–615
34. Happle R (2002) Transposable elements and the lines of Blaschko: a new perspective. Dermatology 204:4–7
35. Happle R (2010) Phylloid hypermelanosis: an unusual form of pigmentary mosaicism. Dermatology 220: 183–185
36. Happle R (2010) Giant melanocytic nevus may be explained as a superimposed patchy manifestation of a polygenic trait. Dermatology 221:30–33
37. Happle R, Franco-Guío MF, Santacoloma-Osorio G (2012) Phylloid hypermelanosis: a cutaneous marker of several different disorders? Pediatr Dermatol. doi:10.1111/pde.12053
38. Jackson R (1976) The lines of Blaschko: a review and reconsideration: observations of the cause of certain unusual linear conditions of the skin. Br J Dermatol 95:349–360
39. Jaenisch R (1985) Mammalian neural crest cells participate in normal embryonic development on microinjection into post-implantation mouse embryos. Nature 318:181–183
40. Jenkins ZA, van Kogelenberg M, Morgan T, Jeffs A, Fukuzawa R, Pearl E, Thaller C, Hing AV, Porteous ME, Garcia-Minaur S, Bohring A, Lacombe D, Stewart F, Fiskerstrand T, Bindoff L, Berland S, Ades LC, Tchan M, David A, Wilson LC, Hennekam RC, Donnai D, Mansour S, Cormier-Daire V, Robertson SP (2009) Germline mutations in *WTX* cause a sclerosing skeletal dysplasia but do not predispose to tumorigenesis. Nat Genet 41:95–100
41. Kerns JA, Cargill EJ, Clark LA, Candille SI, Berryere TG, Olivier M, Lust G, Todhunter RJ, Schmutz SM, Murphy KE, Barsh GS (2007) Linkage and segregation analysis of black and brindle coat color in domestic dogs. Genetics 176:1679–1689
42. Kinsler VA, Thomas AC, Ishida M, Bulstrode NW, Loughlin S, Hing S, Chalker J, McKenzie K, Abu-mero S,

Slater O, Chanudet E, Palmer R, Morrogh D, Stanier P, Healy E, Sebire NJ, Moore GE (2013) Multiple congenital melanocytic nevi and neurocutaneous melanosis are caused by post-zygotic mutations in codon 61 of *NRAS*. J Invest Dermatol. doi:10.1038/jid.2013.70
43. Koniszewski G, Rott HD (1985) The Lyon effect of the lens: findings in the carriers of X chromosome-linked cataract and in Lowe syndrome. Klin Monbl Augenheilkd 187:525–528
44. Kucera GT, Bortner DM, Rosenberg MP (1996) Overexpression of an *Agouti* cDNA in the skin of transgenic mice recapitulates dominant coat color phenotypes of spontaneous mutants. Dev Biol 173:162–173
45. Lapunzina P, Gairí A, Delicado A, Mori MA, Torres ML, Goma A, Navia M, Pajares IL (2004) Macrocephaly-cutis marmorata telangiectatica congenita: report of six new patients and a review. Am J Med Genet A 130A:45–51
46. Lenz W (1970) Medizinische Genetik: Grundlagen, Ergebnisse und Probleme, 2nd edn. Thieme, Stuttgart
47. Lenz W (1975) Half chromatid mutations may explain incontinentia pigmenti in males. Am J Hum Genet 27:690–691
48. Macdonald RH, Sims RT (1969) Linear lesions. Br J Dermatol 81:72–79
49. Mascaró JM, Torres V, Mascaro-Galy C, Botella R (1976) Un cas de fibromatose juvénile linéaire à tendance progressive et ulcéreuse. Bull Soc Fr Dermatol Syph 83:278–279
50. Meyer CH, Freyschmidt-Paul P, Happle R, Kroll P (2004) Unilateral linear hyperpigmentation of the skin with ipsilateral sectorial hyperpigmentation of the retina. Am J Med Genet A 126A:89–92
51. Meyer CH, Rodrigues EB, Mennel S, Schmidt JC, Kroll P (2005) Grouped congenital hypertrophy of the retinal pigment epithelium follows developmental patterns of pigmentary mosaicism. Ophthalmology 112:841–847
52. Mintz B (1967) Gene control of mammalian pigmentary differentiation. I. Clonal origin of melanocytes. Proc Natl Acad Sci U S A 58:344–351
53. Montgomery DW (1901) The cause of the streaks in naevus linearis. J Cutan Genitourin Dis 19:455–464
54. Moss C, Savin J (1995) Dermatology and the new genetics. Blackwell Science Ltd, Osney Mead, Oxford
55. Moss C (1999) Cytogenetic and molecular evidence for cutaneous mosaicism: the ectodermal origin of Blaschko lines. Am J Med Genet 85:330–333
56. Oiso N, Tsuruta D, Imanishi H, Sayasa H, Narita T, Kobayashi H, Ikegami H, Kawada A (2010) Phylloid hypermelanosis and melanocytic nevi with aggregated and disfigured melanosomes: causal relationship between phylloid pigment distribution and chromosome 13 abnormalities. Dermatology 220:169–172
57. Perdu B, de Freitas F, Frints SG, Schouten M, Schrander-Stumpel C, Barbosa M, Pinto-Basto J, Reis-Lima M, de Vernejoul MC, Becker K, Freckmann ML, Keymolen K, Haan E, Savarirayan R, Koenig R, Zabel B, Vanhoenacker FM, Van Hul W (2010) Osteopathia striata with cranial sclerosis due to WTX gene defect. J Bone Miner Res 25:82–90
58. Rott HD, Koniszewski G (1987) Analogy of Blaschko lines in the eye. J Genet Hum 35:19–27
59. Rott HD, Lang GE, Huk W, Pfeiffer RA (1990) Hypomelanosis of Ito (incontinentia pigmenti achromians). Ophthalmological evidence for somatic mosaicism. Ophthalmic Paediatr Genet 11:273–279
60. Ruggieri M (2000) Cutis tricolor: congenital hyper- and hypopigmented lesions in a background of normal skin with and without associated systemic features: further expansion of the phenotype. Eur J Pediatr 159:745–749
61. Ruggieri M, Roggini M, Kennerknecht I, Polizzi A, Distefano A, Pavone V (2011) Spectrum of skeletal abnormalities in a complex malformation syndrome with "cutis tricolor" (Ruggieri-Happle syndrome). Acta Paediatr 100:121–127
62. Siegel DH (2008) Cutaneous mosaicism: a molecular and clinical review. Adv Dermatol 24:223–244
63. Siemens HW (1948) Algemene Dermatologie: Diagnostiek en Therapie. Scheltema & Holtema's Boekhandel en Uitgeversmatschappij, Amsterdam
64. Siemens HW (1952) Allgemeine Diagnostik und Therapie der Hautkrankheiten: als Einführung in die Dermatologie für Studierende und Praktiker. Springer, Berlin
65. Stevanovic D (1977) Multiple, continuous and progressive fibromatosis (author's transl). Ann Dermatol Venereol 104:141–146
66. Tan SS, Breen S (1993) Radial mosaicism and tangential cell dispersion both contribute to mouse neocortical development. Nature 362:638–640
67. Tan SS, Faulkner-Jones B, Breen SJ, Walsh M, Bertram JF, Reese BE (1995) Cell dispersion patterns in different cortical regions studied with an X-inactivated transgenic marker. Development 121:1029–1039
68. Töndury G (1964) Embryologie und Hauttopographie. Arch Klin Exp Dermatol 219:12–24
69. Torrelo A, Baselga E, Nagore E, Zambrano A, Happle R (2005) Delineation of the various shapes and patterns of nevi. Eur J Dermatol 15:439–450
70. Warnke PH, Russo PA, Schimmelpenning GW, Happle R, Harle F, Hauschild A, Sherry E, Luttges J, Terheyden H, Dunsche A, Springer IN (2005) Linear intraoral lesions in the sebaceous nevus syndrome. J Am Acad Dermatol 52:62–64
71. Witkop CJ Jr (1967) Partial expression of sex-linked recessive amelogenesis imperfecta in females compatible with the Lyon hypothesis. Oral Surg Oral Med Oral Pathol 23:174–182
72. Yoshida A, Takahashi K, Maeda F, Akasaka T (2010) Multiple vascular eccrine spiradenomas: a case report and published work review of multiple eccrine spiradenomas. J Dermatol 37:990–994

6 Less Well Defined or So Far Unclassifiable Patterns

Contents

The list of different mosaic patterns as noted in human skin will probably increase in the years to come. In a girl with neurological abnormalities including mental deficiency and epileptic seizures, Ruggieri et al. [10] described cutis tricolor in the form of diffuse hypopigmentation of the trunk leaving a small area of normal skin on the left lateral aspect of the thorax and a segmental hyperpigmentation involving both arms and the trunk, apparently with a midline separation. It is so far unclear whether this arrangement represents a further distinct pattern of pigmentary mosaicism.

6.1 The Pallister-Killian Pattern

Children with Pallister-Killian syndrome, a sporadic phenotype reflecting mosaic tetrasomy 12p (isochromosome 12p syndrome), tend to display streaks and spots of hypopigmentation. In part, these macules are not compatible with the pattern of Blaschko's lines (Fig. 6.1a), whereas other lesions appear to correspond to "atypical" Blaschko lines (Fig. 6.1b) [2, 5, 11–13]. Hyperpigmented macules have also been reported [3, 11]. A comprehensive iconography of the pigmentary anomalies of Pallister-Killian syndrome was published by Wilkens et al. [13].

R. Happle, *Mosaicism in Human Skin*,
DOI 10.1007/978-3-642-38765-4_6, © Springer-Verlag Berlin Heidelberg 2014

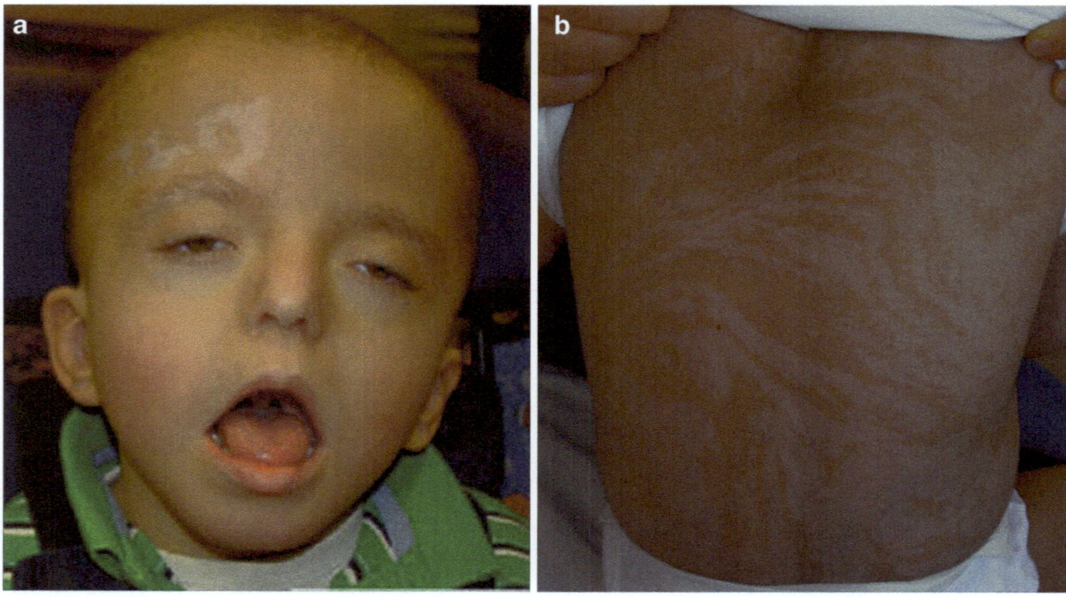

Fig. 6.1 Mosaic hypopigmentation in Pallister-Killian syndrome. (**a**) A facial pattern being incompatible with Blaschko's lines; (**b**) "atypical" Blaschko lines showing a pronounced deviation from the midline (**a**: Reprinted with permission from John Wiley and Sons, USA; **b**: Courtesy of Dr. Hülya Kayserili, Istanbul, Turkey)

6.2 The Mesotropic Facial Pattern

A mesotropic port-wine patch involving the median area of the face is a characteristic feature of macrocephaly-livedo reticularis congenita syndrome ("macrocephaly-capillary malformation") (Fig. 6.2) (see Sect. 7.4.1.5). This pattern has been documented in numerous cases [1, 6–8] and can thus be taken as well defined, but it is so far uncertain whether this capillary malformation reflects mosaicism. One might argue that the spectrum of all other lesions of "macrocephaly-capillary malformation," including livedo reticularis congenita, can best be explained by the action of a lethal mutation surviving by mosaicism [4, 9], but it seems prudent to wait for additional molecular data until the striking affinity of this particular port-wine patch to the median area of the face can be taken as a new mosaic pattern. In other words, until now we cannot be sure that this unusual skin lesion represents a nevus (see Chap. 7).

The mesotropic port-wine patch preponderantly involves the philtrum and the upper lip, but the forehead and the lower lip may likewise be

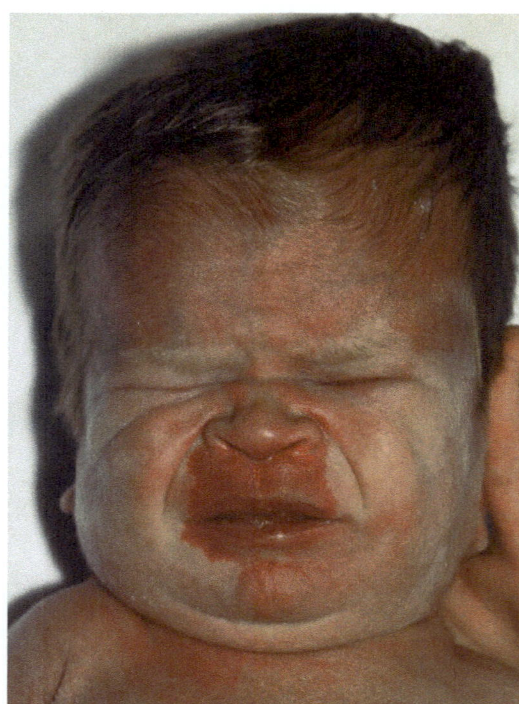

Fig. 6.2 Mesotropic port-wine patch as noted in macrocephaly-livedo reticularis congenita ("macrocephaly-capillary malformation") syndrome

affected [1]. Contrasting with lateralized nevi flammei, this vascular disorder tends to fade during the first years of life [1, 6]. On the other hand, the disorder should be distinguished from the facial salmon patch. This common pale-pink non-nevus is likewise mesotropic and tends to fade during childhood (see Sect. 12.2.3.1).

References

1. Clayton-Smith J, Kerr B, Brunner H, Tranebjaerg L, Magee A, Hennekam RC, Mueller RF, Brueton L, Super M, Steen-Johnsen J, Donnai D (1997) Macrocephaly with cutis marmorata, haemangioma and syndactyly–a distinctive overgrowth syndrome. Clin Dysmorphol 6:291–302
2. Cormier-Daire V, Le Merrer M, Gigarel N, Morichon N, Prieur M, Lyonnet S, Vekemans M, Munnich A (1997) Prezygotic origin of the isochromosome 12p in Pallister-Killian syndrome. Am J Med Genet 69:166–168
3. Geneviève D, Cormier-Daire V, Sanlaville D, Faivre L, Gosset P, Allart L, Picq M, Munnich A, Romana S, de Blois M, Vekemans M (2003) Mild phenotype in a 15-year-old boy with Pallister-Killian syndrome. Am J Med Genet A 116A:90–93
4. Happle R (1990) Nichterbliche Genodermatosen. Hautarzt 41:104–109
5. Horn D, Majewski F, Hildebrandt B, Körner H (1995) Pallister-Killian syndrome: normal karyotype in prenatal chorionic villi, in postnatal lymphocytes, and in slowly growing epidermal cells, but mosaic tetrasomy 12p in skin fibroblasts. J Med Genet 32:68–71
6. Lapunzina P, Gairi A, Delicado A, Mori MA, Torres ML, Goma A, Navia M, Pajares IL (2004) Macrocephaly-cutis marmorata telangiectatica congenita: report of six new patients and a review. Am J Med Genet A 130A:45–51
7. Martínez-Glez V, Romanelli V, Mori MA, Gracia R, Segovia M, González-Meneses A, López-Gutierrez JC, Gean E, Martorell L, Lapunzina P (2010) Macrocephaly-capillary malformation: analysis of 13 patients and review of the diagnostic criteria. Am J Med Genet A 152A:3101–3106
8. Moore CA, Toriello HV, Abuelo DN, Bull MJ, Curry CJ, Hall BD, Higgins JV, Stevens CA, Twersky S, Weksberg R, Dobyns WB (1997) Macrocephaly-cutis marmorata telangiectatica congenita: a distinct disorder with developmental delay and connective tissue abnormalities. Am J Med Genet 70:67–73
9. Riviere JB, Mirzaa GM, O'Roak BJ, Beddaoui M, Alcantara D, Conway RL, St-Onge J, Schwartzentruber JA, Gripp KW, Nikkel SM, Worthylake T, Sullivan CT, Ward TR, Butler HE, Kramer NA, Albrecht B, Armour CM, Armstrong L, Caluseriu O, Cytrynbaum C, Drolet BA, Innes AM, Lauzon JL, Lin AE, Mancini GM, Meschino WS, Reggin JD, Saggar AK, Lerman-Sagie T, Uyanik G, Weksberg R, Zirn B, Beaulieu CL, Majewski J, Bulman DE, O'Driscoll M, Shendure J, Graham JM Jr, Boycott KM, Dobyns WB (2012) De novo germline and postzygotic mutations in *AKT3*, *PIK3R2* and *PIK3CA* cause a spectrum of related megalencephaly syndromes. Nat Genet 44: 934–940
10. Ruggieri M, Iannetti P, Pavone L (2003) Delineation of a newly recognized neurocutaneous malformation syndrome with "cutis tricolor". Am J Med Genet A 120A:110–116
11. Schaefer GB, Jochar A, Muneer R, Sanger WG (1997) Clinical variability of tetrasomy 12p. Clin Genet 51:102–108
12. Schinzel A (1991) Tetrasomy 12p (Pallister-Killian syndrome). J Med Genet 28:122–125
13. Wilkens A, Liu H, Park K, Campbell LB, Jackson M, Kostanecka A, Pipan M, Izumi K, Pallister P, Krantz ID (2012) Novel clinical manifestations in Pallister-Killian syndrome: comprehensive evaluation of 59 affected individuals and review of previously reported cases. Am J Med Genet A 158A:3002–3017

Part III

Mosaic Skin Disorders

7 Nevi

Contents

R. Happle, *Mosaicism in Human Skin*,
DOI 10.1007/978-3-642-38765-4_7, © Springer-Verlag Berlin Heidelberg 2014

The concept of mosaicism has helped develop a reasonable definition of the term nevus. In 1995 the following criteria were proposed: "Nevi are visible, circumscribed, long-lasting lesions of the skin or the neighboring mucosa, reflecting mosaicism. With the exception of melanocytic nevi, they do not show neoplastic growth. They never show malignant neoplasia" [82]. Recent molecular research has shown that this is a workable definition.

The statement that nevi "never show malignant neoplasia" needs a clarifying comment. Of course, some nevi may show secondary malignant growth, but these superimposed tumors do no longer represent nevi.

The new criteria were developed from an earlier tentative definition as proposed by Hermann Pinkus in 1965: "The definition of the two forms of nevi can approximately be given in the following way: (1) The nevus cell nevus, or cellular nevus, is a benign neoplasia of the skin, consisting of specific tumor cells, the nevus cells. (2) Otherwise we can designate as nevi all those circumscribed malformations of the skin that are characterized by a surplus (or, occasionally, a deficit) of one or several mature tissue components, and are relatively stable" [181]. A disadvantage of this definition was that it could be applied to all developmental anomalies of the skin, including supernumerary nipples, rudimentary polydactyly, preauricular tags, and palmoplantar keratoderma. For obvious reasons, such lesions do not represent nevi, which is why they are excluded by the new definition.

7.1 The Theory of Lethal Genes Surviving by Mosaicism

Most nevi reflect the action of a lethal gene that can only survive in proximity to a population of wild-type cells [75]. There are, however, many important exceptions from this rule such as the epidermolytic type of epidermal nevus being caused by nonlethal postzygotic *KRT1* or *KRT10* mutations (see Sect. 7.3.1.7).

7.2 Pigmentary Nevi

This group comprises melanocytic nevi and other nevi reflecting pigmentary mosaicism.

7.2.1 Melanocytic Nevi

Melanocytic nevi represent the only type of nevi showing neoplastic proliferation [181]. They may be present at birth or develop throughout life.

7.2.1.1 Common Small Melanocytic Nevus

Both congenital and acquired small melanocytic nevi have a polygenic background, which means that they are not inherited as a Mendelian trait. In common acquired melanocytic nevi, events of LOH involving various gene loci including *BRAF*, *NRAS*, *BRCA1*, *MC1R*, and *IRF4* were documented [174, 219], providing evidence that these skin lesions reflect mosaicism [157, 194, 217].

Some unusual cases of segmental arrangement of multiple small melanocytic nevi, suggesting an early postzygotic mutational event, have been reported (Fig. 7.1) [47, 227].

7.2.1.2 Common Atypical Melanocytic Nevus

In atypical nevi ("dysplastic nevi"), the number of LOH events appears to be even higher [127, 128, 219]. In the past, a so-called dysplastic nevus

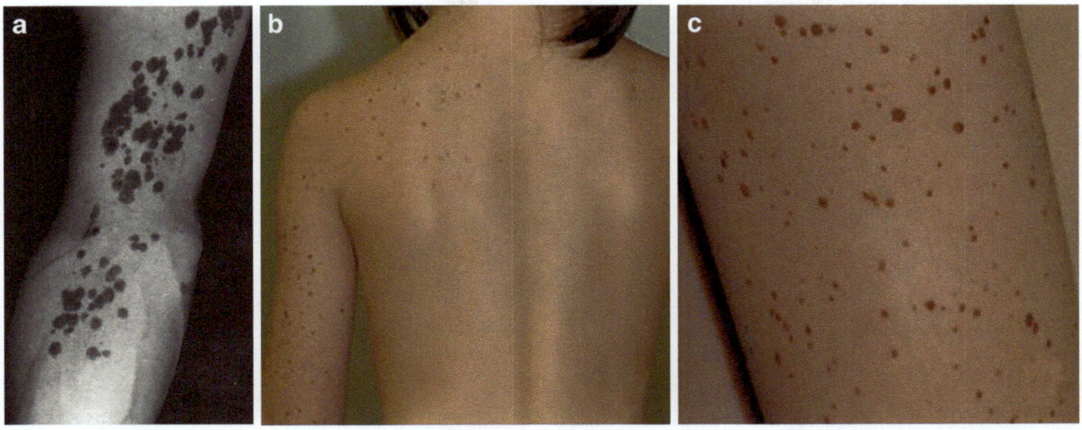

Fig. 7.1 Segmentally arranged small melanocytic nevi. (**a**) Congenital lesions in a 28-year-old man [47]; (**b**, **c**) acquired lesions in a 7-year-old girl [227] (**a**: Reprinted with permission from John Wiley & Sons, USA; **b**, **c**: Reprinted with permission from S. Karger AG, Basel, Switzerland)

syndrome was described and said to be inherited as a Mendelian trait [48]. Other authors, however, argued that dysplastic nevi are a very common polygenic trait and do not constitute a distinct hereditary "syndrome" [78, 119, 218]. Recent molecular research has confirmed this alternative view.

Sometimes the atypical melanocytic nevi may occur in a segmental distribution, representing a particular form of mosaicism [206].

7.2.1.3 Large Congenital Melanocytic Nevus

Large congenital melanocytic nevi including giant lesions usually occur sporadically. The assumption of a very early postzygotic mutation, surviving by mosaicism [77], was supported by several reports on nevus cell aggregates involving the placenta of mothers giving birth to a child with a giant melanocytic nevus [10]. In 2013, Kinsler et al. [136] found that most cases of multiple congenital nevi and neuromelanosis are caused by heterozygosity for a postzygotic *NRAS* codon 61 mutation originating from one single progenitor cell. If these results can be confirmed in larger series, the occurrence of extensive or giant congenital melanocytic nevi may be categorized as a monogenic disorder.

As a consequence, the hypothesis that large congenital melanocytic nevi may be explained as a superimposed mosaic manifestation of a polygenic trait [100] may today either be taken as refuted or it may only be valid for a rather small proportion of patients. In this minority of cases, a polygenic background would explain why the affected individuals often show, in addition, multiple small nevi scattered over their entire body (Fig. 7.2) [57, 192]. At an early developmental stage, LOH would occur at one of the predisposing gene loci, or a postzygotic new mutation would cause heterozygosity at an additional predisposing locus. Future research may show whether this etiological concept can be upheld at least for some cases of giant melanocytic nevi. It would be compatible with the description of several major genes causing large or giant melanocytic nevi, such as *SOX10* [201] or *MC1R* [135].

An unusual linear arrangement of large melanocytic nevi, being superimposed on common acquired small nevi, has also been reported [43].

7.2.1.4 Spitz Nevus

Spitz nevi do apparently not develop from *BRAF* mutations, whereas *HRAS* mutations are often found in these melanocytic lesions [193, 220]. Such molecular findings may be related to the fact that Spitz nevi do usually not develop malignant growth.

Spitz nevi per se may be agminated but do not show a segmental arrangement. However, both macular and papular nevi spili are sometimes found to be covered with multiple Spitz nevi (see Sects. 7.2.1.6 and 7.2.1.7). Moreover, Menni et al. [167] have reported a case of flag-like hypomelanosis covered with Spitz nevi (Fig. 7.3).

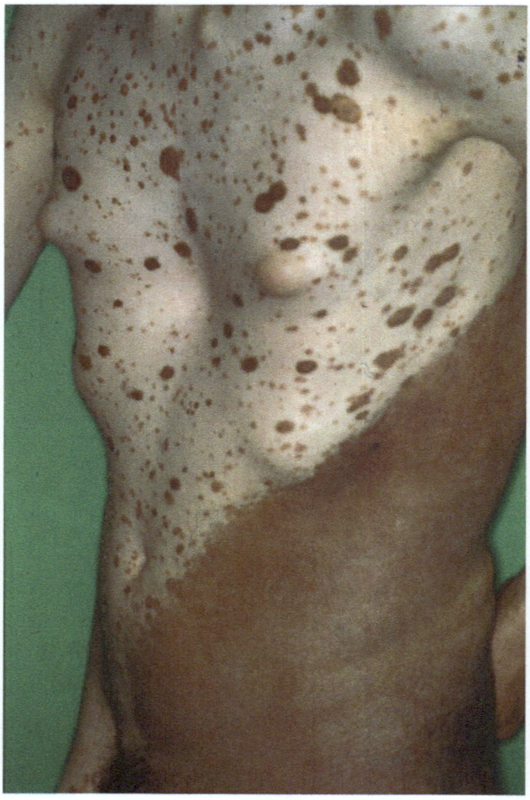

Fig. 7.2 Giant melanocytic nevus associated with multiple small melanocytic nevi, being partly congenital and partly acquired (Courtesy of Dr. Uwe Töllner, Fulda, Germany). Perhaps, the disseminated lesions are not "satellite nevi" but reflect a pronounced polygenic predisposition. According to this theory, the giant nevus would originate from an additional postzygotic mutation that occurred at an early developmental stage

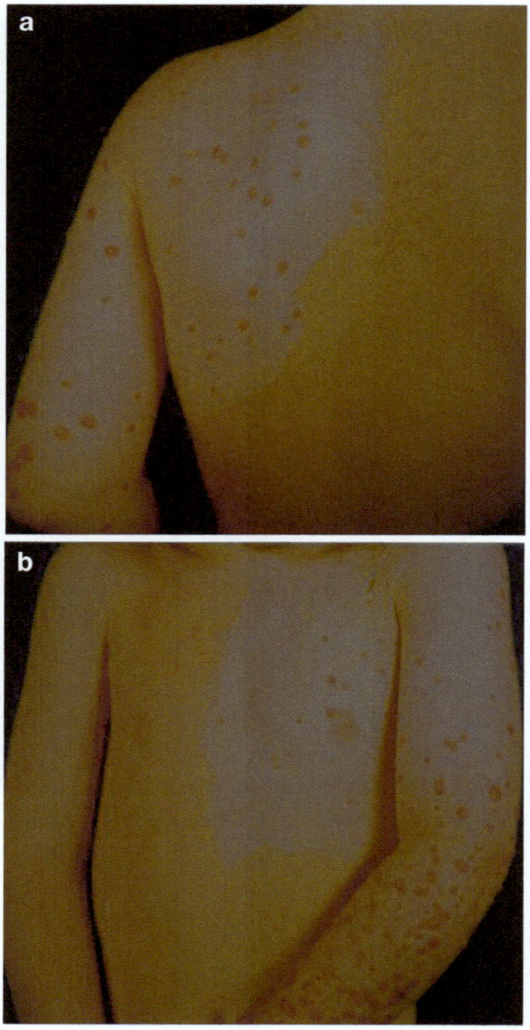

Fig. 7.3 (a, b) Unilateral flag-like nevus achromicus covered with multiple congenital Spitz nevi [167] (Reprinted with permission from John Wiley & Sons, USA)

7.2.1.5 Cellular Blue Nevus

In cellular blue nevi, the spectrum of multistep postzygotic mutations appears to differ from that observed in other melanocytic nevi [196]. A rather characteristic finding appears to be the presence of *GNAG* mutations [193].

7.2.1.6 Papular Nevus Spilus

Papular nevus spilus (speckled lentiginous nevus of the papular type) represents a light-brown macule containing multiple speckles in the form of dark brown papules, in addition to small macules (Fig. 7.4). These speckles are distributed in an uneven way, reminiscent of a star map. On histopathological examination, the papules are found to be melanocytic nevi of a dermal or compound type [97]. This nevus is caused by a postzygotic *HRAS* mutation [64].

Papular nevus spilus syndrome: Large papular nevi spili are arranged in a flag-like pattern and may be associated with neurologic defects that are usually localized ipsilaterally and include hyperhidrosis, muscular weakness, dysesthesia, or thinning of a nerve [88, 184]. In the past this phenotype has been called "speckled lentiginous nevus syndrome," but this term is ambiguous because two different types of SLN

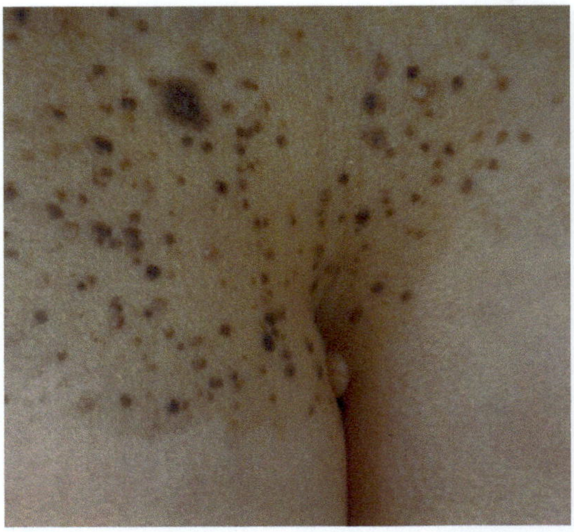

Fig. 7.4 Papular nevus spilus with speckles showing a characteristic star-map pattern

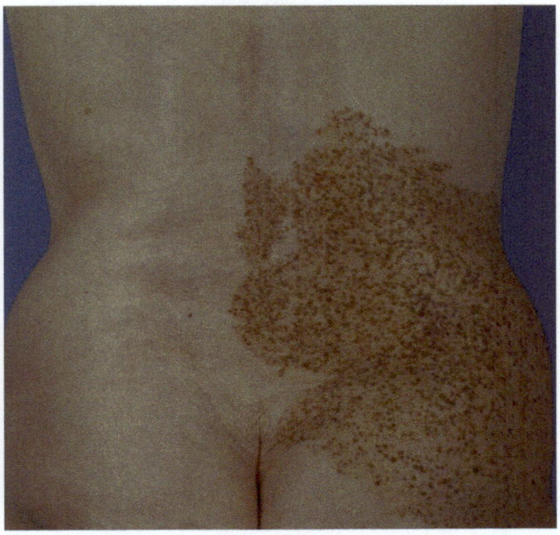

Fig. 7.5 Macular nevus spilus with speckles showing a characteristic polka-dot distribution (Courtesy of Dr. Alexander Stella, Vienna, Austria)

syndromes exist. The designation "papular nevus spilus syndrome" seems more appropriate. The disorder may also occur as a component of phacomatosis pigmentokeratotica that represents a variant Schimmelpenning syndrome (see Sect. 9.1).

7.2.1.7 Macular Nevus Spilus

This disorder is characterized by a light-brown macule containing multiple, small, completely flat, dark speckles that are rather evenly distributed, resembling a polka-dot pattern (Fig. 7.5) [97, 221]. Histopathologically, increased numbers of melanocytes are confined to the elongated rete ridges. At the tips of these structures, the melanocytes tend to form nests at the dermoepidermal junction, giving rise to a so-called "jentigo" pattern [158].

Macular nevus spilus sometimes occurs in combination with nevus roseus, giving rise to a peculiar disorder called "phacomatosis spilorosea" (see Sect. 9.5.2).

7.2.1.8 Linear Lentiginous Nevus

This unusual birthmark is characterized by multiple, tightly packed lentigines arranged along Blaschko's lines. As a distinguishing feature, these speckles do not show any abnormal background in the form of hyper- or hypopigmentation (Fig. 7.6) [114].

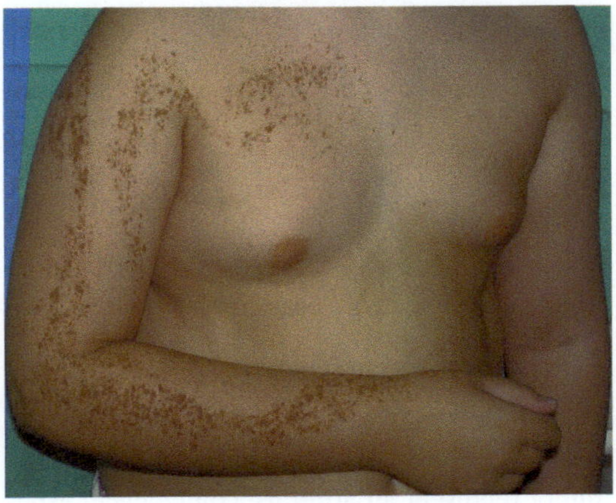

Fig. 7.6 Naevus lentiginosus linearis [114] (Reprinted with permission from the Society for Publication of Acta Dermato-Venereologica, Stockholm, Sweden)

7.2.1.9 Nevus Cesius (Segmental Dermal Melanocytosis)

This lesion (Fig. 7.7) is also called nevus fuscocoeruleus or aberrant Mongolian spot [137]. It should not be confused with the median lumbosacral blue spots being present in many neonates of indigenous populations of many parts of the world. The localization of nevus cesius in the ophthalmo-maxillary region has been called "nevus of Ota," and in the shoulder region some authors use the name "nevus of Ito."

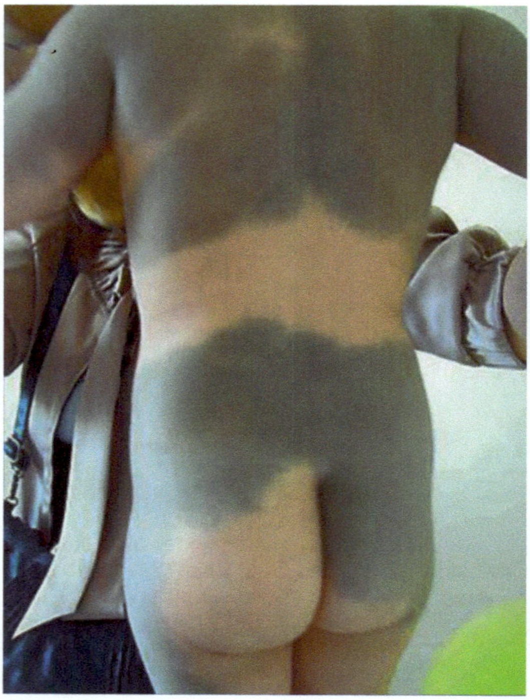

Fig. 7.7 Systematized nevus cesius (segmental dermal melanocytosis) [137] (Reprinted with permission from S. Karger AG, Basel, Switzerland)

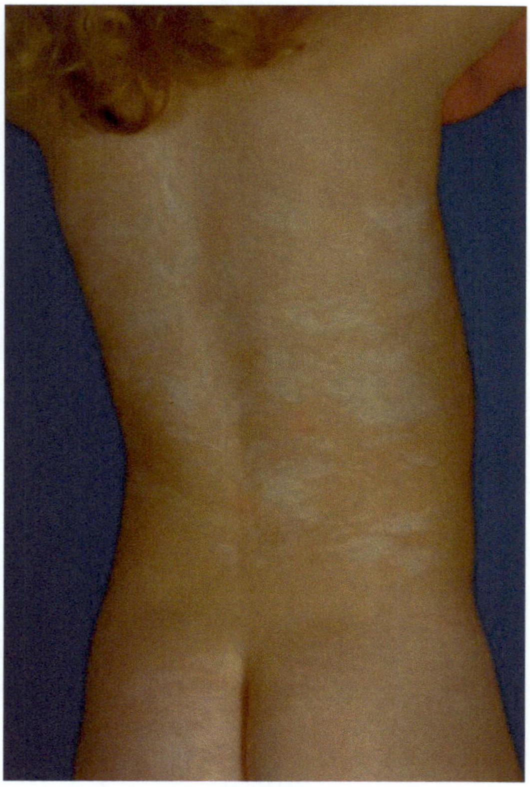

Fig. 7.8 Pigmentary mosaicism of the Ito type

Nevus cesius may be part of a syndrome called "phacomatosis cesioflammea" (see Sect. 9.5.1). At least some of these cases may constitute a binary genodermatosis in the form of nevus cesius coexisting with nevus flammeus. So far, however, it is not clear how many of these cases simply represent an arbitrary coincidence, especially in Latin American or Asian populations.

7.2.2 Other Nevi Reflecting Pigmentary Mosaicism

The following nevi are not characterized by abnormal numbers of melanocytes, but simply by increased or decreased amounts of melanin.

7.2.2.1 Linear Hypomelanosis in Narrow Bands (Pigmentary Mosaicism of the Ito Type)

These disorders are characterized by depigmented skin lesions following Blaschko's lines (Fig. 7.8). If the lesions are not systematized, they are often described under the names *nevus achromicus* or *nevus depigmentosus*. The term "hypomelanosis if Ito" [38] should be avoided because it fosters the erroneous belief that there is a nosological entity with this name. In fact, linear hypomelanosis is an umbrella term that includes many different forms of cellular mosaicism that may or may not be discernible at the cytogenetic level [142].

Linear hypomelanosis is sometimes associated with neurologic or other extracutaneous defects, but again it should be borne in mind that such cases do not represent one distinct phenotype but a group of numerous different mosaic states. When percentages of the frequency of associated extracutaneous anomalies as noted in "hypomelanosis of Ito" are presented, this simply reflects absence of genetic thinking.

Sometimes it may be rather difficult or even impossible to determine whether the light or the dark component is the pathological one in a given case of linear pigmentary mosaicism. A conspicuous example was reported by Thapa et al. [213] (Fig. 7.9).

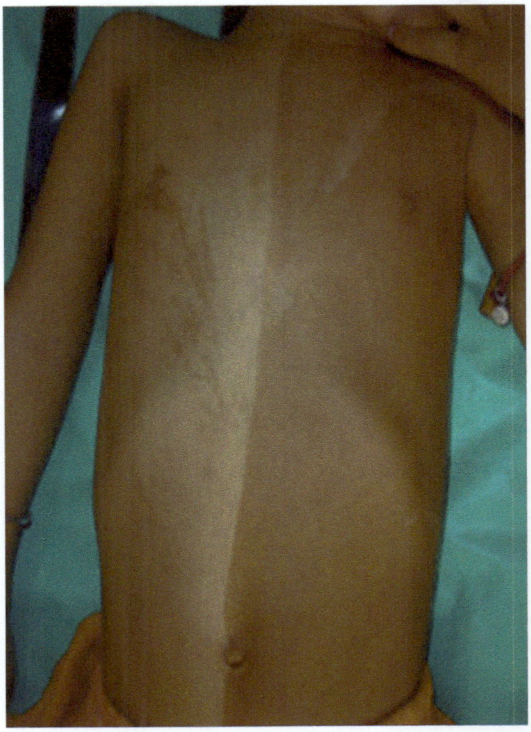

Fig. 7.9 Is this a linear hypo- or hypermelanosis? Sometimes it may be difficult to say which component is the abnormal one, but in this Indian boy it is certainly the light component [213] (Reprinted with permission from John Wiley & Sons, USA)

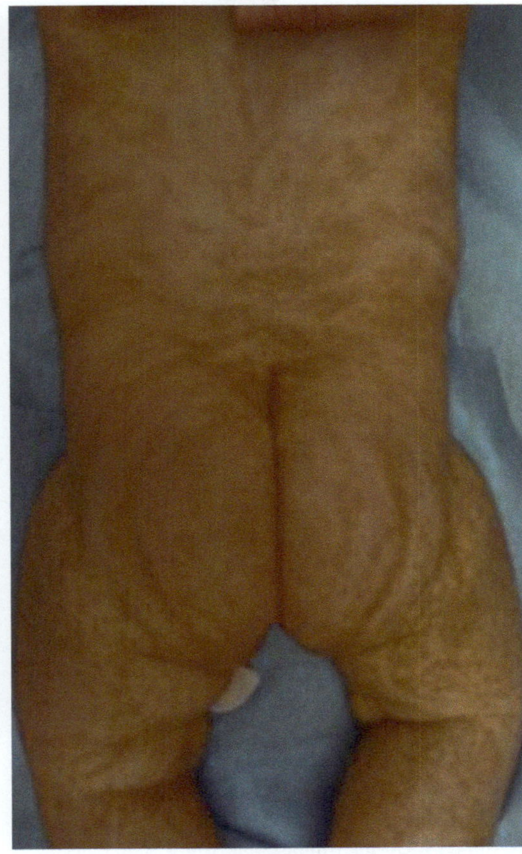

Fig. 7.10 Linear hypermelanosis in narrow bands [162] (Reprinted with permission from Elsevier Limited, UK)

7.2.2.2 Linear Hypermelanosis in Narrow Bands

This is an umbrella term including many different states of pigmentary mosaicism that may or may not be associated with extracutaneous anomalies [5]. The disorder (Fig. 7.10) has often been called "linear and whorled nevoid hypermelanosis" [132, 162], but again it is important to realize that this term does not describe any distinct nosological entity.

7.2.2.3 Linear Hypermelanosis in Broad Bands

The McCune-Albright syndrome is characterized by very broad bands of hypermelanosis (Fig. 7.11), in combination with fibrous dysplasia of bones and endocrinological abnormalities [222]. When many cases are compared, it becomes obvious that these café-au-lait macules are arranged along Blaschko's lines [81, 187]. These bands reflect the clonal outgrowth of cells carrying a postzygotic *GNAS1* mutation [200, 222].

In the past century it was a favorite point of controversy among pediatricians and endocrinologists whether the various hormonal disturbances as noted in patients with McCune-Albright syndrome were of "central" or "peripheral" origin. The concept of randomly distributed mosaicism has rendered this discussion pointless. For obvious reasons both mechanisms are possible [76].

7.2.2.4 Segmental Hypomelanosis Arranged in a Checkerboard Pattern

This is a rarely reported type of hypomelanotic nevus (Fig. 7.12) [167]. Molecular findings are so far not available.

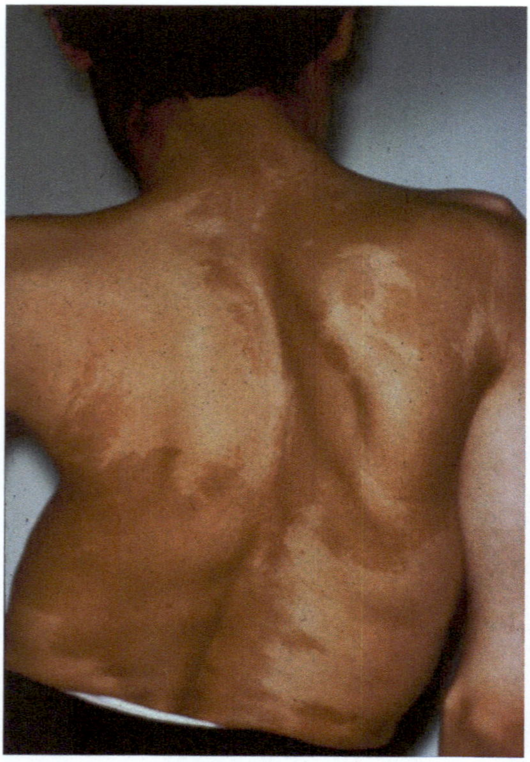

Fig. 7.11 McCune-Albright syndrome characterized by linear hypermelanosis in broad bands (Courtesy of the late Dr. Robert J. Gorlin, Minneapolis, Minnesota, USA)

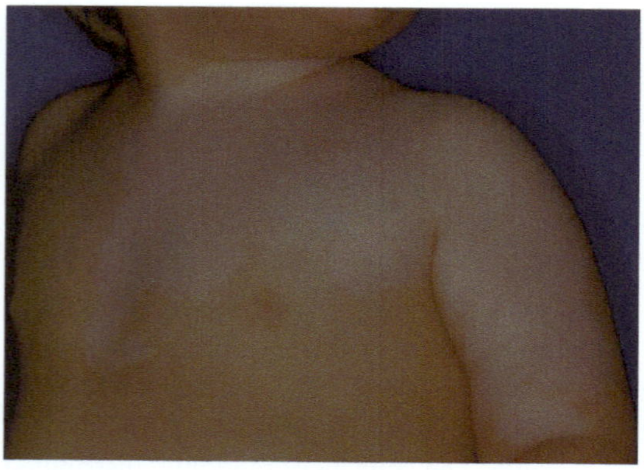

Fig. 7.12 Nevus achromicus arranged in a checkerboard pattern (Courtesy of Dr. Michael Landthaler, Regensburg, Germany)

7.2.2.5 Segmental Hypermelanosis Arranged in a Checkerboard Pattern

Hyperpigmentation arranged in a checkerboard pattern likewise occurs rather rarely (Fig. 7.13).

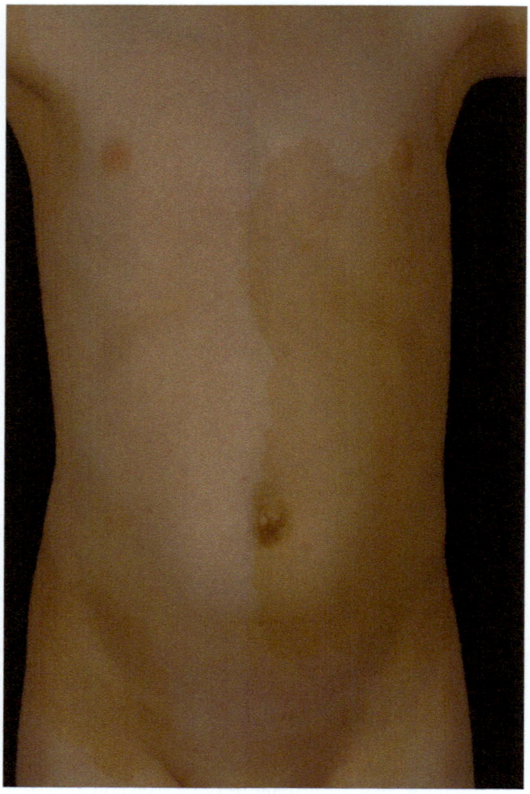

Fig. 7.13 Segmental hypermelanosis arranged in a checkerboard pattern (Courtesy of Dr. Regina Fölster-Holst, Kiel, Germany)

It was reported by Stoll et al. [207] in a boy with developmental delay, facial dysmorphism, and large, low-set ears. Cytogenetic analysis of lesional skin fibroblasts revealed a 12q;14q translocation present in 70 % of cells, whereas no chromosomal abnormality was found in peripheral blood lymphocytes.

7.2.2.6 Phylloid Hypomelanosis

The pattern of this pigmentary disorder (Fig. 7.14) reflects mosaicism but cannot be described as "segmental." Most cases of phylloid hypomelanosis are caused by mosaic trisomy 13q [39, 61, 85] (see Sect. 5.3). Hence, this phenotype cannot simply be taken as a mosaic manifestation of complete trisomy 13 (Patau syndrome) [131], although it is closely related to this severe form of numeric chromosomal aberration.

The associated extracutaneous anomalies show likewise a mosaic distribution. They include

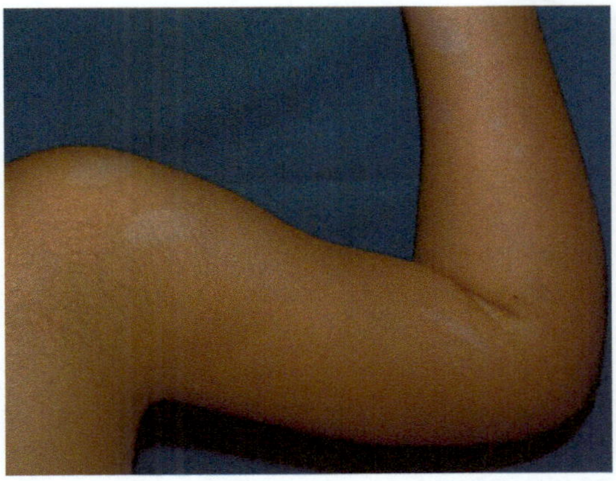

Fig. 7.14 Phylloid hypomelanosis (Courtesy of Dr. Antonia González-Enseñat, Barcelona, Spain)

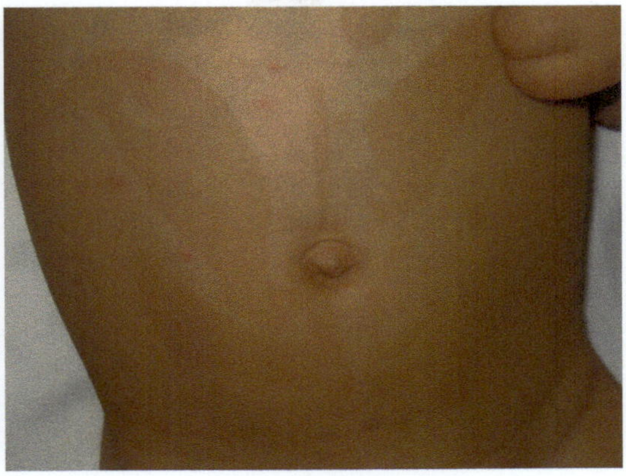

Fig. 7.15 Phylloid hypermelanosis [110] (Courtesy of Dr. Mario F. Franco-Guío, Manizales, Colombia)

absence of corpus callosum and other CNS defects, conductive hearing loss, coloboma, craniofacial defects, brachydactyly, clinodactyly, and camptodactyly [51, 86].

7.2.2.7 Phylloid Hypermelanosis

In 1993, the term "phylloid pattern" was proposed in an article describing an arrangement of leaf-like hyperpigmented macules reminiscent of floral ornaments in a 12-year-old girl who had an arteria subclavia originating from the aorta descendens [81]. Phylloid hypermelanosis (Fig. 7.15) can be taken as a hyperpigmented counterpart of phylloid hypomelanosis. Contrasting with the hypopigmented phenotype, however, it represents a class of heterogeneous disorders [101]. Oiso et al. [175] reported a case of phylloid hypermelanosis and mental deficiency. In this patient, analysis of blood lymphocytes showed three different cell types containing either a dicentric chromosome 13, or monosomy 13, or a ring chromosome (13) (p11.2q34). Remarkably, no normal karyotype was found in 30 cells examined. Already in 1956, Dockx et al. [41] had documented a similar pattern of hypermelanotic macules in a girl with mental deficiency and abnormalities of the bones and joints (Fig. 7.16). Additional cases have recently been reported [23, 110]. Another example of phylloid hypermelanosis has been mistaken for "linear and whorled nevoid hypermelanosis" [150].

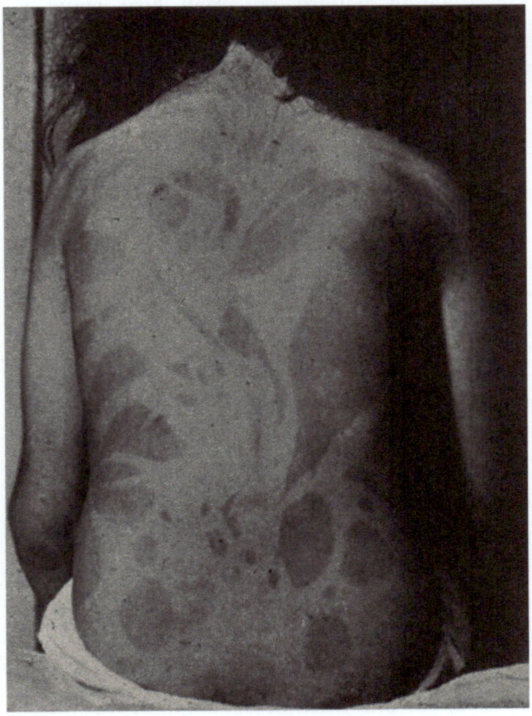

Fig. 7.16 A historical case of phylloid hypermelanosis described by Dockx et al. in 1956 [41] (Reprinted with permission from Elsevier Masson SAS, France)

7.3 Epidermal Nevi

There are two large categories of epidermal nevi. The keratinocytic nevi are "true" epidermal nevi because they are exclusively characterized by epidermal changes, whereas in the group of organoid nevi the essential changes involve the adnexal structures of the skin. In principle, the designation "epithelial nevi" would be a better generic term to denote both categories, but the name "epidermal nevi" has firmly been entrenched in our literature and will be maintained in the following paragraphs.

7.3.1 Keratinocytic Nevi

Most keratinocytic nevi follow the lines of Blaschko. An exception from this rule is the CHILD nevus that is characterized by a peculiar lateralization pattern but may also show an arrangement along Blaschko's lines.

The different epidermal nevus syndromes characterized by particular keratinocytic nevi are listed in Table 7.1.

7.3.1.1 Common Keratinocytic Nevi of the Soft Type, Including Seborrheic Keratoses

These nevi have a soft, velvety surface and are usually of gray or brown color. The earlier the postzygotic mutation has occurred during embryonic life, the more widespread the involvement will be. Some lesions are so small that a linear arrangement is barely discernible. Some form one single band, whereas others show a systematized pattern involving many parts of the body (Fig. 7.17). When the postzygotic mutation occurs later in life, the resulting keratinocytic nevus is small and nonlinear and called, by tradition, "seborrheic keratosis."

In about one third of such cases, mosaicism has been proven at the cellular level by documenting a heterozygous R248C hotspot mutation in the FGFR3 gene within the nevus [71]. A different *FGFR3* mutation has likewise been documented [177]. In other cases such nevi are caused by postzygotic activating *PIK3CA* mutations [72]. Both *FGFR3* and *PIK3CA* mutations are likewise found in seborrheic keratoses [66, 68, 69]. Moreover, many keratinocytic nevi are associated with postzygotic *HRAS* or, more rarely, *KRAS* mutations [21, 69, 70].

Seborrheic Keratoses are Acquired Keratinocytic Nevi

Seborrheic keratoses (Fig. 7.18) have recently turned out to be acquired epidermal nevi. Histopathologically, they very closely resemble linear epidermal nevi of the common type. The spectrum of postzygotic mutations as found in seborrheic keratoses corresponds to that observed in linear keratinocytic nevi of the nonepidermolytic type [66, 68, 72]. In particular, R248C mutations of the FGFR3 gene as well as activating *PIK3CA* mutations are frequently found. The R248C *FGFR3* mutation, when present in the germline, causes thanatophoric dysplasia [71]. This allele can, therefore, be categorized as a lethal mutation surviving by mosaicism.

Activating *PIK3CA* mutations are present in a broad spectrum of malignant tumors [68]. Paradoxically, however, epidermal nevi harboring such mutations do usually not show any proclivity to develop malignant growth.

An Early Postzygotic *FGFR3* Mutation May Cause a Distinct Neurocutaneous Syndrome

In a 5-year-old girl with severe structural CNS defects causing mental deficiency and epileptic seizures, García et al. [54] described a systematized, soft, velvety epidermal nevus of the common, nonorganoid, nonepidermolytic type (Fig. 7.19). A mosaic R248C mutation of the FGFR3 gene was found in the blood and within the nevus, whereas the normal skin contained the wild-type allele only. This disorder, for which the name "FGFR3 epidermal nevus syndrome" was proposed [102], can be categorized as a mosaic manifestation of thanatophoric dysplasia. A similar patient was found to be affected with a different *FGFR3* mutation [177]. A case of FGFR3 epidermal nevus associated with mild facial dysmorphism [30] can be taken as an oligosymptomatic example of the same syndrome.

Early Postzygotic *PIK3CA* Mutations Cause CLOVES Syndrome

The term CLOVE was proposed as an acronym for congenital lipomatous overgrowth with vascular and epidermal anomalies [198]. In 2009 the name was expanded to CLOVES syndrome to

Table 7.1 Epidermal nevus syndromes characterized by keratinocytic nevi

Syndrome	Type of epidermal nevus	Pattern of epidermal nevus	Discriminating clinical criteria	Formal genetic classification	Molecular basis
Proteus syndrome	Linear Proteus nevus (nonepidermolytic keratinocytic nevus of a soft, flat type)	Blaschko lines	Cerebriform connective tissue nevi of palms or soles; asymmetric macrodactyly	Not heritable	Mosaic activating *AKT1* mutation
CLOVES syndrome	Nonepidermolytic keratinocytic nevus	Blaschko lines	Disproportionate overgrowth of fatty tissue, truncal vascular malformations, nonprogressive overgrowth of toes ("ballooning"), "wrinkling" of palms or soles, cerebral defects	Not heritable	Mosaic *PIK3CA* mutations
Type 2 segmental PTEN hamartoma syndrome (including type 2 segmental Cowden disease)	Linear PTEN nevus (nonepidermolytic keratinocytic nevus of a soft and rather thick, papillomatous type)	Blaschko lines	Macrocephaly, overgrowth of a limb, colon polyps, focal segmental glomerulosclerosis; family members with nonsegmental PTEN hamartoma syndrome	Type 2 segmental manifestation of PTEN hamartoma syndrome	*PTEN* germline mutation, with loss of the wild-type allele in the involved segment
FGFR3 epidermal nevus syndrome (García-Hafner-Happle syndrome)	Nonepidermolytic keratinocytic nevus of a soft type	Blaschko lines	Presence of keratinocytic nevus of the common type; absence of disproportionate overgrowth	Not heritable	Mosaic *FGFR3* mutation
CHILD syndrome	CHILD nevus	Two patterns in the form of lateralization and Blaschko lines	Lateralized, inflammatory skin lesions; ptychotropism; ipsilateral limb defects	X-linked dominant inheritance with lethality in male embryos	*NSDHL* mutations
NEVADA syndrome (*n*evus *e*pidermicus with *a*ngio*d*ysplasia and *a*neurysms)	Nonepidermolytic, hystrix-like epidermal nevus	Blaschko lines or lateralization pattern	Dysplasia of large vessels, arteriovenous shunts; dysplasia of retinal vessels	Not heritable	Unknown

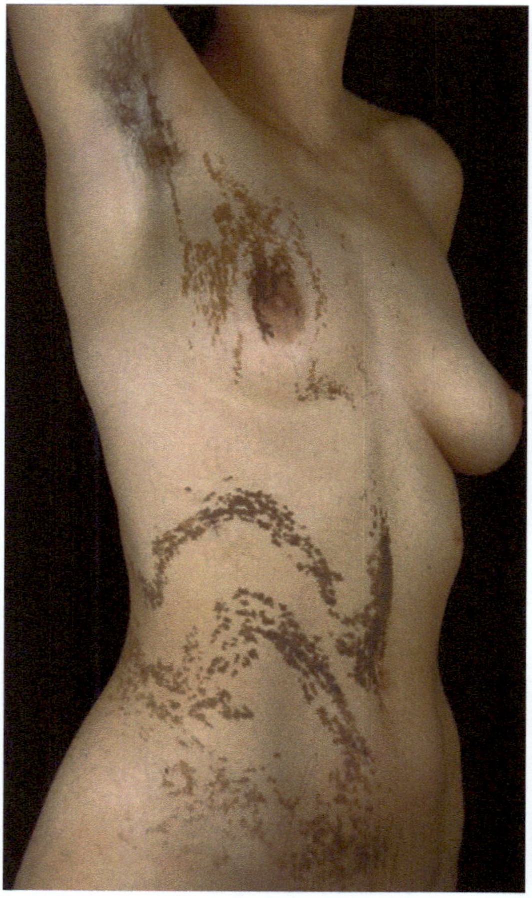

Fig. 7.17 Systematized keratinocytic nevus of the common, soft type

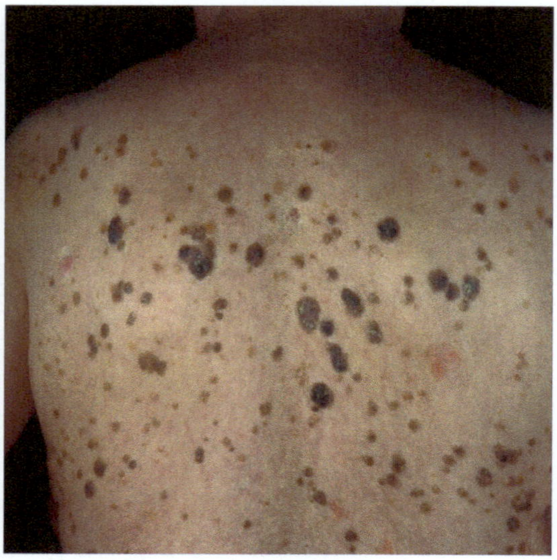

Fig. 7.18 Seborrheic keratoses can today be categorized as acquired keratinocytic nevi of the common type

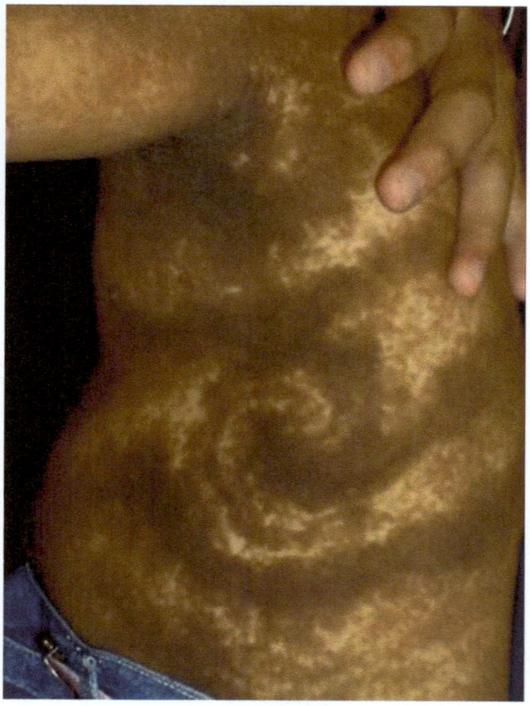

Fig. 7.19 Systematized keratinocytic nevus in a girl with severe cerebral defects (FGFR3 epidermal nevus syndrome) [54] (Courtesy of Dr. Alejandro García Vargas, Guadalajara, Mexico)

denote the additional presence of *s*keletal anomalies [7]. The associated keratinocytic nevus is of a soft, velvety type and may sometimes show a rather broad segmental involvement (Fig. 7.20). In the past the phenotype was confused with Proteus syndrome, but it represents a distinct nosological entity being caused by postzygotic activating mutations in the PIK3CA gene [140].

7.3.1.2 Common Keratinocytic Nevi of the Hard, Verrucous Type

These nevi have a hard and rough surface. Histopathologically they show a more pronounced hyperkeratosis. As an important clinical feature,

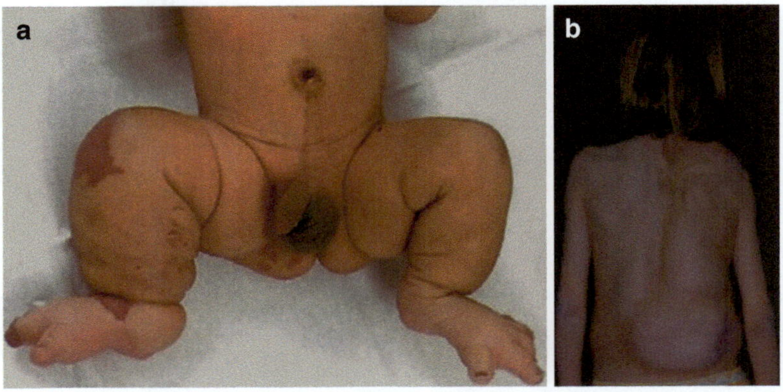

Fig. 7.20 CLOVES syndrome. (**a**) Overgrowth of right leg, vascular nevi, contralateral epidermal nevus, and wide feet with broad first interdigital space [120] (Reprinted with permission from John Wiley & Sons, USA); (**b**) 14-year-old girl with large subcutaneous lipomas and keratinocytic nevus (Courtesy of Dr. Sigrid Tinschert, Berlin, Germany)

the hard type is very difficult to treat, whereas the soft type responds rather well to argon laser or carbon dioxide laser therapy [124, 125].

7.3.1.3 SASKIA Nevus (Segmentally Arranged Seborrheic Keratoses with Impending Atypia): A New Skin Disorder?

Multiple seborrheic keratoses arranged along Blaschko's lines have been suggested to reflect genetic mosaicism [155]. Livingstone et al. [149] described a 76-year-old woman who had since birth multiple seborrheic keratosis-like lesions arranged in a systematized linear pattern on the left side of her body (Fig. 7.21). Some years ago, a squamous cell carcinoma had developed within one of these lesions. Subsequently, carcinomata in situ in association with seborrheic keratosis-like structures were found in four out of seven biopsies obtained from rather hyperkeratotic and erythematous lesions. An *FGFR3* mutation was present in three of the lesions, and an additional *PIK3CA* mutation was found in one of them. No mutation could be documented in specimens from the squamous cell carcinoma. The authors raised the question whether this unusual nevoid disorder may represent a new nosological entity.

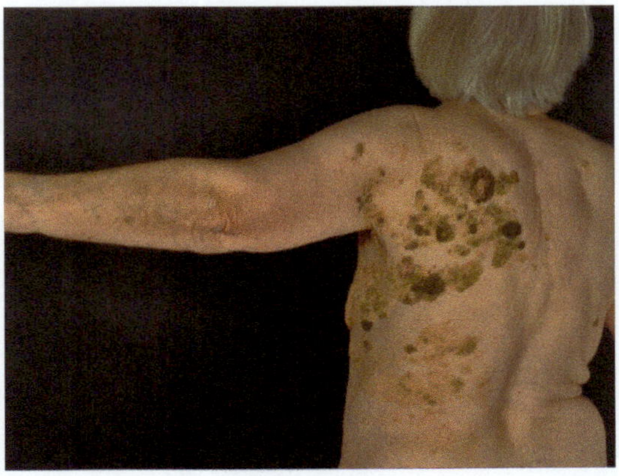

Fig. 7.21 SASKIA nevus (segmentally arranged seborrheic keratoses with impending atypia) [149]

7.3.1.4 Linear PTEN Nevus (Linear Cowden Nevus Included)

This nevus represents a type 2 segmental manifestation of the PTEN hamartoma syndrome (see Table 3.4 and Sect. 10.1.20). It is caused by loss of heterozygosity at the *PTEN* locus [95]. The linear PTEN nevus tends to be thicker and more papillomatous than common keratinocytic nevi (Fig. 7.22).

7.3.1.5 Epidermal Nevus of the Proteus Type

Proteus syndrome is characterized by disproportionate and progressive overgrowth of various tissues including asymmetrical macrodactyly, cerebriform connective tissue nevi of the soles, lipomas, cystic lymphangiomas, and vascular nevi (Fig. 7.23a, b) [15, 102]. The frequently associated epidermal nevus is a soft and rather flat linear lesion (Fig. 7.23c) [94, 102]. The underlying

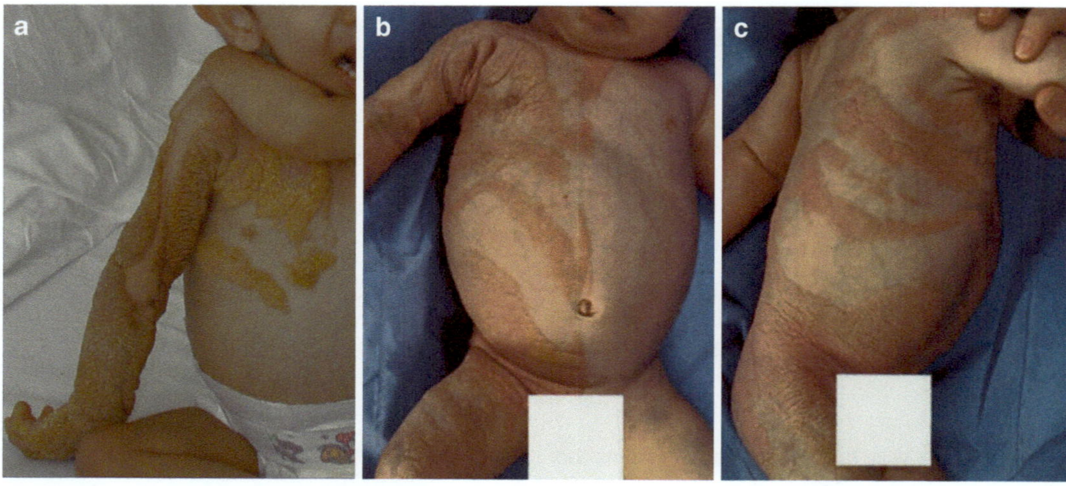

Fig. 7.22 (**a**–**c**) Papillomatous appearance of linear PTEN nevus [153, 211] (Reprinted with permission from John Wiley & Sons, USA; **a**: Color photograph kindly provided by Dr. Mustafa Tekin, Ankara, Turkey)

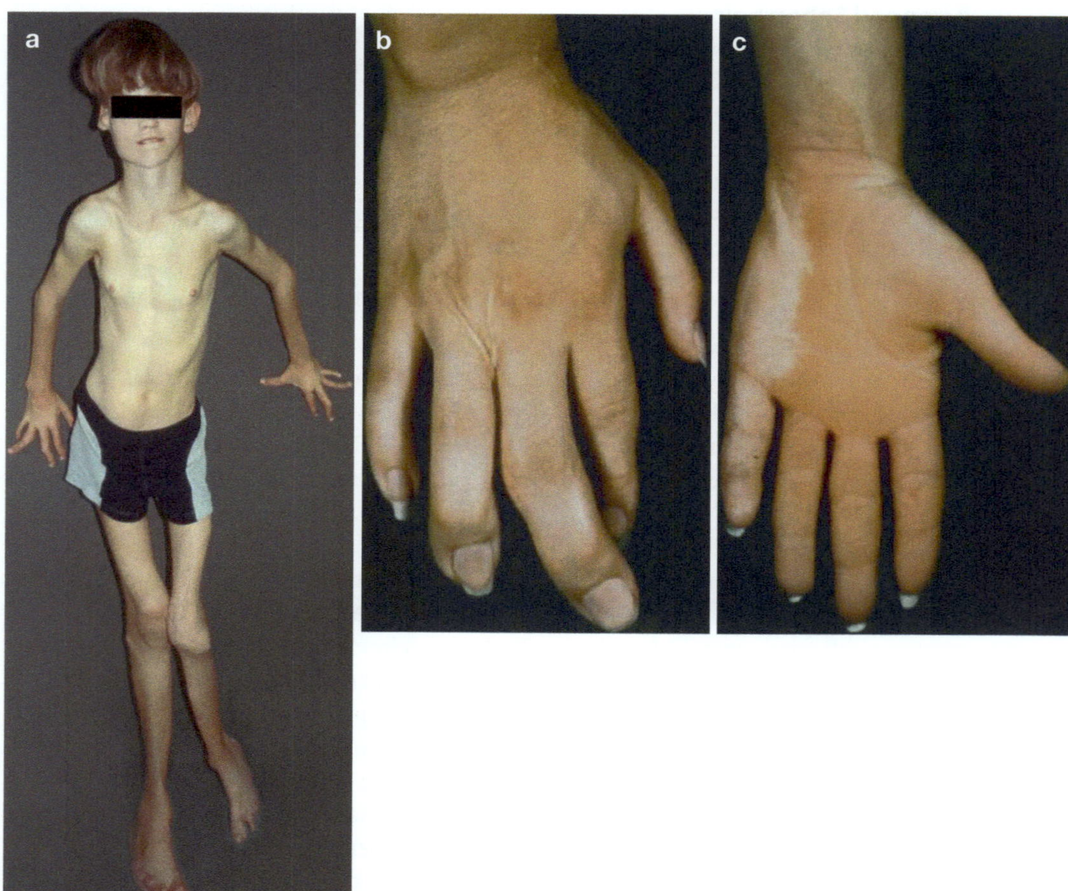

Fig. 7.23 Proteus syndrome (**a**) A 9-year-old boy who had reached a height of 186 cm (Courtesy of Dr. Sigrid Tinschert, Berlin, Germany) (**b**) Macrodactyly of the right hand and (**c**) flat, soft epidermal nevus of the left forearm and hand in a 24-year-old woman [28] (Reprinted with permission from John Wiley & Sons, USA)

mutation involves the AKT1 gene [148]. Remarkably, the mutant gene cannot be found in blood DNA because its lethal action eliminates all lymphocytes carrying the mutation.

7.3.1.6 Hystrix-Like Epidermal Nevus of NEVADA Syndrome

A peculiar keratinocytic nevus characterized by a thick, hystrix-like white or brownish hyperkeratosis has been described in patients with multiple vascular malformations (Fig. 7.24) [160]. The acronym *NEVADA* stands for "*n*evus *e*pidermicus *v*errucosus with *a*ngio*d*ysplasia and *a*neurysms" [103]. It is likely, albeit unproven, that this nevus represents a distinct entity.

7.3.1.7 Epidermolytic Epidermal Nevus

This nevus (Fig. 7.25) represents a mosaic manifestation of epidermolytic ichthyosis of Brocq [178]. Affected individuals run an increased risk to give birth to a child with the diffuse form of this disorder (see Sect. 3.1.1.3).

7.3.1.8 Inflammatory Linear Verrucous Epidermal Nevus (ILVEN)

This nevus is characterized by erythema, scaling, and itching arranged in a linear pattern (Fig. 7.26). In the past, the disorder has often been confused with linear psoriasis [2, 123] or CHILD nevus [58, 171, 172] (see Sect. 7.3.1.9). ILVEN usually occurs sporadically. However, exceptional familial cases have been reported [9, 73] and may be explained as examples of monoallelic autosomal expression, thus reflecting epigenetic mosaicism [89]. So far, a bona fide association of ILVEN with extracutaneous anomalies has not been reported.

7.3.1.9 CHILD Nevus

This skin disorder represents a hallmark of CHILD syndrome that is caused by *NSDHL* mutations and inherited as an X-linked dominant,

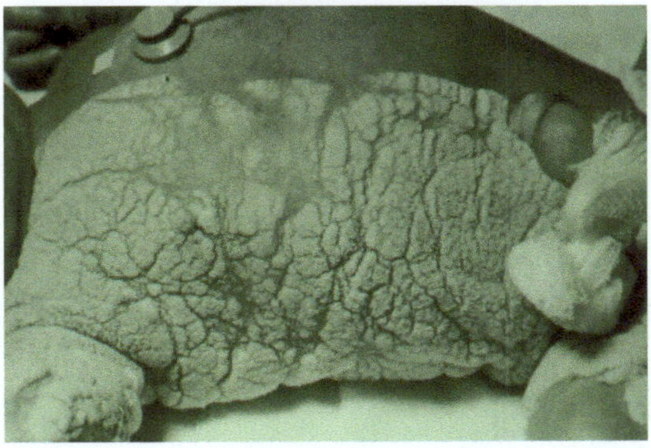

Fig. 7.24 NEVADA syndrome. Hystrix-like epidermal nevus in a newborn who died because of malformation of large vessels [160] (Courtesy of Dr. Wolfgang Marsch, Halle [Saale], Germany)

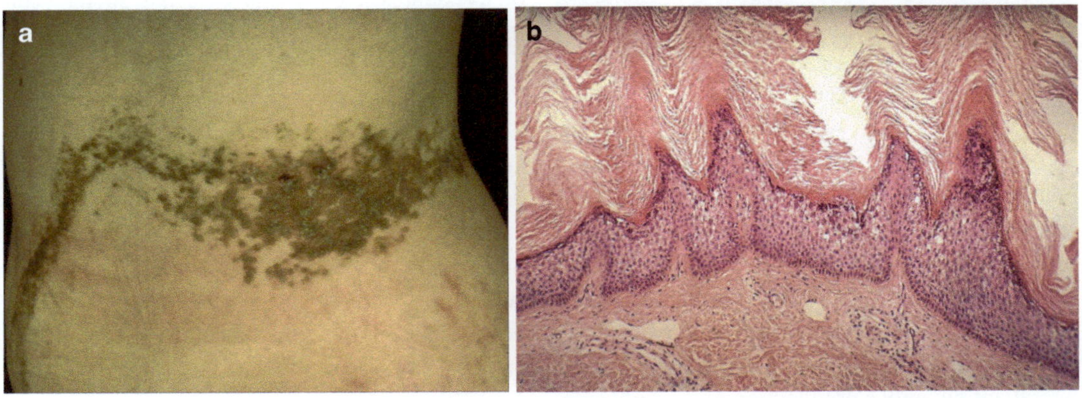

Fig. 7.25 Epidermolytic epidermal nevus showing (**a**) a rather hard surface and (**b**) the characteristic histopathological feature of acanthokeratolysis

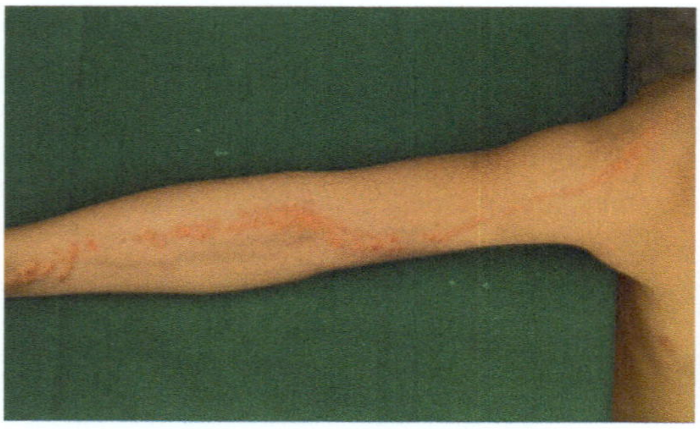

Fig. 7.26 Inflammatory linear verrucous epidermal nevus (ILVEN)

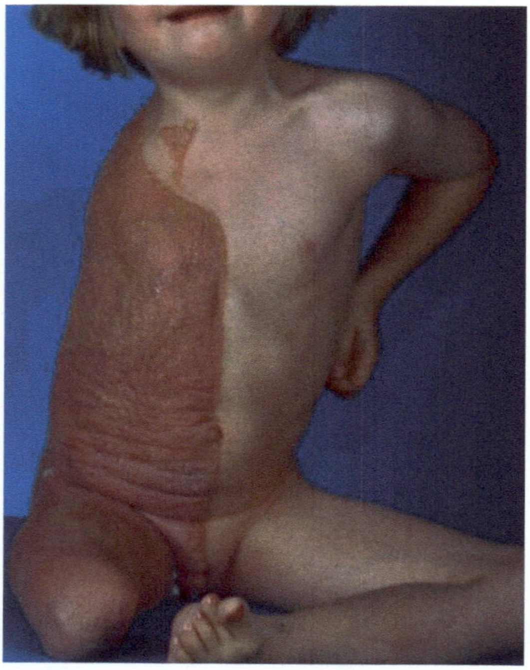

Fig. 7.27 CHILD nevus showing a nonlinear arrangement

male-lethal trait (Table 7.1). It occurs almost exclusively in females and is usually present at birth. Strictly unilateral areas of erythema are covered with yellowish waxy scales [115]. The CHILD nevus displays two different patterns of distribution that are often intermingled. A diffuse lateralization, with a predilection of the right side, is highly characteristic (Fig. 7.27), whereas lesions arranged along Blaschko's lines may be noted both ipsilaterally and contralaterally (Fig. 7.28). Most of the histopathological features

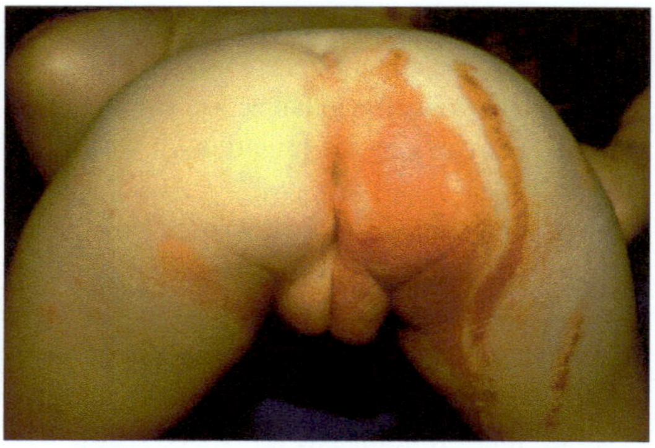

Fig. 7.28 CHILD nevus showing a linear arrangement (Courtesy of Dr. Adelaide A. Hebert, Houston, Texas, USA)

are indistinguishable from those of psoriasis, which is why CHILD nevus has sometimes been misdiagnosed as psoriasis [108, 185, 202]. However, the phenomenon of verruciform xanthoma is typical of CHILD nevus and not found in psoriasis [115].

Practical note: The best treatment for CHILD nevus is topical application of a lotion or ointment containing lovastatin or simvastatin in combination with cholesterol [179].

7.3.1.10 Nevus Corniculatus

This disorder is characterized by cutaneous horns and comedo-like lesions arranged along Blaschko's lines (Fig. 7.29) and showing the histopathological feature of acantholysis without any sign of dyskeratosis. Only one case has so far been reported [118].

7.3.1.11 Nevus Kerinokeratoticus

This nevus is a mosaic manifestation of kerinokeratosis papulosa ("waxy keratoses of childhood"), a disorder that may be noted, by way of exception, also in adulthood [42]. A type 1 segmental manifestation was documented by Mehrabi et al. [164], and a type 2 segmental involvement, being superimposed on bilaterally disseminated papules, was also reported (Fig. 7.30) [109].

7.3.1.12 Other Keratinocytic Nevi

Cases of type 2 segmental manifestation of autosomal dominant defects of keratinocytes can likewise be categorized as keratinocytic nevi. Examples

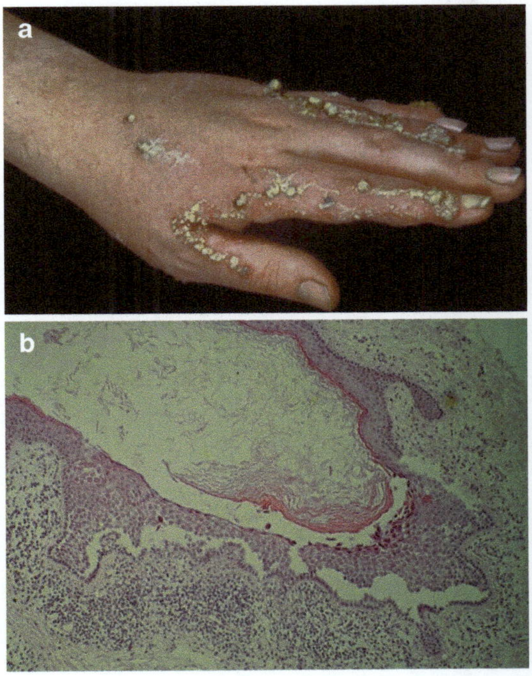

Fig. 7.29 Nevus corniculatus. (**a**) Linear arrangement of horn-like hyperkeratoses [118]; (**b**) comedo-like lesion showing suprabasal acantholysis (**a**: Reprinted with permission from John Wiley & Sons, USA)

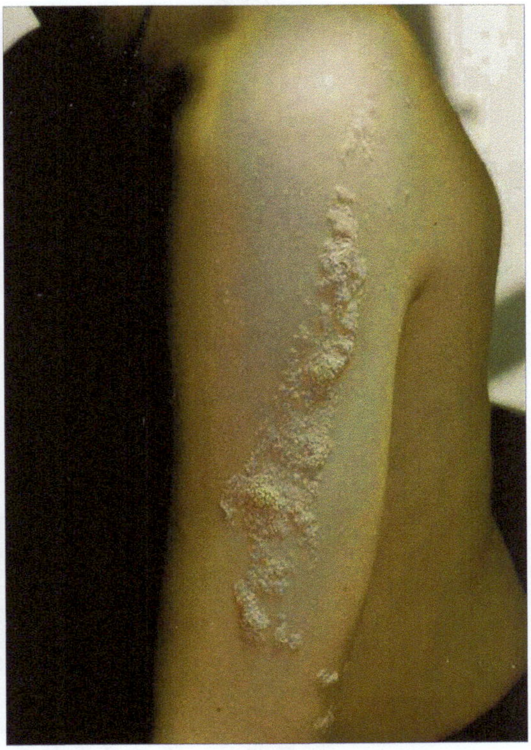

Fig. 7.30 Nevus kerinokeratoticus, representing a type 2 segmental manifestation of kerinokeratosis papulosa [109] (Reprinted with permission from Elsevier, Oxford, UK)

include pronounced and rather stable linear forms of Darier disease, Hailey-Hailey disease, disseminated superficial actinic porokeratosis, plaque-like porokeratosis of Mibelli, and epidermolytic ichthyosis of Brocq [98] (see Sect. 10.2).

7.3.2 Organoid Epidermal Nevi

The group of organoid epidermal nevi is rather large, and many of them are diagnostic markers of distinct multisystem birth defects (Table 7.2). The molecular basis of most of these disorders is still unknown, but the cause of nevus sebaceus and Schimmelpenning syndrome has already been found [63].

7.3.2.1 Nevus Sebaceus

In principle, this nevus (Fig. 7.31) is characterized by hyperplasia of sebaceous glands. It should be borne in mind, however, that during childhood these glands are usually underdeveloped. A typical finding in young children is the presence of less differentiated structures reminiscent of embryonic hair follicles [165, 205]. Moreover, the sebaceous component may be minimal or absent in parts of the nevus involving areas outside of the head and neck [83]. Nevus sebaceus is a mosaic RASopathy being caused by postzygotic *HRAS* or *KRAS* mutations [63].

Several familial cases have been reported. Such familial aggregation is difficult to explain because nevus sebaceus reflects heterozygosity for a *HRAS* or *KRAS* mutation [63, 107, 146], which means that this type of organoid nevus can no longer be taken as a paradominant trait [107, 112] (see also Sect. 3.1.1.4).

Nevus sebaceus is a hallmark of Schimmelpenning syndrome that includes cerebral, ocular, and skeletal defects [102]. The hypothesis that the disorder reflects the action of a lethal gene surviving by mosaicism [75, 188] has now been confirmed by

Table 7.2 Epidermal nevus syndromes characterized by organoid nevi

Syndrome	Type of epidermal nevus	Pattern of epidermal nevus	Discriminating clinical criteria	Formal genetic classification	Molecular basis
Schimmelpenning syndrome, including phacomatosis pigmentokeratotica	Nevus sebaceus	Blaschko lines	Epibulbar lipodermoid	A lethal gene surviving by mosaicism	*HRAS* and *KRAS* mutations
Nevus comedonicus syndrome	Nevus comedonicus	Blaschko lines	Ipsilateral cataract; absence deformities of bones	Not heritable	Unknown
Angora hair nevus syndrome (Schauder syndrome)	Angora hair nevus	Blaschko lines	Broad bands covered with long, soft, white hair; porencephaly	Not heritable	Unknown
Becker nevus syndrome	Becker nevus	Checkerboard arrangement	No Blaschko-linear pattern; in women, ipsilateral breast hypoplasia is often present	Paradominant (?)	Unknown
Castori syndrome	Complex nevus with proliferation of partly immature eccrine structures	Band-like, but dissimilar from Blaschko's lines	Pox-like facial and intraoral scars; gingival synechiae; dental crowding; dysmorphic facial appearance; mental deficiency; EEG anomalies; conductive hearing loss	Probably not heritable	Unknown
Nevus trichilemmocysticus syndrome	Nevus trichilemmocysticus	Blaschko lines	Large or small cystic lesions; filiform hyperkeratoses; comedo-like lesions; bone defects	Not heritable	Unknown
Porokeratotic eccrine nevus syndrome	Porokeratotic eccrine nevus	Blaschko lines	Hyperparakeratosis of eccrine infundibula; conductive hearing loss	Type 1 mosaic manifestation of KID syndrome	*GJB2* mutations

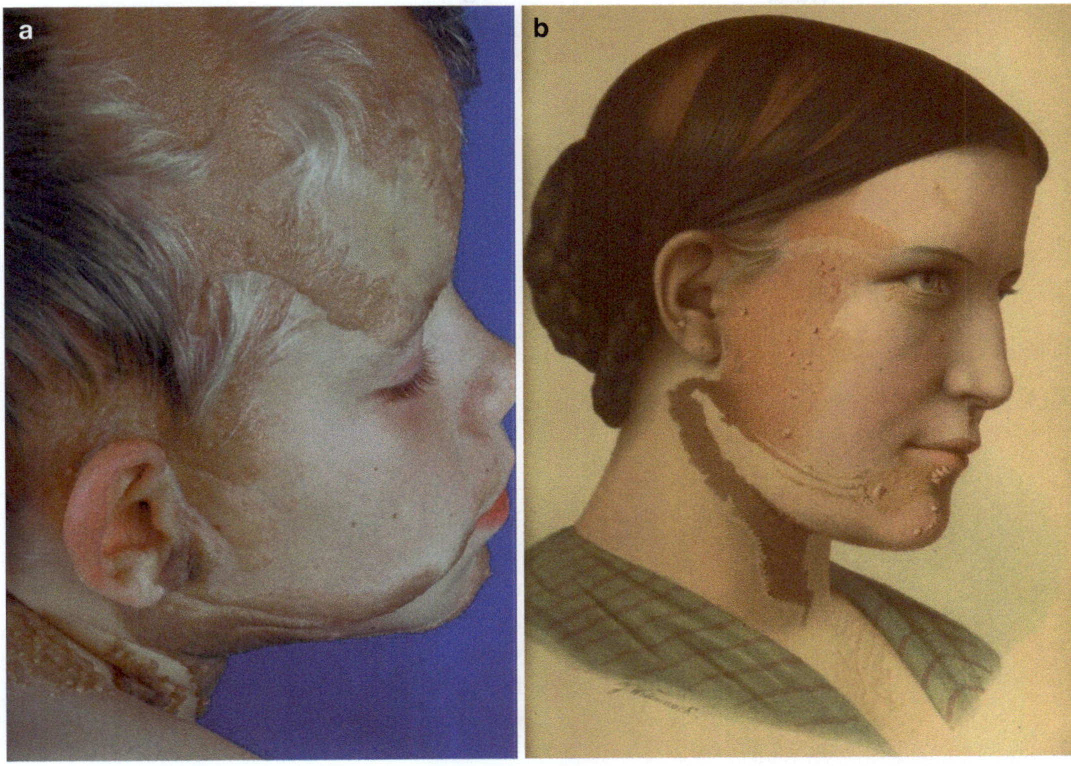

Fig. 7.31 Systematized sebaceous nevus (**a**) in a young boy and (**b**) in a 16-year-old girl [50]

molecular data [63, 209]. Apparently, these mutations are lethal to such degree that they cannot be found in the patients' blood [63, 146].

This syndrome may be associated with papular nevus spilus in the form of phacomatosis pigmentokeratotica that represents a pseudodidymosis [106] (see Sect. 9.1), or with aplasia cutis congenita (see Sect. 9.3).

Nevus Marginatus: A Peculiar Variant of Nevus Sebaceus

An unusual birthmark was described by Hafner et al. [67] in a 35-year-old man. A congenital linear lesion involving one side of his trunk was conspicuously bordered by dark brown papules (Fig. 7.32). On histopathological examination, the flat central area showed sebaceous hyperplasia indicative of a sebaceous nevus, whereas the elevated margin showed features of a nonorganoid keratinocytic nevus. Subsequent molecular analysis revealed, in both components of the lesion, the presence of a *HRAS* p.G13R mutation, which is known to be the most frequently occurring mutation in nevus sebaceus [65].

In several other reports, nevus marginatus can be found on photographs [22, 212]. It may best be categorized, according to present knowledge, as a particular clinical variant of nevus sebaceus. Apparently, the epithelial progenitor cells of nevus sebaceus possess a variable developmental potential to differentiate into sebocytes or keratinocytes [65].

7.3.2.2 Nevus Comedonicus

The disorder is characterized by multiple comedones showing a linear arrangement. The keratin-filled structures are often surrounded by markedly atrophic skin (Fig. 7.33), and sometimes they are associated with large epithelial cysts.

Nevus comedonicus syndrome includes skeletal, ocular, and neurologic abnormalities [102]. The presence of an ipsilateral cataract can be taken as a diagnostic clue.

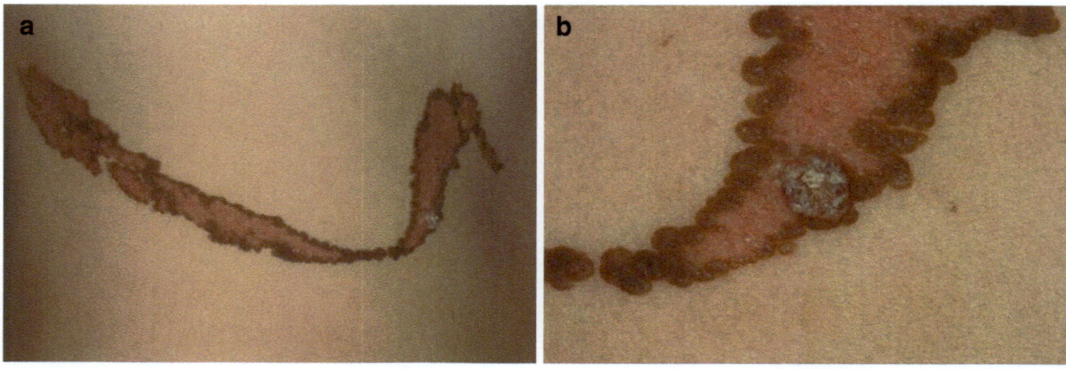

Fig. 7.32 Nevus marginatus [67]. (**a**) General view; (**b**) close-up (Reprinted with permission from S. Karger AG, Basel, Switzerland)

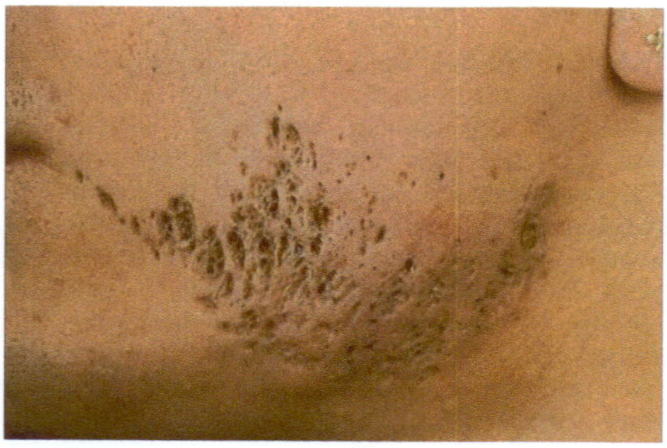

Fig. 7.33 Nevus comedonicus (Courtesy of Dr. Wolf I. Worret, Munich, Germany)

7.3.2.3 Linear Epidermolytic Comedones

This disorder has unduly been conflated with nevus comedonicus [6, 154]. It should be separated as a distinct entity. The rather large and elevated comedones are arranged along Blaschko's lines (Fig. 7.34a, b) and show, as a distinguishing feature, the histopathological phenomenon of epidermolytic hyperkeratosis (Fig. 7.34c) [6, 182]. For unknown reasons, a man affected with this mosaic disorder had a child with epidermolytic ichthyosis of Brocq [154].

7.3.2.4 Angora Hair Nevus and Schauder Syndrome

The disorder consists of band-like areas covered with soft, white hair that grows out from dilated follicular infundibula (Fig. 7.35). This nevus was first described by Schauder [199] in a man with multiple mosaic CNS defects giving rise to mental deficiency and epileptic seizures (Table 7.2). Moreover, facial dysmorphism and severe ocular defects were present. In another case the associated abnormalities were rather mild [18].

7.3.2.5 Becker Nevus and Becker Nevus Syndrome

Within the group of organoid epidermal nevi, this disorder has the particularity not to follow Blaschko's lines but to be arranged in a checkerboard pattern. Becker nevus is an androgen-dependent lesion. Therefore, women and prepubertal boys merely show a hyperpigmented macule with bizarre outlines and mild or absent hairiness, whereas in adult men the pigmented patch is darker and covered by a pronounced hypertrichosis (Fig. 7.36). This explains why isolated Becker nevi are reported more frequently in men. Adolescents may develop acne within the Becker nevus [1, 197]. Histopathologically, the lesion usually shows increased numbers of smooth muscles, resembling a "smooth muscle hamartoma."

On the other hand, Becker nevus syndrome is reported more frequently in women because of the more noticeable feature of ipsilateral breast hypoplasia (Fig. 7.37) [31]. Other syndromic associations include asymmetric patchy hypoplasia of subcutaneous fatty tissue, supernumerary nipples, absence or hypoplasia of ipsilateral muscles

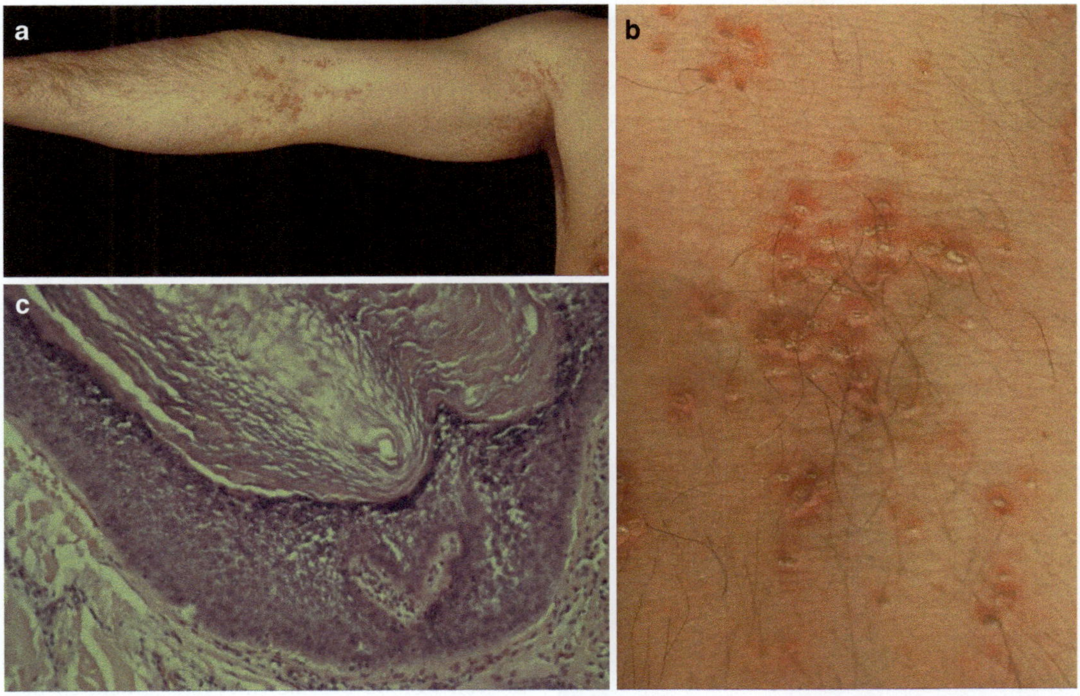

Fig. 7.34 Linear epidermolytic comedones. (**a**) Systematized unilateral involvement; (**b**) close-up; (**c**) comedo showing the histopathological feature of epidermolytic hyperkeratosis

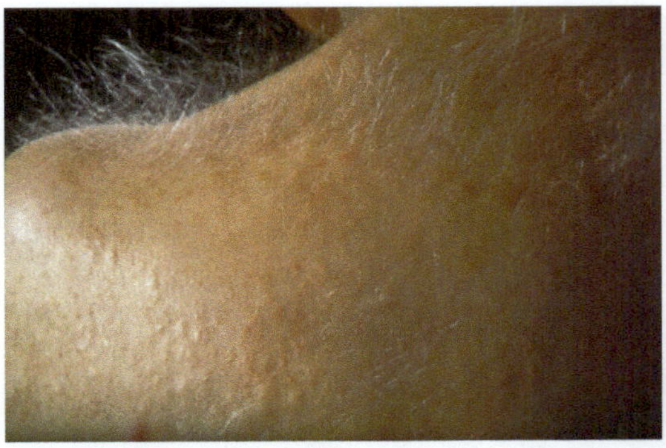

Fig. 7.35 Angora hair nevus, being a hallmark of Schauder syndrome [102] (Reprinted with permission from Elsevier Limited, Oxford, UK)

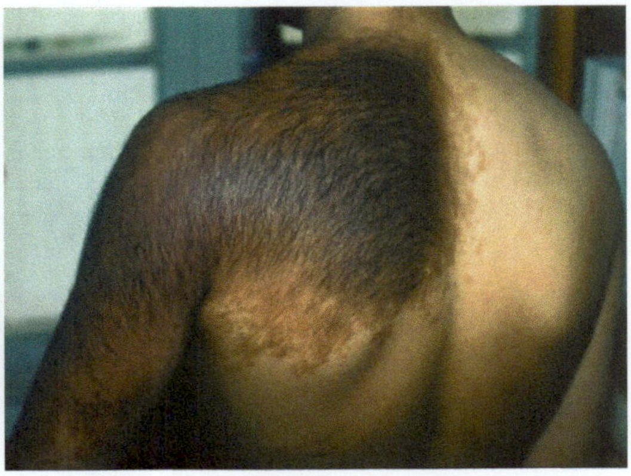

Fig. 7.36 Becker nevus (Courtesy of Dr. Aïcha Salhi, Algiers, Algeria)

of the shoulder girdle, scoliosis, and skeletal anomalies of the thorax, as well as segmental odontomaxillary hypoplasia or dysplasia [102].

Both isolated Becker nevus and Becker nevus syndrome usually occur sporadically. Some exceptional familial cases [156] can probably not be explained by the mechanism of paradominant inheritance [80] (see Sect. 3.1.1.4).

7.3.2.6 Porokeratotic Eccrine Nevus: A Mosaic Manifestation of KID Syndrome

The long-winded synonym "porokeratotic eccrine ostial and dermal duct nevus" (PEODDN) is redundant because an eccrine nevus always involves the dermal ducts, and its porokeratotic component is always ostial [116]. The infundibular parakeratotic plugs (Fig. 7.38a, b) are most prominent in lesions involving the palms and soles. In other parts of the body, they may be far less conspicuous or even absent. The disorder has sometimes been mistaken for "linear porokeratosis" or "unilateral punctate porokeratosis" [11, 183, 190].

Titeux et al. [214] described systematized linear lesions in the mother of a child with KID syndrome. In several cases of porokeratotic eccrine nevi, Easton et al. [45] found mosaic *GJB2* mutations resulting in a defective connexin 26, and they suggested that such nevi represent a mosaic manifestation of KID syndrome. Remarkably, Jamora and Celis [130] observed a widespread porokeratotic eccrine nevus in a boy who had gross hearing loss (Fig. 7.38c). This case may also be taken as an example of "porokeratotic eccrine nevus syndrome".

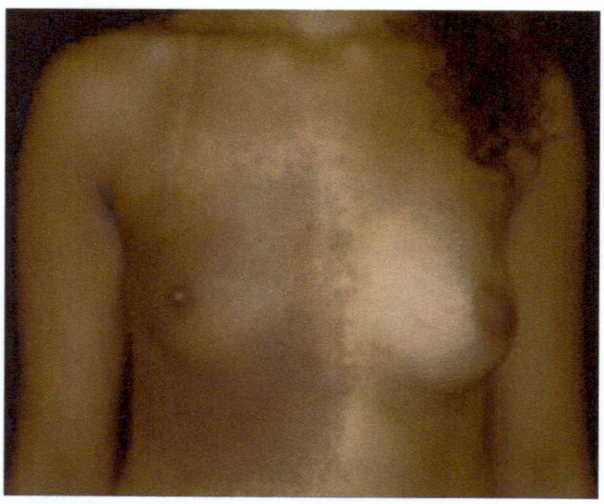

Fig. 7.37 A typical feature of Becker nevus syndrome is hypoplasia of the ipsilateral breast [31] (Reprinted under license of Creative Commons)

Masferrer [163] described linear lesions of porokeratotic eccrine nevus in a mother and her daughter. Because the daughter had, in addition, "a few solitary units," it is tempting to speculate that the mother had a type 1 segmental manifestation of KID syndrome heralding gonadal mosaicism, whereas her daughter had a type 2 segmental involvement (see Sect. 10.2.6).

7.3.2.7 Eccrine Nevus of the Castori Type

Castori et al. [26] described a complex organoid nevus characterized by proliferations of immature to well-formed eccrine duct-like structures that were localized in the deep dermis, being scattered within an abundant fibrous stroma. The keratinocytes were hyperpigmented, especially at the base of the rete ridges. Clinically, the lesions formed large depressed plaques with a brownish-violaceous hue (Fig. 7.39). Their arrangement was segmental, but did not appear to follow Blaschko's lines. The systematized nevus was found to be associated with several unusual extracutaneous

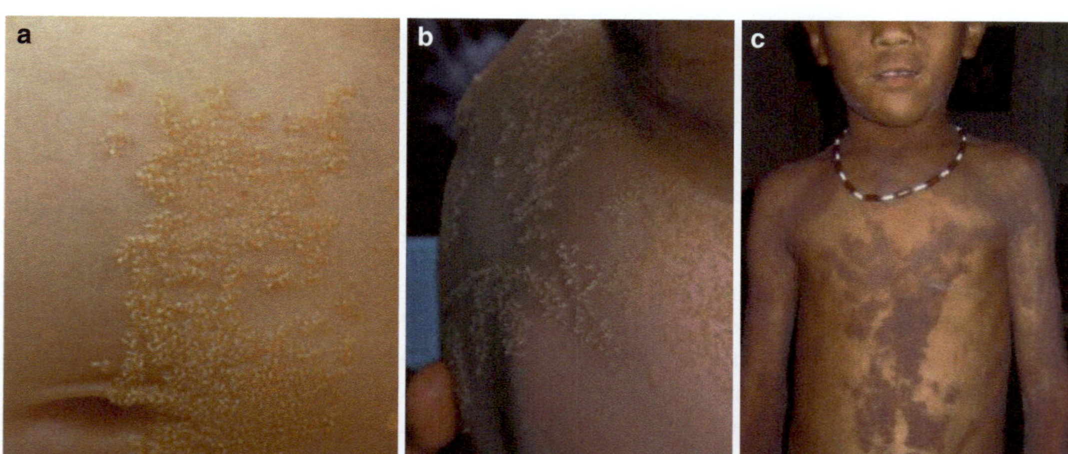

Fig. 7.38 Porokeratotic eccrine nevus. (**a**, **b**) Linear arrangement of filiform hyperkeratoses in an 8-month-old girl [24]; (**c**) systematized involvement in a 7-year-old boy with bilateral hearing loss [130] (**a**, **b**: Reprinted with permission from John Wiley & Sons, USA; **c**: Reprinted with permission from Elsevier Limited, Oxford, UK)

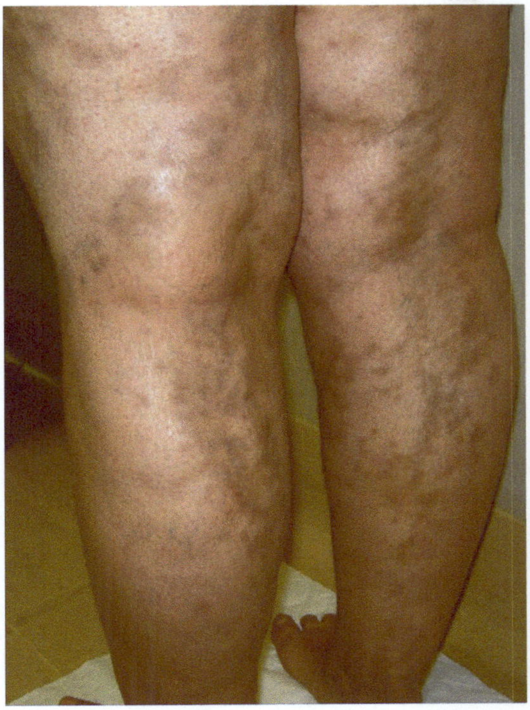

Fig. 7.39 Eccrine nevus of the Castori type [26] (Courtesy of Dr. Marco Castori, Rome, Italy)

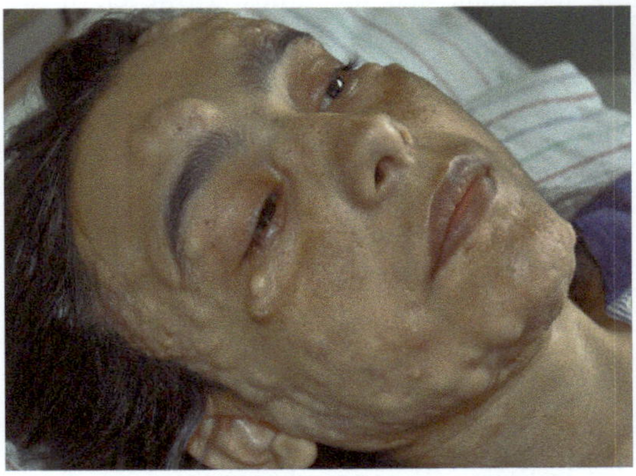

Fig. 7.40 Nevus trichilemmocysticus

defects (see Table 7.2). Most likely it represents a new mosaic skin disorder.

7.3.2.8 Nevus Trichilemmocysticus

The hallmark of this organoid nevus is the presence of multiple trichilemmal cysts being arranged along Blaschko's lines (Fig. 7.40) and usually intermingled with multiple filiform hyperkeratosis [210]. Comedo-like plugs may likewise be present. Additional cases have recently been described [52, 143, 144]. A coexistence with bone lesions as noted in two cases [52, 210] may represent a distinct epidermal nevus syndrome [103].

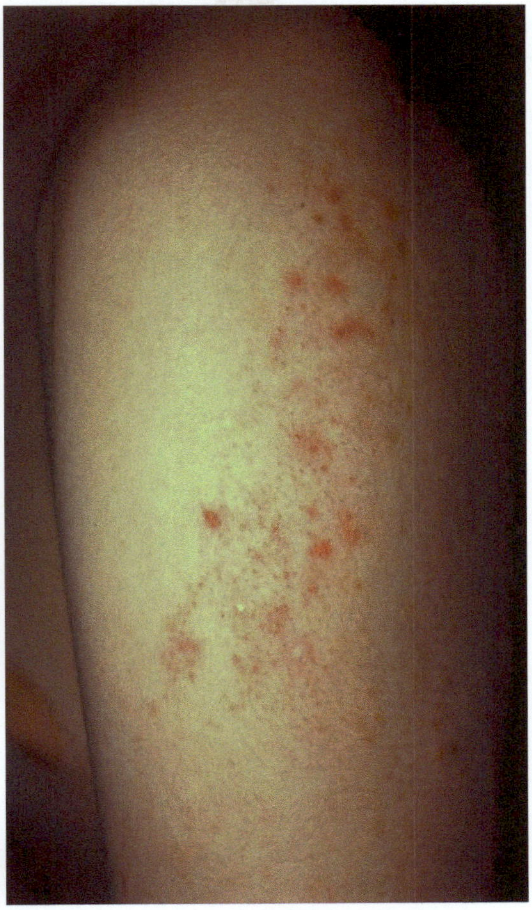

Fig. 7.41 Acne nevus of Munro [173] (Reprinted with permission from Elsevier Limited, Oxford, UK)

7.3.2.9 Acne Nevus of Munro

Munro's acne nevus (Fig. 7.41) is a mosaic manifestation of Apert syndrome, an autosomal dominant trait caused by *FGFR2* mutations [173]. The inflammatory acne lesions that are arranged along Blaschko's lines have been conflated with nevus comedonicus [166, 208]. It should be borne in mind that nevus comedonicus does usually *not* show any inflammatory acne lesions and is *not* caused by *FGFR2* mutations.

7.4 Vascular Nevi

The group of vascular nevi includes many different entities. It is important to realize that the presently fashionable name "capillary malformation" is ambiguous and unsuitable to categorize a specific vascular disorder like nevus flammeus or nevus roseus. "Capillary malformation" is an umbrella term that includes, in addition to the vascular nevi, some skin lesions that cannot be categorized as nevi such as the nuchal, glabellar, or sacral salmon patches or the vascular lesions of Osler-Rendu-Weber disease [96].

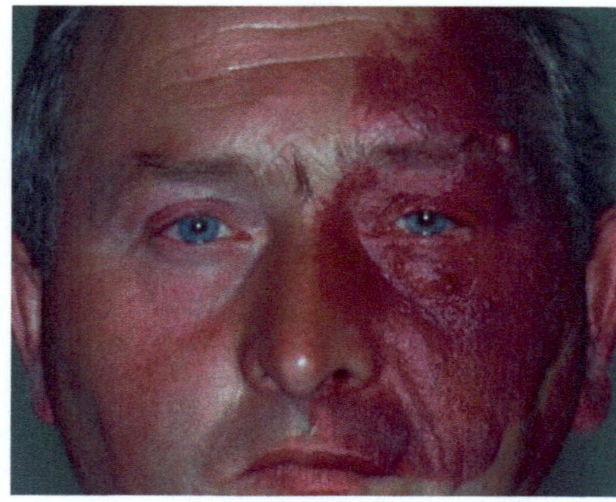

Fig. 7.42 Nevus flammeus

7.4.1 Capillary Nevi

This group comprises nevus flammeus, nevus roseus, rhodoid nevus, cutis marmorata telangiectatica congenita, livedo reticularis congenita, angiokeratoma circumscriptum, angioma serpiginosum, and nevus anemicus.

7.4.1.1 Nevus Flammeus

This is a dark red vascular patch arranged in a lateralized flag-like or checkerboard pattern (Fig. 7.42). The disorder is caused by a lethal mutation in *GNAQ* [204]. Hence, its mosaic origin [77] can now be taken as proven.

Trigeminal Arrangement of Facial Nevi Flammei?

Since more than 100 years our textbooks say that facial nevi flammei correspond to the dermatomal segments of trigeminal innervation. This dogma, however, has no scientific basis. When large numbers of photographs are studied, there is ample evidence that facial nevi flammei display a flag-like arrangement that does not correspond to the cutaneous representation of the three trigeminal branches [93]. This iconoclastic view is not entirely new. Similar arguments against the prevailing belief of neural origin of these nevi were presented by Alexander already in 1972 [4]. Today the concept of mosaicism implies that the assumption of a neurological origin is unnecessary. Because the mosaic nature of nevus flammeus was proven at the molecular level [204], the doctrine of trigeminal arrangement of facial nevi flammei will most likely be abandoned.

Sturge-Weber-Klippel-Trenaunay Syndrome as a Unifying Concept

According to our scholastic tradition, we distinguish a Sturge-Weber syndrome involving the face from a Klippel-Trenaunay syndrome involving the trunk and the limbs. However, if we accept the concept of mosaicism, it becomes obvious that the two syndromes merely represent geographical variants of one clinicogenetic entity for which the name "Sturge-Weber-Klippel-Trenaunay syndrome" has been proposed [90].

Sturge-Weber syndrome is defined by a nevus flammeus involving either unilaterally or bilaterally the upper half of the face in combination with vascular lesions of the brain giving rise to seizures and vascular anomalies of the eye that may cause glaucoma. The generally held belief that in Sturge-Weber syndrome the nevus flammeus is located in a V1 or V1–V2 neurotomal distribution is probably not true.

Klippel-Trenaunay syndrome includes nevus flammeus outside the head and neck region and enlargement of a limb that usually shows varicosities. In addition, dysplasia of lymph vessels may be found. The disorder may be unilateral or bilateral, and the enlargement may involve some fingers or toes or an entire limb. By way of exception, hypoplasia of the involved limb may be noted [35]. Because Parkes Weber was the first to describe arteriovenous anastomoses in the hypertrophic limb, some authors prefer the term "Klippel-Trenaunay-Weber syndrome," whereas others consider the "Parkes Weber syndrome" to represent a separate disorder [62]. Most authors, however, agree that Klippel-Trenaunay and Parkes Weber syndromes are clinical variants of the same disease.

For the following reasons it seems justified to take Sturge-Weber and Klippel-Trenaunay syndromes as a single clinicogenetic entity. Firstly, the two disorders sometimes occur together (Fig. 7.43) [37, 84]. Deutsch [37] collected 39 reports of such combination and described an additional case. Remarkably and revealingly, some authors who categorize the syndromes as separate entities are presenting photographs showing vascular lesions of both syndromes in the same patient [59, 62]. Of 14 patients with Sturge-Weber syndrome treated for glaucoma, 5 also suffered from Klippel-Trenaunay syndrome [12]. Conversely, in patients with Klippel-Trenaunay syndrome, the presently available imaging techniques may reveal intracranial vascular lesions characteristic of Sturge-Weber syndrome [129]. The concept of mosaicism implies that the responsible mutation may involve various regions of the body (Fig. 7.44). When the nevus flammeus involves exclusively the trunk with a corresponding vascular malformation of the spinal cord, the disorder has also been called "Cobb syndrome" [203].

At the time when this book appears, we will almost certainly know whether the Klippel-Trenaunay phenotype is likewise caused by a postzygotic mutation in *GNAQ*, which seems rather likely.

7.4.1.2 Nevus Roseus

This is a lateralized telangiectatic birthmark of light-red or pale-pink color (Fig. 7.45). The disorder has so far been mistaken as a nevus flammeus from which it should be distinguished for the

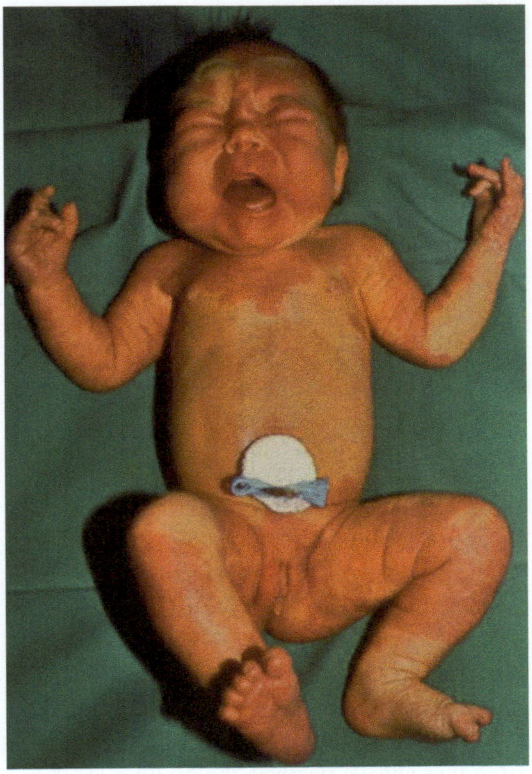

Fig. 7.43 Sturge-Weber-Klippel-Trenaunay syndrome

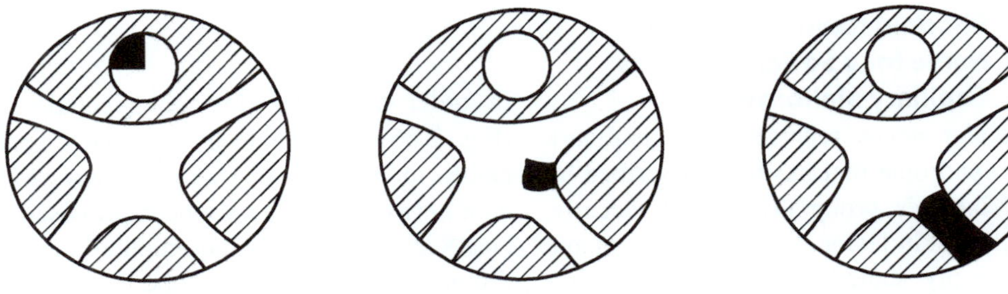

Fig. 7.44 The underlying mutation may involve various regions of the body, but from a genetic point of view these are three names for the same entity

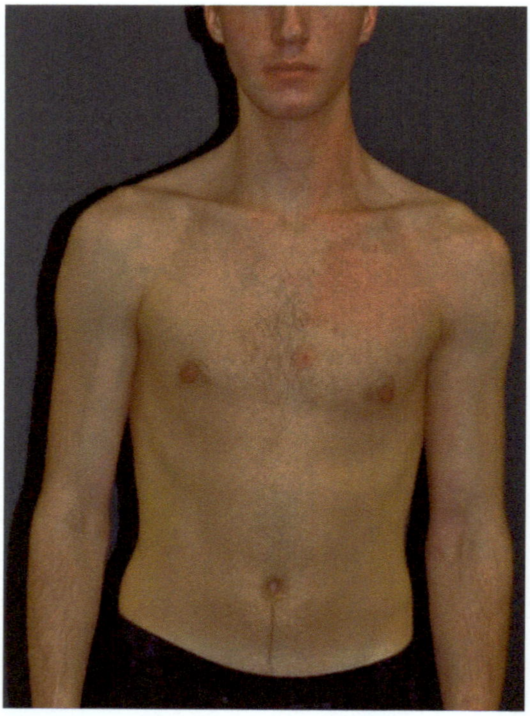

Fig. 7.45 Nevus roseus

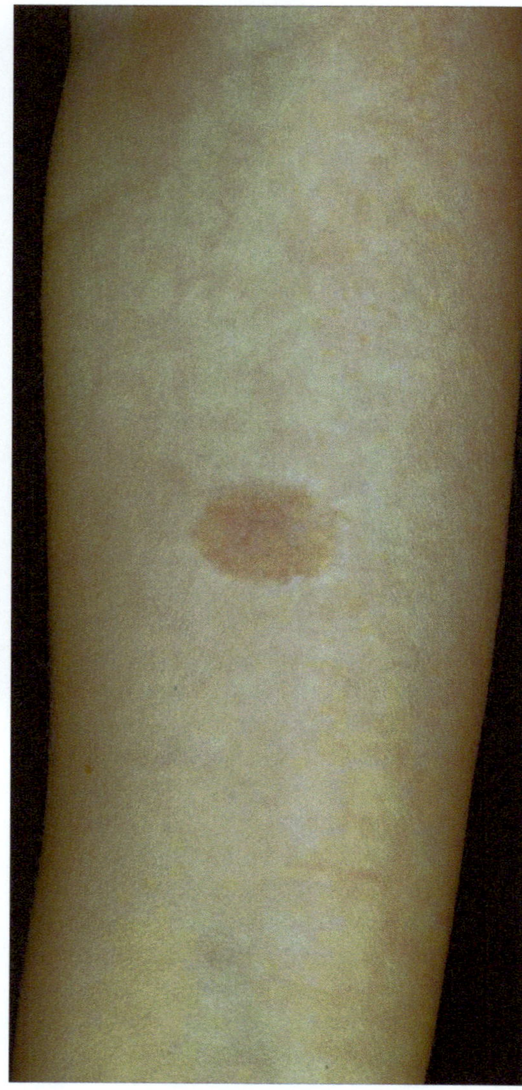

Fig. 7.46 Rhodoid nevus. Note the characteristic anemic halo (Courtesy of Dr. Regina Fölster-Holst, Kiel, Germany)

following reasons. Nevus roseus is consistently found in phacomatosis spilorosea, whereas nevus flammeus is a regular component of phacomatosis cesioflammea [91, 92]. It should be noted, however, that during early infancy a nevus flammeus may also show a pink color and may thus be indistinguishable from nevus roseus. Moreover, nevus roseus should be distinguished from the salmon patch that always involves the midline, does not reflect mosaicism, and can, therefore, not be categorized as a nevus [82] (see Sect. 12.2.3.1).

7.4.1.3 Rhodoid Nevus: A Specific Name for a So Far Nameless Capillary Malformation

This vascular nevus represents the hallmark of a new syndrome that has been described by Miikka Vikkula's group [46]. Until recently, however, the skin lesion had no specific name. It was simply called "capillary malformation," representing the cutaneous component of the phenotype "capillary malformation-arteriovenous malformation" and being caused by *RASA1* mutations [20, 46, 122, 186]. In order to render this telangiectatic nevus (Fig. 7.46) identifiable by clinicians, separate names in the form of *rhodoid nevus* and *rhodoid nevus syndrome* have been proposed [104] (see Sect. 10.4.2). Notably, rhodoid nevus syndrome includes the phenotype "capillary malformation without arteriovenous malformation" (OMIM 608354) [176].

Rhodoid nevi are lighter than nevus flammeus but darker than nevus roseus. The lesions tend to be rather small and their shape is round or oval. In contrast to nevus flammeus and nevus roseus, rhodoid nevi are transmitted as an autosomal dominant trait and they show a haphazard, nonsegmental distribution. As a highly characteristic feature, they are often surrounded by a pale anemic halo.

7.4.1.4 Cutis Marmorata Telangiectatica Congenita (Van Lohuizen Syndrome)

This nevus is characterized by a segmentally arranged, rather coarse-meshed erythema (Fig. 7.47) that blanches with pressure. It is accentuated by cold, but does not disappear with rewarming. Both telangiectasia and phlebectasia are noted, and ulceration may also occur. Because even in cases of most severe involvement some small unaffected areas are found, it has been assumed that the disorder is caused by a lethal mutation surviving by mosaicism [79].

The Van Lohuizen syndrome includes extracutaneous defects such as hyper- or hypoplasia of limbs, syndactyly, cleft palate, glaucoma, and mental deficiency [34].

Perhaps, cutis marmorata telangiectatica congenita represents a paradominant trait. This would explain why the phenotype usually occurs sporadically and why the exceptional familial cases do not show any consistent mode of transmission [34].

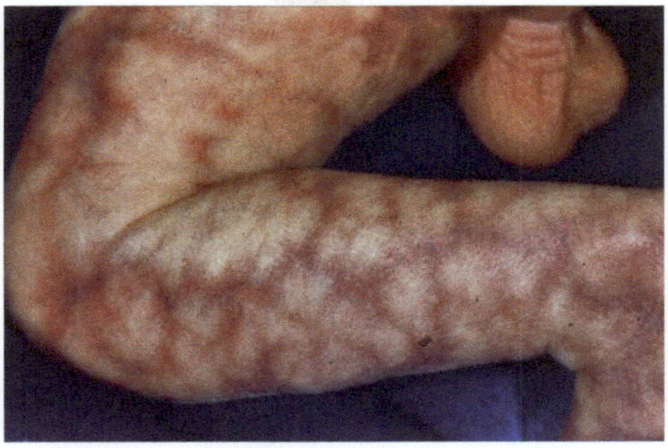

Fig. 7.47 Cutis marmorata telangiectatica congenita

7.4.1.5 Livedo Reticularis Congenita, a Hallmark of Megalencephaly-Livedo Reticularis Congenita Syndrome ("Macrocephaly-Capillary Malformation Syndrome")

Contrasting with cutis marmorata telangiectatica congenita, this type of vascular nevus is characterized by a rather close-meshed erythema (Fig. 7.48) [60, 225]. Megalencephaly-livedo reticularis congenita syndrome is a distinct multisystem birth defect caused by postzygotic *PIK3CA* mutations [189]. Associated neurologic defects include mental deficiency, brain asymmetry, megalencephaly, polymicrogyria, and hydrocephalus [168]. Asymmetric overgrowth and syndactyly are likewise present. As a highly characteristic additional cutaneous feature, a

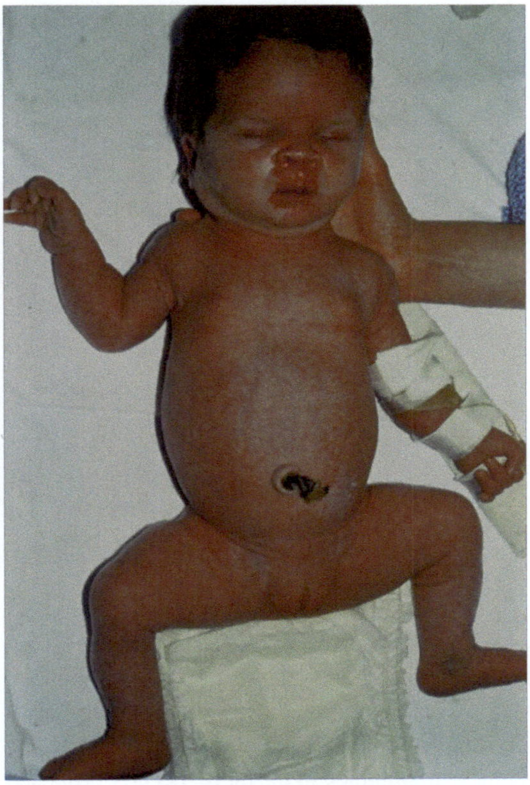

Fig. 7.48 Megalencephaly-livedo reticularis congenita syndrome. Note mesotropic port-wine stain of the forehead and upper philtrum

mesotropic port-wine patch with a predilection for the philtrum and upper lip may be noted (Fig. 6.2) [53, 79, 161] (see Sect. 6.2).

This particular syndrome was delineated in 1997 by Moore et al. [170] under the inappropriate designation "macrocephaly-cutis marmorata telangiectatica congenita." Later on, the term was

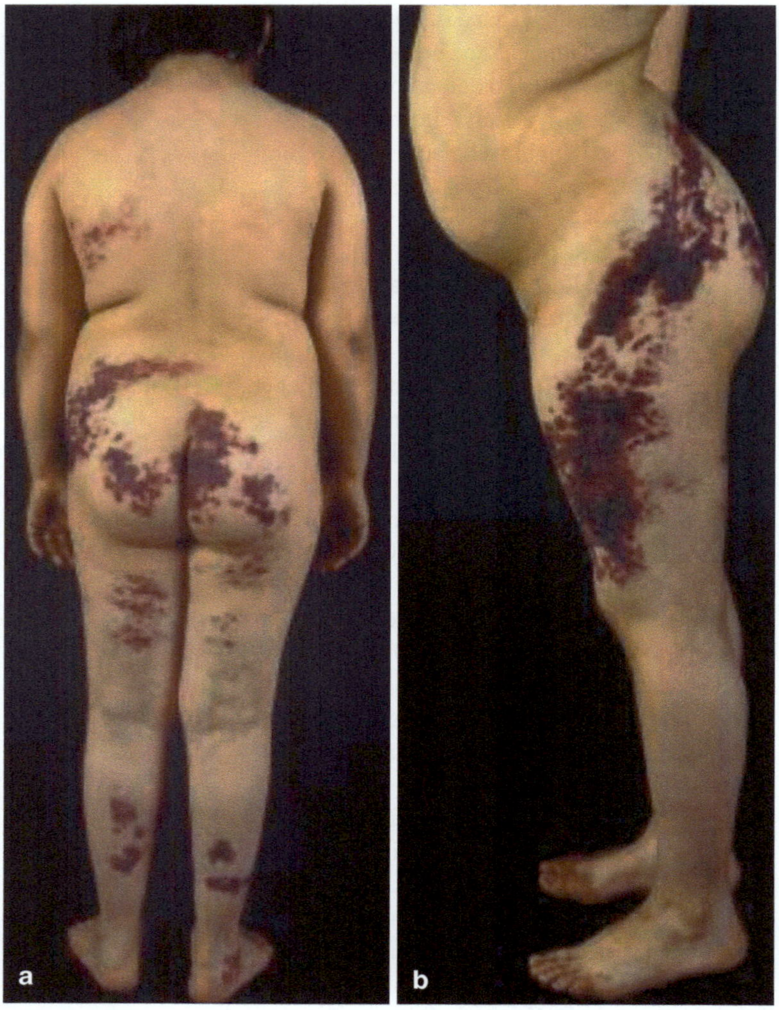

Fig. 7.49 Angiokeratoma circumscriptum showing a systematized band-like arrangment. (**a**) Dorsal view; (**b**) lateral view [14]. (Reprinted with permission from John Wiley & Sons, USA)

replaced by the even more inaccurate name "macrocephaly-capillary malformation" [180, 215, 225]. Clinicians should bear in mind that "capillary malformation" is an empty notion [96]. The name is maintained in the more recently proposed synonyms "megalencephaly-capillary malformation syndrome" and "megalencephaly-capillary malformation-polymicrogyria syndrome" [168, 176]. Here, the term *livedo reticularis congenita* is proposed to distinguish this vascular nevus from all other capillary malformations.

7.4.1.6 Angiokeratoma Circumscriptum

The disorder does not represent a benign neoplasia but a nevus, which is why "angiokeratoma" is a misnomer [33]. The nevus is arranged in a band-like pattern but does not follow Blaschko's lines [14, 141]. Cases of systematized involvement (Fig. 7.49) strongly suggest mosaicism resulting from a postzygotic mutation [14].

7.4.1.7 Segmentally Arranged Angioma Serpiginosum

"Angioma serpiginosum" does not show neoplastic proliferation and does, therefore, not represent an angioma [27, 139]. A more appropriate designation would be "telangiectasia serpiginosa," but the incorrect name has firmly been entrenched in our terminology. The disorder is characterized by bright-red punctate telangiectatic lesions that are often grouped to form a patchy or segmental arrangement [13, 27]. Histopathologically,

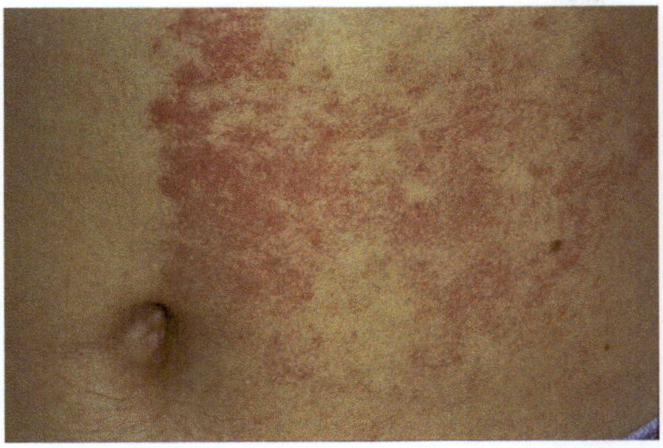

Fig. 7.50 Angioma serpiginosum showing a segmental distribution [27] (Reprinted with permission from S. Karger AG, Basel, Switzerland)

dilated and tortuous capillaries are noted in the uppermost part of the dermis.

Nonsegmental angioma serpiginosum sometimes shows a familial aggregation, suggesting transmission as an autosomal dominant trait [159]. On the other hand, cases of segmentally arranged angioma serpiginosum, suggesting mosaicism (Fig. 7.50), have so far always been sporadic [27]. It is as yet unclear whether such cases can be taken as a type 1 or type 2 segmental manifestation of the nonsegmental phenotype.

The arrangement of segmental angioma serpiginosum can so far not be assigned to any of the established archetypical patterns of mosaicism. Some authors assumed a distribution along Blaschko's lines [3, 55], but in most cases the lesions do not follow this pattern.

Segmental angioma serpiginosum has been confused with focal dermal hypoplasia and was, therefore, erroneously described as an X-linked trait [16, 126]. According to present knowledge, X-linked angioma serpiginosum (OMIM 300652) does most likely not exist [99].

7.4.1.8 Nevus Anemicus

This nevus represents a vasoconstrictive counterpart of telangiectatic nevi. It is a pale macule with a highly irregular margin (Fig. 7.51). If the skin is rubbed, there is no erythematous reflex within the nevus. Multiple lesions are not arranged in any archetypical mosaic pattern but rather show a patchy, haphazard distribution [216].

Nevus anemicus may occur as part of several binary genetic skin disorders such as phacomatosis cesioflammea [91], "phacomatosis cesioanemica" [105, 121], and mixed vascular nevus (see Sect. 8.1.1.2 and 9.5).

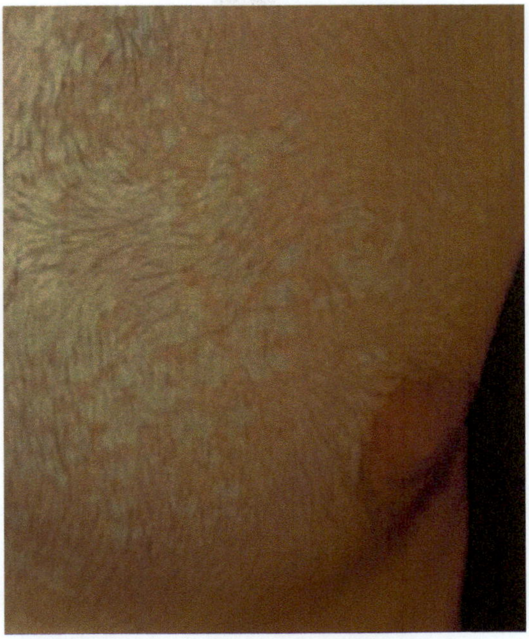

Fig. 7.51 Nevus anemicus (Courtesy of Dr. Helena de las Heras, Madrid, Spain)

7.4.1.9 Nevus Vascularis Mixtus

This disorder represents an admixture of anemic and reticular telangiectatic skin changes [74]. Possibly, the dermatosis is a particular example of allelic twin spotting. It represents a hallmark of a novel neurocutaneous phenotype, mixed vascular nevus syndrome [145, 195] (see Sect. 8.1.1.2).

7.4.2 Venous Nevi

According to present knowledge, three types of venous nevi can be distinguished.

7.4.2.1 Large Venous Nevus

This venous malformation occurs sporadically and is characterized by the presence of enlarged and tortuous veins involving the skin or the neighboring mucosa in a segmental pattern with a strict midline separation (Fig. 7.52) [228]. The lesions almost certainly reflect mosaicism and

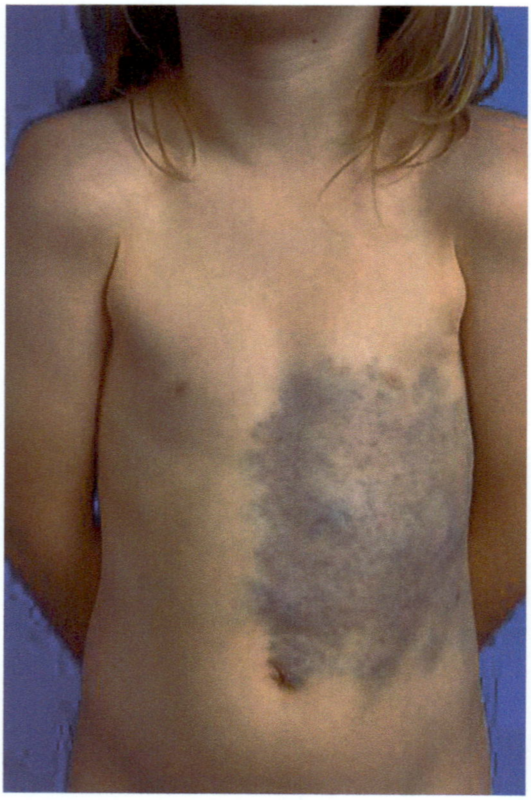

Fig. 7.52 Large venous nevus [228] (Reprinted with permission from S. Karger AG, Basel, Switzerland)

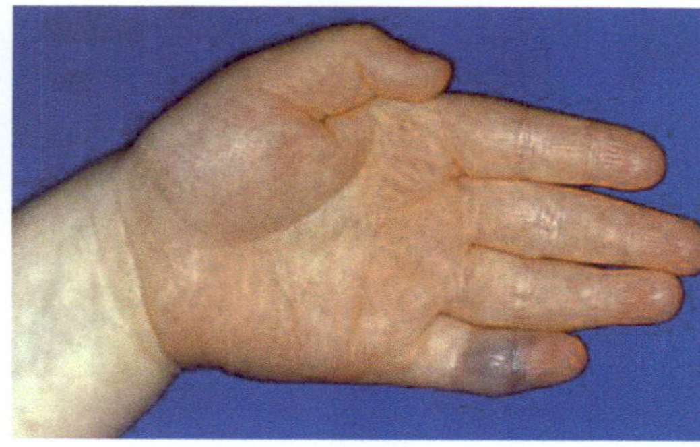

Fig. 7.53 Small venous nevus [228] (Reprinted with permission from S. Karger AG, Basel, Switzerland)

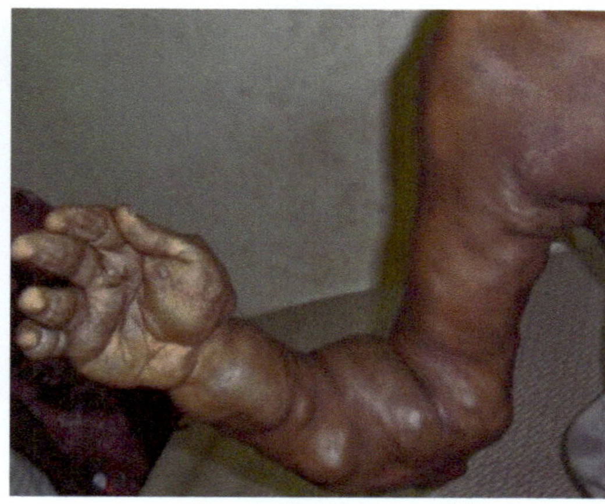

Fig. 7.54 Venous nevus of the Servelle-Martorell type [133] (Reprinted with license of Creative Commons)

appear to fulfill all criteria of a true nevus. The name "venous malformation," as presently used by many authors [49], is an umbrella term unsuitable to appropriately denote this disorder.

7.4.2.2 Small Venous Nevi ("Hereditary Cutaneomucosal Venous Malformations")

Small venous nevi (Fig. 7.53) that have been called "hereditary cutaneomucosal venous malformations" are inherited as an autosomal dominant trait [147]. They are caused by *TIE2/TEK* mutations [224] and fulfill all criteria of true nevi. Their relationship to large venous nevi is so far unclear.

7.4.2.3 Venous Nevus of the Servelle-Martorell Type

This malformation is characterized by a segmental area of huge phlebectasia with hyperplasia of the involved limb (Fig. 7.54) [133, 223]. The underlying bones may sometimes be hypotrophic.

7.5 Connective Tissue Nevi

These skin lesions are sometimes noted as part of various multisystem birth defects.

7.5.1 Collagen Nevi of Tuberous Sclerosis

Such nevi often represent a type 2 segmental manifestation of the disorder (see Sect. 10.3.1).

7.5.2 Linear Collagen Nevus

This peculiar connective tissue nevus (Fig. 7.55) that follows Blaschko's lines [44] has also been called "papulolinear collagenoma" [56, 152, 191]. A rather pronounced form of the disorder

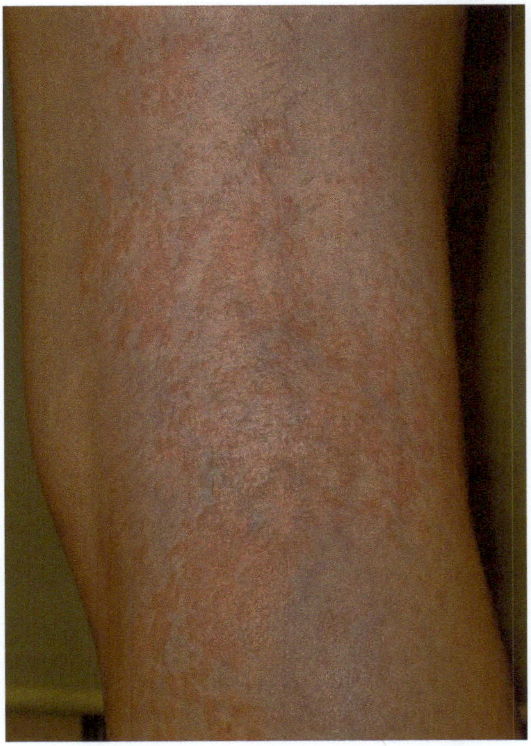

Fig. 7.55 Linear collagen nevus [44] (Reprinted with permission from the Society for Publication of Acta Dermato-Venereologica)

was described in a boy with phacomatosis pigmentokeratotica [17].

7.5.3 Elastin-Rich Nevus

Small, skin-colored or yellowish, firm papules as noted in Buschke-Ollendorff syndrome can be categorized as elastin-rich nevi. Large plaques showing an asymmetrical arrangement have been called "juvenile elastoma" [87]. They probably represent a type 2 segmental manifestation of the syndrome (see Sect. 10.3.2).

7.5.4 Segmental Manifestation of Ehlers-Danlos Syndromes

Type 1 segmental forms of autosomal dominant types of Ehlers-Danlos syndrome have been documented by some authors [32, 138], and a type 2

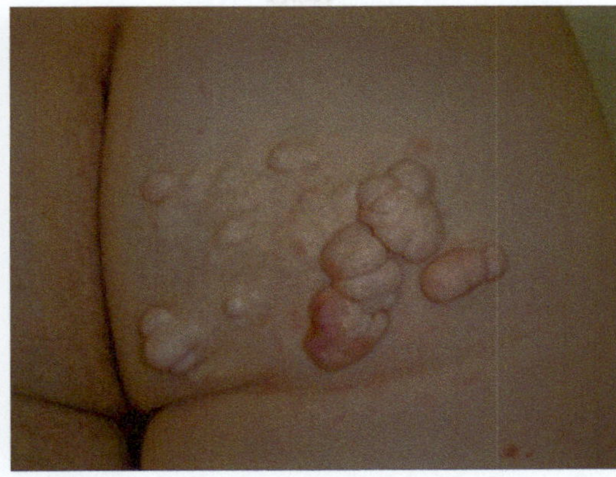

Fig. 7.56 Nevus lipomatosus superficialis [25] (Reprinted with permission from John Wiley & Sons, USA)

segmental involvement has also been described (see Sect. 10.3.3).

7.6 Fatty Tissue Nevi

Two different types of fatty tissue nevi are so far known.

7.6.1 Nevus Lipomatosus Superficialis

This malformation consists of multiple soft papules and nodules involving the trunk in a linear arrangement (Fig. 7.56) [29, 226]. The arrangement is linear, but presumably not Blaschko-linear as assumed by some authors [19]. Histopathologically, hyperplastic and lobulated fatty tissue is noted.

7.6.2 Nevus Psiloliparus

This nevus is a flat and hairless lesion involving the scalp [113]. In the past it has been mistaken for sebaceous nevus and sometimes even for alopecia areata. The hair follicles are either absent or present in the form of embryonic anlagen [111]. The nevus shows a patchy, nonlinear arrangement (Fig. 7.57). It may occur either as a hallmark of encephalocraniocutaneous lipomatosis or as an isolated lesion [111, 151]. The etiological concept

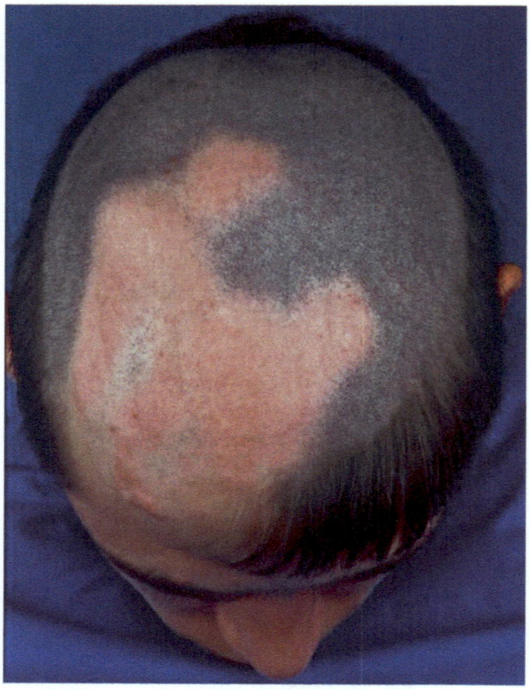

Fig. 7.57 Nevus psiloliparus

of a lethal mutation surviving by mosaicism [117] has now been accepted [36, 40, 134, 169], although molecular proof is still lacking.

Nevus psiloliparus should be distinguished from nevus lipomatosus superficialis of the Hoffmann-Zurhelle type that is characterized by nodular lesions (see Sect. 7.6.1). It is usually localized on the trunk but may, by way of exception, also involve the scalp [8].

References

1. Agrawal S, Garg VK, Sah SP, Agarwalla A (2001) Acne in Becker's nevus. Int J Dermatol 40:583–585
2. Akelma AZ, Cizmeci MN, Kanburoglu MK, Mete E (2012) A diagnostic dilemma: inflammatory linear verrucous epidermal nevus versus linear psoriasis. J Pediatr 162:879
3. Al Hawsawi K, Al Aboud K, Al Aboud D, Al Githami A (2003) Linear angioma serpiginosum. Pediatr Dermatol 20:167–168
4. Alexander GL (1972) Sturge-Weber syndrome. In: Vinken PJ, Bruyn GW (eds) Handbook of clinical neurology, vol 14. North Holland Publishing Company, Amsterdam, pp 223–240
5. Alimurung FM, Lapenas D, Willis I, Lang P (1979) Zebra-like hyperpigmentation in an infant with multiple congenital defects. Arch Dermatol 115:878–881
6. Aloi FG, Molinero A (1987) Nevus comedonicus with epidermolytic hyperkeratosis. Dermatologica 174:140–143
7. Alomari AI (2009) Characterization of a distinct syndrome that associates complex truncal overgrowth, vascular, and acral anomalies: a descriptive study of 18 cases of CLOVES syndrome. Clin Dysmorphol 18:1–7
8. Al-Oudah N, Al Rujaii A, Al-Rikabi AC (2009) Nevus lipomatosus of the scalp mimicking sebaceous nevus: report of an unusual case. Eur J Pediatr Dermatol 19:135–138
9. Alsaleh QA, Nanda A, Hassab-el-Naby HM, Sakr MF (1994) Familial inflammatory linear verrucous epidermal nevus (ILVEN). Int J Dermatol 33: 52–54
10. Antaya RJ, Keller RA, Wilkerson JA (1995) Placental nevus cells associated with giant congenital pigmented nevi. Pediatr Dermatol 12: 260–262
11. Apel HP, Kuhlwein A, Jänner M (1986) Porokeratosis linearis unilateralis (linear porokeratosis). Hautarzt 37:284–286
12. Azuma I (1984) Congenital glaucoma. Klin Monbl Augenheilkd 184:287–289
13. Bayramgurler D, Filinte D, Kiran R (2008) Angioma serpiginosum with sole involvement. Eur J Dermatol 18:708–709
14. Bechara FG, Happle R, Altmeyer P, Grabbe S, Jansen T (2006) Angiokeratoma circumscriptum arranged in a systematized band-like pattern suggesting mosaicism. J Dermatol 33:489–491
15. Biesecker LG, Happle R, Mulliken JB, Weksberg R, Graham JM Jr, Viljoen DL, Cohen MM Jr (1999) Proteus syndrome: diagnostic criteria, differential diagnosis, and patient evaluation. Am J Med Genet 84:389–395
16. Blinkenberg EO, Brendehaug A, Sandvik AK, Vatne O, Hennekam RC, Houge G (2007) Angioma serpiginosum with oesophageal papillomatosis is an X-linked dominant condition that maps to Xp11.3-Xq12. Eur J Hum Genet 15:543–547
17. Boente Mdel C, Asial RA, Happle R (2008) Phacomatosis pigmentokeratotica: a follow-up report documenting additional cutaneous and extracutaneous anomalies. Pediatr Dermatol 25:76–80
18. Boente MC, Asial R, Primc NB, Happle R (2013) Angora hair nevus: a further case of an unusual epidermal nevus representing a hallmark of angora hair nevus syndrome (Schauder syndrome). J Dermatol Case Rep 7:49–51
19. Bolognia JL, Orlow SJ, Glick SA (1994) Lines of Blaschko. J Am Acad Dermatol 31:157–190, quiz 190–152
20. Boon LM, Mulliken JB, Vikkula M (2005) *RASA1*: variable phenotype with capillary and arteriovenous malformations. Curr Opin Genet Dev 15:265–269

21. Bourdeaut F, Hérault A, Gentien D, Pierron G, Ballet S, Reynaud S, Paris R, Schleiermacher G, Baumann C, Philippe-Chomette P, Gauthier-Villars M, Peuchmaur M, Radvanyi F, Delattre O (2010) Mosaicism for oncogenic G12D *KRAS* mutation associated with epidermal nevus, polycystic kidneys and rhabdomyosarcoma. J Med Genet 47:859–862
22. Bouthors J, Vantyghem MC, Manouvrier-Hanu S, Soudan B, Proust E, Happle R, Piette F (2006) Phacomatosis pigmentokeratotica associated with hypophosphataemic rickets, pheochromocytoma and multiple basal cell carcinomas. Br J Dermatol 155:225–226
23. Bygum A, Petkov Y, Graakjaer J, Jensen UB, Fagerberg C (2012) Phylloid hypermelanosis in a child with psychomotor delay, cicatricial alopecia, hearing loss and polythelia. Acta Derm Venereol 92:191–192
24. Cambiaghi S, Gianotti R, Caputo R (2007) Widespread porokeratotic eccrine ostial and dermal duct nevus along Blaschko lines. Pediatr Dermatol 24:162–167
25. Cardot-Leccia N, Italiano A, Monteil MC, Basc E, Perrin C, Pedeutour F (2007) Naevus lipomatosus superficialis: a case report with a 2p24 deletion. Br J Dermatol 156:380–381
26. Castori M, Annessi G, Castiglia D, Buffa V, Paradisi A, Cascone P, Zambruno G, Grammatico P, Paradisi M (2010) Systematized organoid epidermal nevus with eccrine differentiation, multiple facial and oral congenital scars, gingival synechiae, and blepharophimosis: a novel epidermal nevus syndrome. Am J Med Genet A 152A:25–31
27. Chen W, Liu TJ, Yang YC, Happle R (2006) Angioma serpiginosum arranged in a systematized segmental pattern suggesting mosaicism. Dermatology 213:236–238
28. Child FJ, Werring DJ, Vivier AW (1998) Proteus syndrome: diagnosis in adulthood. Br J Dermatol 139:132–136
29. Cil Y, Yildiz H, Simsek HA (2011) Sacral nevus lipomatosus cutaneous superficialis. J Cutan Aesthet Surg 4:161–162
30. Collin B, Taylor IB, Wilkie AO, Moss C (2007) Fibroblast growth factor receptor 3 (*FGFR3*) mutation in a verrucous epidermal naevus associated with mild facial dysmorphism. Br J Dermatol 156:1353–1356
31. Cosendey FE, Martinez NS, Bernhard GA, Dias MF, Azulay DR (2010) Becker nevus syndrome. An Bras Dermatol 85:379–384
32. Cullen SI (1979) Localized Ehlers-Danlos syndrome. Arch Dermatol 115:332–333
33. Dammert K (1965) Angiokeratosis naeviformis – a form of naevus telangiectaticus lateralis (naevus flammeus). Dermatologica 130:17–39
34. Danarti R, Happle R, König A (2001) Paradominant inheritance may explain familial occurrence of cutis marmorata telangiectatica congenita. Dermatology 203:208–211
35. Danarti R, König A, Bittar M, Happle R (2007) Inverse Klippel-Trenaunay syndrome: review of cases showing deficient growth. Dermatology 214:130–132
36. Delfino LN, Fariello G, Quattrocchi CC, Aiello C, Menchini L, Devito R, Zama M, Claps D, Vigevano F, Longo D (2011) Encephalocraniocutaneous lipomatosis (ECCL): neuroradiological findings in three patients and a new association with fibrous dysplasia. Am J Med Genet A 155A:1690–1696
37. Deutsch J, Weissenbacher G, Widhalm K, Wolf G, Barsegar B (1976) Combination of the syndrome of Sturge-Weber and the syndrome of Klippel-Trenaunay (author's transl). Klin Pädiatr 188:464–471
38. Devillers C, Quatresooz P, Hermanns-Lê T, Szepetiuk G, Lemaire R, Piérard-Franchimont C, Piérard GE (2011) Hypomelanosis of Ito: pigmentary mosaicism with immature melanosome in keratinocytes. Int J Dermatol 50:1234–1239
39. Dhar SU, Robbins-Furman P, Levy ML, Patel A, Scaglia F (2009) Tetrasomy 13q mosaicism associated with phylloid hypomelanosis and precocious puberty. Am J Med Genet A 149A:993–996
40. Dhouib A, Hanquinet S, La Scala GC (2013) Encephalocraniocutaneous lipomatosis: magnetic resonance imaging findings in a child. J Pediatr 163:297–297.e1
41. Dockx L, Lowenthal A, Van Bogaert L (1956) A congenital pigmentary syndrome close to incontinentia pigmenti with oligophrenia and multiple osteoarticular malformations. Rev Neurol (Paris) 95:48–54
42. Donati P, Trincone S, Cota C, Paolini F, Venuti A (2011) Kerinokeratosis papulosa-like lesions in an adult man with homozygous polymorphism of EVER 2. Eur J Dermatol 21:288–289
43. Droitcourt C, Adenis-Lamarre E, Ezzedine K, Taïeb A, Jouary T (2009) Blaschkolinear congenital melanocytic nevus. Dermatology 219:182–183
44. Dubiel W, Kiehlmann I, Happle R (2013) Linear collagen nevus: an unusual connective tissue nevus. Acta Derm Venereol (Epub ahead of print) doi: 10.2340/00015555-1747
45. Easton JA, Donnelly S, Kamps MA, Steijlen PM, Martin PE, Tadini G, Janssens R, Happle R, van Geel M, van Steensel MA (2012) Porokeratotic eccrine nevus may be caused by somatic connexin26 mutations. J Invest Dermatol 132:2184–2191
46. Eerola I, Boon LM, Mulliken JB, Burrows PE, Dompmartin A, Watanabe S, Vanwijck R, Vikkula M (2003) Capillary malformation-arteriovenous malformation, a new clinical and genetic disorder caused by *RASA1* mutations. Am J Hum Genet 73:1240–1249
47. Effendy I, Happle R (1992) Linear arrangement of multiple congenital melanocytic nevi. J Am Acad Dermatol 27:853–854
48. Elder DE, Green MH, Dt G, Kraemer KH, Clark WH Jr (1982) The dysplastic nevus syndrome: our definition. Am J Dermatopathol 4:455–460
49. Enjolras O, Ciabrini D, Mazoyer E, Laurian C, Herbreteau D (1997) Extensive pure venous

malformations in the upper or lower limb: a review of 27 cases. J Am Acad Dermatol 36:219–225
50. Esmarch F, Kulenkampff D (1885) Die elephantiastischen Formen: eine umfassende Darstellung der angeborenen und erworbenen Elephantiasis sowie aller verwandten Leiden. J. F. Richter, Hamburg, pp 235–236, Table XXV
51. Faletra F, Berti I, Tommasini A, Pecile V, Cleva L, Alberini E, Bruno I, Gasparini P (2012) Phylloid pattern of hypomelanosis closely related to chromosomal abnormalities in the 13q detected by SNP array analysis. Dermatology 225:294–297
52. Ferreira PS, Valente NY, Nico MM (2013) Multiple filiform keratoses and nodules in a 10-year-old girl. Pediatr Dermatol 30:261–262
53. Franceschini P, Licata D, Di Cara G, Guala A, Franceschini D, Genitori L (2000) Macrocephaly-cutis marmorata telangiectatica congenita without cutis marmorata? Am J Med Genet 90:265–269
54. García-Vargas A, Hafner C, Pérez-Rodríguez AG, Rodríguez-Rojas LX, González-Esqueda P, Stoehr R, Hernández-Torres M, Happle R (2008) An epidermal nevus syndrome with cerebral involvement caused by a mosaic *FGFR3* mutation. Am J Med Genet A 146A:2275–2279
55. Gerbig AW, Zala L, Hunziker T (1995) Angioma serpiginosum, a skin change along Blaschko lines? Hautarzt 46:847–849
56. Girard C, Bessis D (2006) Papulolinear collagenoma. J Am Acad Dermatol 54:S240
57. Goldstone S, Samitz MH, Carter DM (1967) 1. Giant cerebriform nevus of the scalp with malignant melanoma and metastases. 2. Multiple intradermal nevi with macular atrophy of many lesions. Arch Dermatol 95:137
58. Golitz LE, Weston WL (1979) Inflammatory linear verrucous epidermal nevus: association with epidermal nevus syndrome. Arch Dermatol 115: 1208–1209
59. Gomez MR, Bebin EM (1987) Sturge-Weber syndrome. In: Gomez MR (ed) Neurocutaneous diseases: a practical approach. Butterworths, Boston, pp 356–167
60. Gonzalez ME, Burk CJ, Barbouth DS, Connelly EA (2009) Macrocephaly-capillary malformation: a report of three cases and review of the literature. Pediatr Dermatol 26:342–346
61. González-Enseñat MA, Vicente A, Poo P, Catalá V, Pérez-Iribarne MM, Fuster C, Geán E, Happle R (2009) Phylloid hypomelanosis and mosaic partial trisomy 13: two cases that provide further evidence of a distinct clinicogenetic entity. Arch Dermatol 145:576–578
62. Gorlin RJ, Cohen MMJ, Hennekam RCM (eds) (2001) Syndromes of the head and neck. Oxford University Press, Oxford
63. Groesser L, Herschberger E, Ruetten A, Ruivenkamp C, Lopriore E, Zutt M, Langmann T, Singer S, Klingseisen L, Schneider-Brachert W, Toll A, Real FX, Landthaler M, Hafner C (2012) Postzygotic *HRAS* and *KRAS* mutations cause nevus sebaceous and Schimmelpenning syndrome. Nat Genet 44:783–787
64. Groesser L, Herschberger E, Sagrera A, Shwayder T, Flux K, Ehmann L, Wollenberg A, Torrelo A, Bagazgoitia L, Diaz-Ley B, Tinschert S, Oschlies I, Siner S, Mickler M, Toll A, Landthaler M, Real FX, Hafner C (2013) Phacomatosis pigmentokeratotica is caused by a postzygotic *HRAS* mutation in a multipotent progenitor cell. J Invest Dermatol 133:1998–2003
65. Groesser L, Vogt T, Happle R, Herschberger E, Mickler M, Landthaler M, Hafner C (2012) Nevus marginatus revisited: a combined organoid and non-organoid epidermal nevus caused by *HRAS* mutation. Br J Dermatol 168:892–894
66. Hafner C, Hartmann A, van Oers JM, Stoehr R, Zwarthoff EC, Hofstaedter F, Landthaler M, Vogt T (2007) *FGFR3* mutations in seborrheic keratoses are already present in flat lesions and associated with age and localization. Mod Pathol 20:895–903
67. Hafner C, Landthaler M, Happle R, Vogt T (2008) Nevus marginatus: a distinct type of epidermal nevus or merely a variant of nevus sebaceus? Dermatology 216:236–238
68. Hafner C, López-Knowles E, Luis NM, Toll A, Baselga E, Fernández-Casado A, Hernández S, Ribé A, Mentzel T, Stoehr R, Hofstaedter F, Landthaler M, Vogt T, Pujol RM, Hartmann A, Real FX (2007) Oncogenic *PIK3CA* mutations occur in epidermal nevi and seborrheic keratoses with a characteristic mutation pattern. Proc Natl Acad Sci U S A 104:13450–13454
69. Hafner C, Toll A, Gantner S, Mauerer A, Lurkin I, Acquadro F, Fernandez-Casado A, Zwarthoff EC, Dietmaier W, Baselga E, Parera E, Vicente A, Casanova A, Cigudosa J, Mentzel T, Pujol RM, Landthaler M, Real FX (2012) Keratinocytic epidermal nevi are associated with mosaic *RAS* mutations. J Med Genet 49:249–253
70. Hafner C, Toll A, Real FX (2011) *HRAS* mutation mosaicism causing urothelial cancer and epidermal nevus. N Engl J Med 365:1940–1942
71. Hafner C, van Oers JM, Vogt T, Landthaler M, Stoehr R, Blaszyk H, Hofstaedter F, Zwarthoff EC, Hartmann A (2006) Mosaicism of activating *FGFR3* mutations in human skin causes epidermal nevi. J Clin Invest 116:2201–2207
72. Hafner C, Vogt T, Landthaler M, Müsebeck J (2008) Somatic *FGFR3* and *PIK3CA* mutations are present in familial seborrhoeic keratoses. Br J Dermatol 159:214–217
73. Hamm H, Happle R (1986) Inflammatory linear verrucous epidermal nevus (ILVEN) in a mother and her daughter. Am J Med Genet 24:685–690
74. Hamm H, Happle R (1986) Mixed vascular nevus: report of 4 cases. Hautarzt 37:388–392
75. Happle R (1986) Cutaneous manifestation of lethal genes. Hum Genet 72:280
76. Happle R (1986) The McCune-Albright syndrome: a lethal gene surviving by mosaicism. Clin Genet 29:321–324

77. Happle R (1987) Lethal genes surviving by mosaicism: a possible explanation for sporadic birth defects involving the skin. J Am Acad Dermatol 16:899–906
78. Happle R (1989) Gregor Mendel and the dysplastic nevi. Hautarzt 40:70–76
79. Happle R (1990) Nichterbliche Genodermatosen. Hautarzt 41:104–109
80. Happle R (1992) Paradominant inheritance: a possible explanation for Becker's pigmented hairy nevus. Eur J Dermatol 2:39–40
81. Happle R (1993) Pigmentary patterns associated with human mosaicism: a proposed classification. Eur J Dermatol 3:170–174
82. Happle R (1995) What is a nevus? A proposed definition of a common medical term. Dermatology 191:1–5
83. Happle R (1995) Epidermal nevus syndromes. Semin Dermatol 14:111–121
84. Happle R (1999) Neurocutaneous diseases. In: Freedberg IM, Eisen AZ, Wolff K et al (eds) Dermatology in general medicine. Mc-Graw-Hill, New York, pp 2131–2148
85. Happle R (2000) Phylloid hypomelanosis is closely related to mosaic trisomy 13. Eur J Dermatol 10:511–512
86. Happle R (2001) Phylloid hypomelanosis and mosaic trisomy 13: a new etiologically defined neurocutaneous syndrome. Hautarzt 52:3–5
87. Happle R (2001) Buschke-Ollendorff syndrome: early, unilateral and pronounced involvement may be explained as a type 2 segmental manifestation. Eur J Dermatol 11:505
88. Happle R (2002) Speckled lentiginous nevus syndrome: delineation of a new distinct neurocutaneous phenotype. Eur J Dermatol 12:133–135
89. Happle R (2002) Transposable elements and the lines of Blaschko: a new perspective. Dermatology 204:4–7
90. Happle R (2003) Sturge-Weber-Klippel-Trenaunay syndrome: what's in a name? Eur J Dermatol 13: 223
91. Happle R (2005) Phacomatosis pigmentovascularis revisited and reclassified. Arch Dermatol 141:385–388
92. Happle R (2005) Nevus roseus: a distinct vascular birthmark. Eur J Dermatol 15:231–234
93. Happle R (2006) Viewpoint 2 in: What is the biological basis of pattern formation in skin lesions? Exp Dermatol 15:553–557
94. Happle R (2007) Type 2 segmental Cowden disease vs Proteus syndrome. Br J Dermatol 156:1089–1090
95. Happle R (2007) Linear Cowden nevus: a new distinct epidermal nevus. Eur J Dermatol 17:133–136
96. Happle R (2008) What is a capillary malformation? J Am Acad Dermatol 59:1077–1079
97. Happle R (2009) Speckled lentiginous naevus: which of the two disorders do you mean? Clin Exp Dermatol 34:133–135
98. Happle R (2009) Superimposed segmental manifestation of both rare and common cutaneous disorders: a new paradigm. Actas Dermosifiliogr 100:77–85
99. Happle R (2009) Angioma serpiginosum is not caused by *PORCN* mutations. Eur J Hum Genet 17:881–882; author reply 882
100. Happle R (2010) Giant melanocytic nevus may be explained as a superimposed patchy manifestation of a polygenic trait. Dermatology 221:30–33
101. Happle R (2010) Phylloid hypermelanosis: an unusual form of pigmentary mosaicism. Dermatology 220:183–185
102. Happle R (2010) The group of epidermal nevus syndromes Part I. Well defined phenotypes. J Am Acad Dermatol 63:1–22, quiz 23–24
103. Happle R (2010) The group of epidermal nevus syndromes Part II. Less well defined phenotypes. J Am Acad Dermatol 63:25–30, quiz 31–22
104. Happle R (2010) The rhodoid nevus: a proposed term for a so far unnamed capillary malformation. Dermatology 221:317–319
105. Happle R (2011) Didymosis cesioanemica: an unusual counterpart of phacomatosis cesioflammea. Eur J Dermatol 21:471
106. Happle R (2013) Phacomatosis pigmentokeratotica is a "pseudodidymosis". J Invest Dermatol 133:1923–1925
107. Happle R (2013) Nevus sebaceus is a mosaic RASopathy. J Invest Dermatol 133:597–600
108. Happle R, Danarti R (2010) CHILD syndrome vs. unilateral psoriasis. Int J Dermatol 49:847–848
109. Happle R, Fleiner J, Loskamp U (2004) Kerinokeratosis papulosa with a type 2 segmental manifestation. J Am Acad Dermatol 50: S84–S85
110. Happle R, Franco-Guío MF, Santacoloma-Osorio G (2012) Phylloid hypermelanosis: a cutaneous marker of several different disorders? Pediatr Dermatol. doi: 10.1111/pde.12053 [E-pub ahead of print]
111. Happle R, Hörster S (2004) Nevus psiloliparus: report of two nonsyndromic cases. Eur J Dermatol 14:314–316
112. Happle R, König A (1999) Familial naevus sebaceus may be explained by paradominant transmission. Br J Dermatol 141:377
113. Happle R, Küster W (1998) Nevus psiloliparus: a distinct fatty tissue nevus. Dermatology 197:6–10
114. Happle R, Metze D, Vera Casaño A (2010) Naevus lentiginosus linearis: a distinct skin disorder. Acta Derm Venereol 90:210–211
115. Happle R, Mittag H, Küster W (1995) The CHILD nevus: a distinct skin disorder. Dermatology 191:210–216
116. Happle R, Rogers M (2002) Epidermal nevi. Adv Dermatol 18:175–201
117. Happle R, Steijlen PM (1993) Encephalocraniocutaneous lipomatosis: a non-hereditary mosaic phenotype. Hautarzt 44:19–22
118. Happle R, Steijlen PM, Kolde G (1990) Naevus corniculatus: a new acantholytic disorder. Br J Dermatol 122:107–112
119. Happle R, Traupe H, Vakilzadeh F, Macher E (1982) Arguments in favor of a polygenic inheritance of precursor nevi. J Am Acad Dermatol 6:540–543

120. Harit D, Aggarwal A (2010) CLOVE syndrome with nevus unis lateris: report of a case. Pediatr Dermatol 27:311–312
121. Hernández-Núñez A, Borbujo J, Córdoba S, Martínez-Morán C, Martínez-Sánchez D, Romero-Maté A (2011) Nevus anemicus of the cheek with ipsilateral melanosis bulbi: an unusual example of didymosis. Eur J Dermatol 21:597–598
122. Hershkovitz D, Bergman R, Sprecher E (2008) A novel mutation in *RASA1* causes capillary malformation and limb enlargement. Arch Dermatol Res 300:385–388
123. Hofer T (2006) Does inflammatory linear verrucous epidermal nevus represent a segmental type 1/type 2 mosaic of psoriasis? Dermatology 212:103–107
124. Hohenleutner U, Landthaler M (1993) Laser therapy of verrucous epidermal naevi. Clin Exp Dermatol 18:124–127
125. Hohenleutner U, Wlotzke U, Konz B, Landthaler M (1995) Carbon dioxide laser therapy of a widespread epidermal nevus. Lasers Surg Med 16:288–291
126. Houge G, Oeffner F, Grzeschik KH (2008) An Xp11.23 deletion containing *PORCN* may also cause angioma serpiginosum, a cosmetic skin disease associated with extreme skewing of X-inactivation. Eur J Hum Genet 16:1027–1028
127. Hussein MR, Roggero E, Tuthill RJ, Wood GS, Sudilovsky O (2003) Identification of novel deletion loci at 1p36 and 9p22-21 in melanocytic dysplastic nevi and cutaneous malignant melanomas. Arch Dermatol 139:816–817
128. Hussein MR, Wood GS (2002) Molecular aspects of melanocytic dysplastic nevi. J Mol Diagn 4:71–80
129. Ishimoto S, Ito K, Matsuzaki M, Kimura M (2002) Sensorineural hearing loss with intracranial venous malformations in Klippel-Trenaunay syndrome. Ann Otol Rhinol Laryngol 111:558–562
130. Jamora MJ, Celis MA (2008) Generalized porokeratotic eccrine ostial and dermal duct nevus associated with deafness. J Am Acad Dermatol 59:S43–S45
131. Jinawath N, Zambrano R, Wohler E, Palmquist MK, Hoover-Fong J, Hamosh A, Batista DA (2011) Mosaic trisomy 13: understanding origin using SNP array. J Med Genet 48:323–326
132. Kalter DC, Griffiths WA, Atherton DJ (1988) Linear and whorled nevoid hypermelanosis. J Am Acad Dermatol 19:1037–1044
133. Karuppal R, Raman RV, Valsalan BP, Gopakumar T, Kumaran CM, Vasu CK (2008) Servelle-Martorell syndrome with extensive upper limb involvement: a case report. J Med Case Reports 2:142
134. Kim DH, Park SB, Lee Y, Im M, Seo YJ, Choi SH, Lee JH (2012) Encephalocraniocutaneous lipomatosis without neurologic anomalies. Ann Dermatol 24:476–478
135. Kinsler VA, Abu-Amero S, Budd P, Jackson IJ, Ring SM, Northstone K, Atherton DJ, Bulstrode NW, Stanier P, Hennekam RC, Sebire NJ, Moore GE, Healy E (2012) Germline melanocortin-1-receptor genotype is associated with severity of cutaneous phenotype in congenital melanocytic nevi: a role for *MC1R* in human fetal development. J Invest Dermatol 132:2026–2032
136. Kinsler VA, Thomas AC, Ishida M, Bulstrode NW, Loughlin S, Hing S, Chalker J, McKenzie K, Abu-Amero S, Slater O, Chanudet E, Palmer R, Morrogh D, Stanier P, Healy E, Sebire NJ, Moore GE (2013) Multiple congenital melanocytic nevi and neurocutaneous melanosis are caused by post-zygotic mutations in codon 61 of *NRAS*. J Invest Dermatol 133: 2229–2236
137. Köse O, Huseynov S, Demiriz M (2012) Giant Mongolian macules with bilateral ocular involvement: case report and review. Dermatology 224:126–129
138. Krieg T, Landthaler M, Ihme A, Wolff HH, Braun-Falco O (1984) Cutis hyperelastica circumscripta: a case report with a morphologic study. Hautarzt 35:415–417
139. Kumakiri M, Katoh N, Miura Y (1980) Angioma serpiginosum. J Cutan Pathol 7:410–421
140. Kurek KC, Luks VL, Ayturk UM, Alomari AI, Fishman SJ, Spencer SA, Mulliken JB, Bowen ME, Yamamoto GL, Kozakewich HP, Warman ML (2012) Somatic mosaic activating mutations in *PIK3CA* cause CLOVES syndrome. Am J Hum Genet 90:1108–1115
141. Kuske H, Baumgartner P (1962) Angiokeratoma corporis naeviforme. Dermatologica 124:303
142. Küster W, König A (1999) Hypomelanosis of Ito: no entity, but a cutaneous sign of mosaicism. Am J Med Genet 85:346–350
143. Lang SC, Bauer B, Bröcker EB, Hamm H (2010) Naevus trichilemmocysticus - the first paediatric case of a newly delineated organoid naevus. J Eur Acad Dermatol Venereol 24:1485–1487
144. Larralde M, Boggio P, Abad ME, Corbella C, Happle R (2011) Nevus trichilemmocysticus: report of a new case of a recently recognized entity. Pediatr Dermatol 28:286–289
145. Leech SN, Taylor AE, Ramesh V, Birchall D, Lynch SA (2004) Widespread capillary malformation associated with global developmental delay and megalencephaly. Clin Dysmorphol 13:169–172
146. Levinsohn JL, Tian LC, Boyden LM, McNiff JM, Narayan D, Loring ES, Yun D, Sugarman JL, Overton JD, Mane SM, Lifton RP, Paller AS, Wagner AM, Antaya RJ, Choate KA (2013) Whole-exome sequencing reveals somatic mutations in *HRAS* and *KRAS*, which cause nevus sebaceus. J Invest Dermatol 133:827–830
147. Limaye N, Boon LM, Vikkula M (2009) From germline towards somatic mutations in the pathophysiology of vascular anomalies. Hum Mol Genet 18:R65–R74
148. Lindhurst MJ, Sapp JC, Teer JK, Johnston JJ, Finn EM, Peters K, Turner J, Cannons JL, Bick D, Blakemore L, Blumhorst C, Brockmann K, Calder P, Cherman N, Deardorff MA, Everman DB, Golas G, Greenstein RM, Kato BM, Keppler-Noreuil KM,

Kuznetsov SA, Miyamoto RT, Newman K, Ng D, O'Brien K, Rothenberg S, Schwartzentruber DJ, Singhal V, Tirabosco R, Upton J, Wientroub S, Zackai EH, Hoag K, Whitewood-Neal T, Robey PG, Schwartzberg PL, Darling TN, Tosi LL, Mullikin JC, Biesecker LG (2011) A mosaic activating mutation in *AKT1* associated with the Proteus syndrome. N Engl J Med 365:611–619
149. Livingstone E, Happle R, Hafner C, Piel S, Schadendorf D, Hillen U (2010) SASKIA-Nävus: ein neuer epidermaler Nävus? Poster presented at the 20th Annual Meeting of the Arbeitsgemeinschaft Onkologische Dermatologie, Kiel, Germany, 23–25 September
150. Llamas-Velasco M, Eguren C, Arranz E, Renedo M, de Argila D, García-Díez A (2010) Linear and whorled nevoid hypermelanosis and Axenfeld-Rieger anomaly: a novel association. Acta Derm Venereol 90:317–318
151. Llamas-Velasco M, Hernandez A, Colmenero I, Torrelo A (2011) Nevus psiloliparus in a child with encephalocraniocutaneous lipomatosis. Actas Dermosifiliogr 102:303–305
152. Lo LK, Tsai TF, Chen YF, Hung CM, Ko WC (2009) Papulolinear collagenoma with arborizing arrangement: report of a case. Pediatr Dermatol 26:111–112
153. Loffeld A, McLellan NJ, Cole T, Payne SJ, Fricker D, Moss C (2006) Epidermal naevus in Proteus syndrome showing loss of heterozygosity for an inherited *PTEN* mutation. Br J Dermatol 154:1194–1198
154. Lookingbill DP, Ladda RL, Cohen C (1984) Generalized epidermolytic hyperkeratosis in the child of a parent with nevus comedonicus. Arch Dermatol 120:223–226
155. Mabuchi T, Akasaka E, Kondoh A, Umezawa Y, Matsuyama T, Ozawa A (2008) Seborrheic keratosis that follows Blaschko's lines. J Dermatol 35:301–303
156. Maessen-Visch MB, Hulsmans RF, Hulsmans FJ, Neumann HA (1997) Melanosis naeviformis of Becker and scoliosis: a coincidence? Acta Derm Venereol 77:135–136
157. Maitra A, Gazdar AF, Moore TO, Moore AY (2002) Loss of heterozygosity analysis of cutaneous melanoma and benign melanocytic nevi: laser capture microdissection demonstrates clonal genetic changes in acquired nevocellular nevi. Hum Pathol 33:191–197
158. Marchesi L, Naldi L, Di Landro A, Cavalieri d'Oro L, Brevi A, Cainelli T (1992) Segmental lentiginosis with "jentigo" histologic pattern. Am J Dermatopathol 14:323–327
159. Marriott PJ, Munro DD, Ryan T (1975) Angioma serpiginosum: familial incidence. Br J Dermatol 93:701–706
160. Marsch WC, Taube KC, Käsemann B (1981) Ein Solitärfall von Ichthyosis hystrix gravior unilateralis: klinische und morphologische Befunde. Z Hautkr 56:1073–1080
161. Martínez-Glez V, Romanelli V, Mori MA, Gracia R, Segovia M, González-Meneses A, López-Gutierrez JC, Gean E, Martorell L, Lapunzina P (2010) Macrocephaly-capillary malformation: analysis of 13 patients and review of the diagnostic criteria. Am J Med Genet A 152A:3101–3106
162. Maruani A, Khallouf R, Machet MC, Lorette G (2012) Diffuse linear and whorled nevoid hypermelanosis in a newborn. J Pediatr 160:171
163. Masferrer E, Vicente MA, Bassas-Vila J, Rovira C, González-Enseñat MA (2010) Porokeratotic eccrine ostial and dermal duct naevus: report of 10 cases. J Eur Acad Dermatol Venereol 24:847–851
164. Mehrabi D, Thomas JE, Selim MA, Prose NS (2001) Waxy keratoses of childhood in a segmental distribution. Pediatr Dermatol 18:415–416
165. Mehregan AH, Pinkus H (1965) Life history of organoid nevi: special reference to nevus sebaceus of Jadassohn. Arch Dermatol 91:574–588
166. Melnik BC, Vakilzadeh F, Aslanidis C, Schmitz G (2008) Unilateral segmental acneiform naevus: a model disorder towards understanding fibroblast growth factor receptor 2 function in acne? Br J Dermatol 158:1397–1399
167. Menni S, Betti R, Boccardi D, Gualandri L (2001) Both unilateral naevus achromicus and congenital agminated Spitz naevi in a checkerboard mosaic pattern. Br J Dermatol 144:187–188
168. Mirzaa GM, Conway RL, Gripp KW, Lerman-Sagie T, Siegel DH, deVries LS, Lev D, Kramer N, Hopkins E, Graham JM Jr, Dobyns WB (2012) Megalencephaly-capillary malformation (MCAP) and megalencephaly-polydactyly-polymicrogyria-hydrocephalus (MPPH) syndromes: two closely related disorders of brain overgrowth and abnormal brain and body morphogenesis. Am J Med Genet A 158A:269–291
169. Moog U, Roelens F, Mortier GR, Sijstermans H, Kelly M, Cox GF, Robson CD, Kimonis VE (2007) Encephalocraniocutaneous lipomatosis accompanied by the formation of bone cysts: harboring clues to pathogenesis? Am J Med Genet A 143A:2973–2980
170. Moore CA, Toriello HV, Abuelo DN, Bull MJ, Curry CJ, Hall BD, Higgins JV, Stevens CA, Twersky S, Weksberg R, Dobyns WB (1997) Macrocephaly-cutis marmorata telangiectatica congenita: a distinct disorder with developmental delay and connective tissue abnormalities. Am J Med Genet 70:67–73
171. Moss C, Burn J (1990) CHILD + ILVEN = PEN or PENCIL. J Med Genet 27:390–391
172. Moss C, Savin J (1995) Dermatology and the new genetics. Blackwell Science Ltd, Osney Mead, Oxford
173. Munro CS, Wilkie AO (1998) Epidermal mosaicism producing localised acne: somatic mutation in *FGFR2*. Lancet 352:704–705
174. Newton-Bishop JA, Chang YM, Iles MM, Taylor JC, Bakker B, Chan M, Leake S, Karpavicius B, Haynes S, Fitzgibbon E, Elliott F, Kanetsky PA, Harland M,

Barrett JH, Bishop DT (2010) Melanocytic nevi, nevus genes, and melanoma risk in a large case–control study in the United Kingdom. Cancer Epidemiol Biomarkers Prev 19:2043–2054

175. Oiso N, Tsuruta D, Imanishi H, Sayasa H, Narita T, Kobayashi H, Ikegami H, Kawada A (2010) Phylloid hypermelanosis and melanocytic nevi with aggregated and disfigured melanosomes: causal relationship between phylloid pigment distribution and chromosome 13 abnormalities. Dermatology 220:169–172
176. Online Mendelian Inheritance in Man (OMIM). Available from: http://www3.ncbi.nlm.nih.gov/Omim/. Accessed 28 Dec 2012
177. Ousager LB, Bygum A, Hafner C (2012) Identification of a novel S249C *FGFR3* mutation in a keratinocytic epidermal nevus syndrome. Br J Dermatol 167:202–204
178. Paller AS, Syder AJ, Chan YM, Yu QC, Hutton E, Tadini G, Fuchs E (1994) Genetic and clinical mosaicism in a type of epidermal nevus. N Engl J Med 331:1408–1415
179. Paller AS, van Steensel MA, Rodriguez-Martin M, Sorrell J, Heath C, Crumrine D, van Geel M, Cabrera AN, Elias PM (2011) Pathogenesis-based therapy reverses cutaneous abnormalities in an inherited disorder of distal cholesterol metabolism. J Invest Dermatol 131:2242–2248
180. Panigrahi I, Bhushan M, Yadav M, Khandelwal N, Singhi P (2012) Macrocephaly-capillary malformation syndrome: three new cases. J Neurol Sci 313:178–181
181. Pinkus H (1965) On the definition of nevi, organ nevi and nevoid tumors. Hautarzt 16:184–190
182. Plewig G, Christophers E (1975) Nevoid follicular epidermolytic hyperkeratosis. Arch Dermatol 111:223–226
183. Rahbari H, Cordero AA, Mehregan AH (1974) Linear porokeratosis: a distinctive clinical variant of porokeratosis of Mibelli. Arch Dermatol 109:526–528
184. Ramolia P, Treadwell P, Haggstrom A (2009) Speckled lentiginous nevus syndrome associated with musculoskeletal abnormalities. Pediatr Dermatol 26:298–301
185. Rawal RC, Goyal P, Rastogi S (2006) Naevoid psoriasis? Unilateral psoriasis? Int J Dermatol 45:1001–1002
186. Revencu N, Boon LM, Mulliken JB, Enjolras O, Cordisco MR, Burrows PE, Clapuyt P, Hammer F, Dubois J, Baselga E, Brancati F, Carder R, Quintal JM, Dallapiccola B, Fischer G, Frieden IJ, Garzon M, Harper J, Johnson-Patel J, Labreze C, Martorell L, Paltiel HJ, Pohl A, Prendiville J, Quere I, Siegel DH, Valente EM, Van Hagen A, Van Hest L, Vaux KK, Vicente A, Weibel L, Chitayat D, Vikkula M (2008) Parkes Weber syndrome, vein of Galen aneurysmal malformation, and other fast-flow vascular anomalies are caused by *RASA1* mutations. Hum Mutat 29:959–965
187. Rieger E, Kofler R, Borkenstein M, Schwingshandl J, Soyer HP, Kerl H (1994) Melanotic macules following Blaschko's lines in McCune-Albright syndrome. Br J Dermatol 130:215–220
188. Rijntjes-Jacobs EG, Lopriore E, Steggerda SJ, Kant SG, Walther FJ (2010) Discordance for Schimmelpenning-Feuerstein-Mims syndrome in monochorionic twins supports the concept of a postzygotic mutation. Am J Med Genet A 152A:2816–2819
189. Rivière JB, Mirzaa GM, O'Roak BJ, Beddaoui M, Alcantara D, Conway RL, St-Onge J, Schwartzentruber JA, Gripp KW, Nikkel SM, Worthylake T, Sullivan CT, Ward TR, Butler HE, Kramer NA, Albrecht B, Armour CM, Armstrong L, Caluseriu O, Cytrynbaum C, Drolet BA, Innes AM, Lauzon JL, Lin AE, Mancini GM, Meschino WS, Reggin JD, Saggar AK, Lerman-Sagie T, Uyanik G, Weksberg R, Zirn B, Beaulieu CL, Majewski J, Bulman DE, O'Driscoll M, Shendure J, Graham JM Jr, Boycott KM, Dobyns WB (2012) De novo germline and postzygotic mutations in *AKT3*, *PIK3R2* and *PIK3CA* cause a spectrum of related megalencephaly syndromes. Nat Genet 44:934–940
190. Roberts LC, DeVillez RL (1984) Congenital unilateral punctate porokeratosis. Am J Dermatopathol 6:57–61
191. Romiti R, Romiti N (2004) Papulolinear collagenoma. J Am Acad Dermatol 50:797–798
192. Rompel R, Meyer W, Petres J (1993) Cerebriform intradermal nevus. Hautarzt 44:238–241
193. Ross AL, Sanchez MI, Grichnik JM (2011) Molecular nevogenesis. Dermatol Res Pract 2011: Article ID 463184
194. Rübben A, Bogdan I, Grussendorf-Conen EI, Burg G, Böni R (2002) Loss of heterozygosity and microsatellite instability in acquired melanocytic nevi: towards a molecular definition of the dysplastic nevus. Recent Results Cancer Res 160: 100–110
195. Ruggieri M, Milone P, Pavone P, Falsaperla R, Polizzi A, Caltabiano R, Fichera M, Gabriele AL, Distefano A, De Pasquale R, Salpietro V, Micali G, Pavone L (2012) Nevus vascularis mixtus (cutaneous vascular twin nevi) associated with intracranial vascular malformation of the Dyke-Davidoff-Masson type in two patients. Am J Med Genet 158A:2870–2880
196. Saldanha G, Purnell D, Fletcher A, Potter L, Gillies A, Pringle JH (2004) High *BRAF* mutation frequency does not characterize all melanocytic tumor types. Int J Cancer 111:705–710
197. Santos-Juanes J, Galache C, Curto JR, Carrasco MP, Ribas A, Sánchez del Río J (2002) Acneiform lesions in Becker's nevus and breast hypoplasia. Int J Dermatol 41:699–700
198. Sapp JC, Turner JT, van de Kamp JM, van Dijk FS, Lowry RB, Biesecker LG (2007) Newly delineated syndrome of congenital lipomatous overgrowth, vascular malformations, and epidermal nevi (CLOVE

syndrome) in seven patients. Am J Med Genet A 143A:2944–2958

199. Schauder S, Hanefeld F, Noske UM, Zoll B (2000) Depigmented hypertrichosis following Blaschko's lines associated with cerebral and ocular malformations: a new neurocutaneous, autosomal lethal gene syndrome from the group of epidermal naevus syndromes? Br J Dermatol 142:1204–1207
200. Schwindinger WF, Francomano CA, Levine MA (1992) Identification of a mutation in the gene encoding the alpha subunit of the stimulatory G protein of adenylyl cyclase in McCune-Albright syndrome. Proc Natl Acad Sci U S A 89: 5152–5156
201. Shakhova O, Zingg D, Schaefer SM, Hari L, Civenni G, Blunschi J, Claudinot S, Okoniewski M, Beermann F, Mihic-Probst D, Moch H, Wegner M, Dummer R, Barrandon Y, Cinelli P, Sommer L (2012) *Sox10* promotes the formation and maintenance of giant congenital naevi and melanoma. Nat Cell Biol 14:882–890
202. Shear CS, Nyhan WL, Frost P, Weinstein GD (1970) Syndrome of unilateral ectromelia, psoriasis and central nervous system anomalies. In: Bergsma D (ed) The Third Conference on the Clinical Delineation of Birth Defects, Baltimore, Maryland, 1971, Birth defects: original article series. Williams and Wilkins, Baltimore, pp 197–203
203. Shim JH, Lee DW, Cho BK (1996) A case of Cobb syndrome associated with lymphangioma circumscriptum. Dermatology 193:45–47
204. Shirley MD, Tang H, Gallione CJ, Baugher JD, Frelin LP, Cohen B, North PE, Marchuk DA, Comi AM, Pevsner J (2013) Sturge-Weber syndrome and port-wine stains caused by somatic mutation in *GNAQ*. N Engl J Med 368:1971–1979
205. Steigleder GK, Cortes Cortes A (1971) The activity of sebaceous glands in sebaceous nevi during childhood. Arch Klin Exp Dermatol 239:323–328
206. Sterry W, Christophers E (1988) Quadrant distribution of dysplastic nevus syndrome. Arch Dermatol 124:926–929
207. Stoll C, Alembik Y, Grosshans E, de Saint MA (2002) An unusual human mosaic for skin pigmentation. Genet Couns 13:281–287
208. Sugarman JL (2008) Nevus comedonicus syndrome. In: Ruggieri M, Pascual-Castroviejo I, Di Rocco C (eds) Neurocutaneous diseases: the phacomatoses and hamartoneoplastic syndromes. Springer, Vienna, pp 581–587
209. Sun BK, Saggini A, Sarin KY, Kim J, Benjamin L, Leboit PE, Khavari PA (2013) Mosaic activating *RAS* mutations in nevus sebaceus and nevus sebaceus syndrome. J Invest Dermatol 133:824–827
210. Tantcheva-Poor I, Reinhold K, Krieg T, Happle R (2007) Trichilemmal cyst nevus: a new complex organoid epidermal nevus. J Am Acad Dermatol 57:S72–S77
211. Tekin M, Hişmi BO, Fitoz S, Yalçinkaya F, Ekim M, Kansu A, Ertem M, Deda G, Tutar E, Arsan S, Zhou XP, Pilarski R, Eng C, Akar N (2006) A germline PTEN mutation with manifestations of prenatal onset and verrucous epidermal nevus. Am J Med Genet A 140:1472–1475
212. Tévar E, Torrelo A, Contreras F, Colmenero I, Zambrano A (2006) Multiple basal cell carcinomas on phacomatosis pigmentokeratotica. Actas Dermosifiliogr 97:518–521
213. Thapa R, Dhar S, Malakar R, Chakrabartty S (2007) Hypomelanosis of Ito-whorled hyperpigmentation combination: a mirror image presentation. Pediatr Dermatol 24:572–573
214. Titeux M, Mendonça V, Décha A, Moreira E, Magina S, Maia A, Lacaze-Buzy L, Mejía JE, Torrão L, Carvalho F, Eça-Guimarães J, Hovnanian A (2009) Keratitis-ichthyosis-deafness syndrome caused by GJB2 maternal mosaicism. J Invest Dermatol 129:776–779
215. Toriello HV, Mulliken JB (2007) Accurately renaming macrocephaly-cutis marmorata telangiectatica congenita (M-CMTC) as macrocephaly-capillary malformation (M-CM). Am J Med Genet A 143A:3009
216. Torrelo A, Baselga E, Nagore E, Zambrano A, Happle R (2005) Delineation of the various shapes and patterns of nevi. Eur J Dermatol 15:439–450
217. Tran TP, Titus-Ernstoff L, Perry AE, Ernstoff MS, Newsham IF (2002) Alteration of chromosome 9p21 and/or p16 in benign and dysplastic nevi suggests a role in early melanoma progression (United States). Cancer Causes Control 13:675–682
218. Traupe H, Macher E, Hamm H, Happle R (1989) Mutation rate estimates are not compatible with autosomal dominant inheritance of the dysplastic nevus "syndrome". Am J Med Genet 32:155–157
219. Uribe P, Wistuba II, Gonzalez S (2009) Allelotyping, microsatellite instability, and *BRAF* mutation analyses in common and atypical melanocytic nevi and primary cutaneous melanomas. Am J Dermatopathol 31:354–363
220. van Dijk MC, Bernsen MR, Ruiter DJ (2005) Analysis of mutations in B-RAF, N-RAS, and H-RAS genes in the differential diagnosis of Spitz nevus and spitzoid melanoma. Am J Surg Pathol 29:1145–1151
221. Vidaurri-de la Cruz H, Happle R (2006) Two distinct types of speckled lentiginous nevi characterized by macular versus papular speckles. Dermatology 212:53–58
222. Weinstein LS (2006) G(s)alpha mutations in fibrous dysplasia and McCune-Albright syndrome. J Bone Miner Res 21:P120–P124
223. Weiss T, Mädler U, Oberwittler H, Kahle B, Weiss C, Kübler W (2000) Peripheral vascular malformation (Servelle-Martorell). Circulation 101:E82–E83
224. Wouters V, Limaye N, Uebelhoer M, Irrthum A, Boon LM, Mulliken JB, Enjolras O, Baselga E, Berg J, Dompmartin A, Ivarsson SA, Kangesu L, Lacassie Y, Murphy J, Teebi AS, Penington A, Rieu P, Vikkula M (2010) Hereditary cutaneomucosal venous

malformations are caused by *TIE2* mutations with widely variable hyper-phosphorylating effects. Eur J Hum Genet 18:414–420
225. Wright DR, Frieden IJ, Orlow SJ, Shin HT, Chamlin S, Schaffer JV, Paller AS (2009) The misnomer "macrocephaly-cutis marmorata telangiectatica congenita syndrome": report of 12 new cases and support for revising the name to macrocephaly-capillary malformations. Arch Dermatol 145:287–293
226. Yap FB (2009) Nevus lipomatosus superficialis. Singapore Med J 50:e161–e162
227. Yu X, Nagai H, Nishigori C, Horikawa T (2008) Acquired unilateral melanocytic nevi in otherwise normal skin. Dermatology 217:63–65
228. Zietz S, Happle R, Hohenleutner U, Landthaler M (2008) The venous nevus: a distinct vascular malformation suggesting mosaicism. Dermatology 216: 31–36

Didymotic Skin Disorders

8

Contents

The two dissimilar alleles causing didymotic skin disorders (see Sect. 3.1.1.5) may involve the same gene locus (allelic didymosis), or they may affect different loci (nonallelic didymosis) on either of a pair of homologous chromosomes.

8.1 Allelic Didymosis

The following disorders are suggestive of allelic twin spotting, but molecular proof is so far lacking.

8.1.1 Capillary Didymosis

This form of didymosis occurs rather frequently. From a clinical point of view, different types can be distinguished.

8.1.1.1 Nevus Flammeus Twinned with Nevus Anemicus

This type of possible twin spotting (Fig. 8.1) is noted rather frequently [8]. In part, the two nevi may overlap each other.

8.1.1.2 Nevus Vascularis Mixtus, a Hallmark of Mixed Vascular Nevus Syndrome (Ruggieri-Leech Syndrome)

This phenotype represents an admixture of reticular telangiectatic lesions and angiospastic spots of nevus anemicus [7]. It occurs as an isolated skin disease (Fig. 8.2) or as a hallmark of

R. Happle, *Mosaicism in Human Skin*,
DOI 10.1007/978-3-642-38765-4_8, © Springer-Verlag Berlin Heidelberg 2014

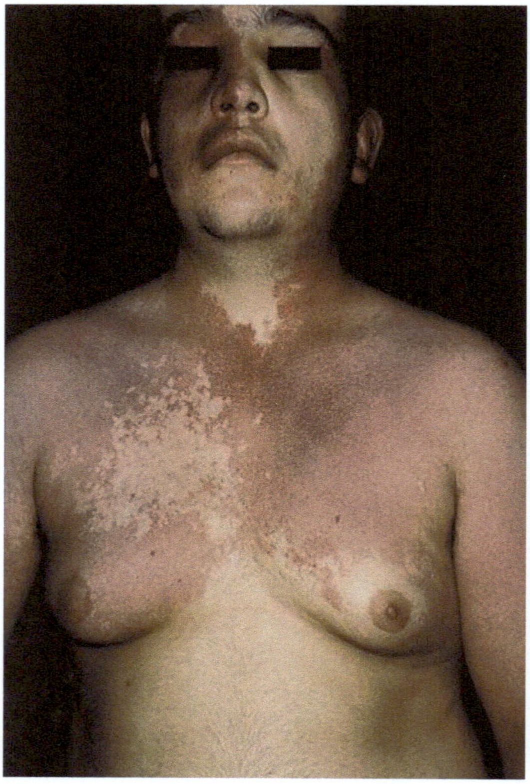

Fig. 8.1 Nevus flammeus twinned with nevus anemicus [3] (Reprinted with permission from John Libbey Eurotext, Montrouge, France)

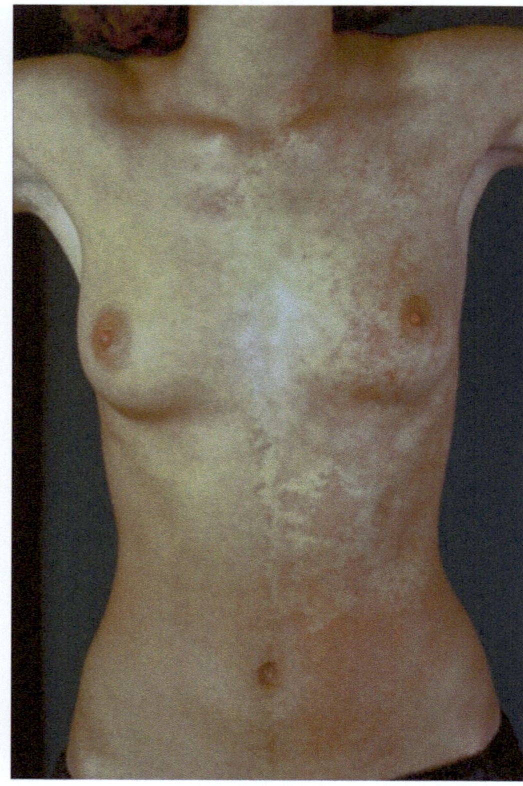

Fig. 8.2 Nevus vascularis mixtus (Courtesy of Dr. Henning Hamm, Würzburg, Germany)

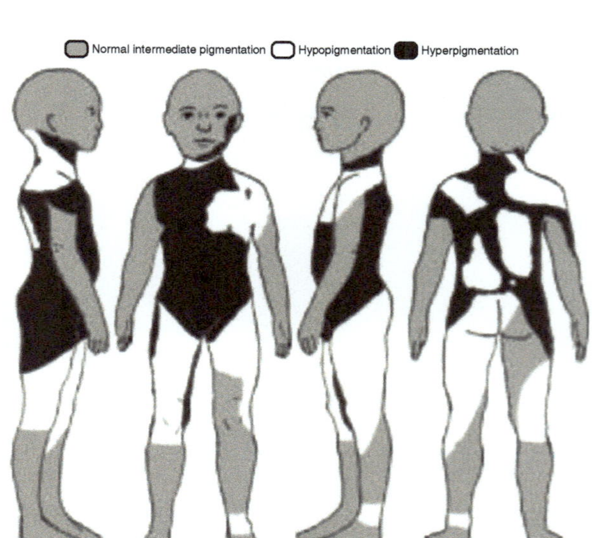

Fig. 8.3 Cutis tricolor arranged in large patches without midline separation as noted in the Ruggieri-Happle syndrome [20]

a neurocutaneous syndrome characterized by distinct brain malformations (mixed vascular nevus syndrome, Ruggieri-Leech syndrome) [16, 22].

8.1.2 Cutis Tricolor

This group of disorders is characterized by hyper- and hypomelanotic macules that are often localized in close proximity to each other.

8.1.2.1 Ruggieri-Happle Syndrome

A hallmark of this disorder is the presence of rather large light and dark macules on a background skin of intermediate color (Fig. 8.3), sometimes being arranged in a distinct sash-like pattern (Fig. 5.26) (see Sect. 5.6). Extracutaneous mosaic defects involve the brain [10, 17, 20] and the bones [21].

8.1.2.2 Cutis Tricolor Parvimaculata

The trait is characterized by rather small hyper- and hypomelanotic spots on a background of intermediate skin (Fig. 8.4) [6]. Rather often it is associated with cerebral defects [1, 2, 15].

8.1.2.3 Cutis Tricolor of the Blaschko-Linear Type

Niessen et al. [18] described hyper- and hypopigmented bands following Blaschko's lines in a 6-year-old girl with double aneuploidy mosaicism 47,XX+7/45,X.

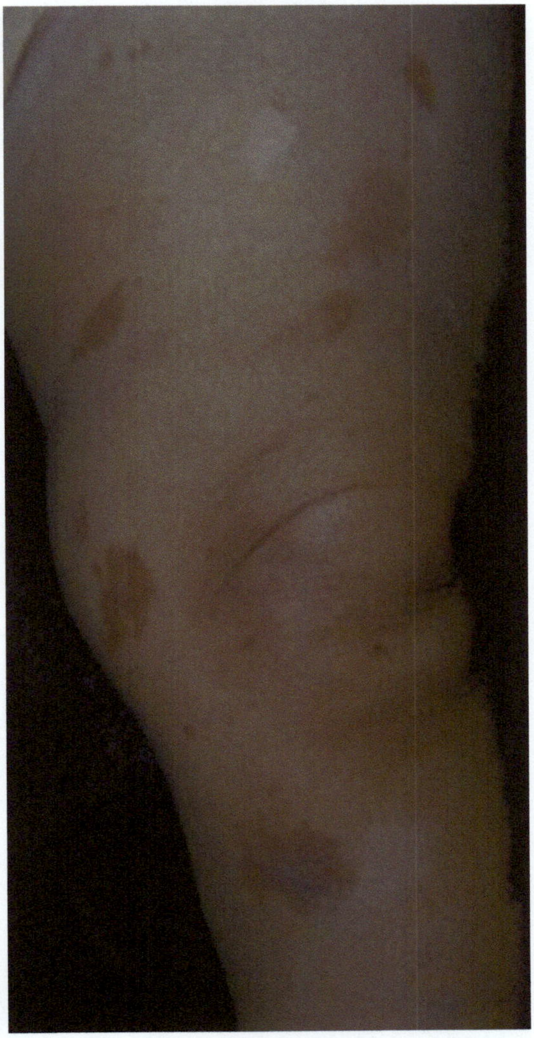

Fig. 8.4 Cutis tricolor parvimaculata [1] (Reprinted with permission from John Wiley & Sons, USA)

8.1.3 Didymosis in Epidermolytic Ichthyosis of Brocq

In 1991, Eng et al. [4] described a boy with "congenital ichthyosiform erythroderma and epidermal nevus." Happle and König [11] reclassified this case as an example of epidermolytic ichthyosis of Brocq showing twin spotting in the form of paired bands of either excessive or absent involvement.

8.1.4 Didymosis in Darier Disease

Patients with Darier disease may likewise have paired bands of either pronounced or absent involvement, which can be interpreted as cases of allelic twin spotting (Fig. 8.5). Three clinical examples suggestive of such mechanism have so far been reported [14, 19, 26], but molecular evidence is still lacking.

8.2 The Theoretical Concept of Nonallelic Didymosis

In nonallelic didymosis, the two components would involve different tissues. In the past, this concept has been proposed to explain several binary genodermatoses such as phacomatosis

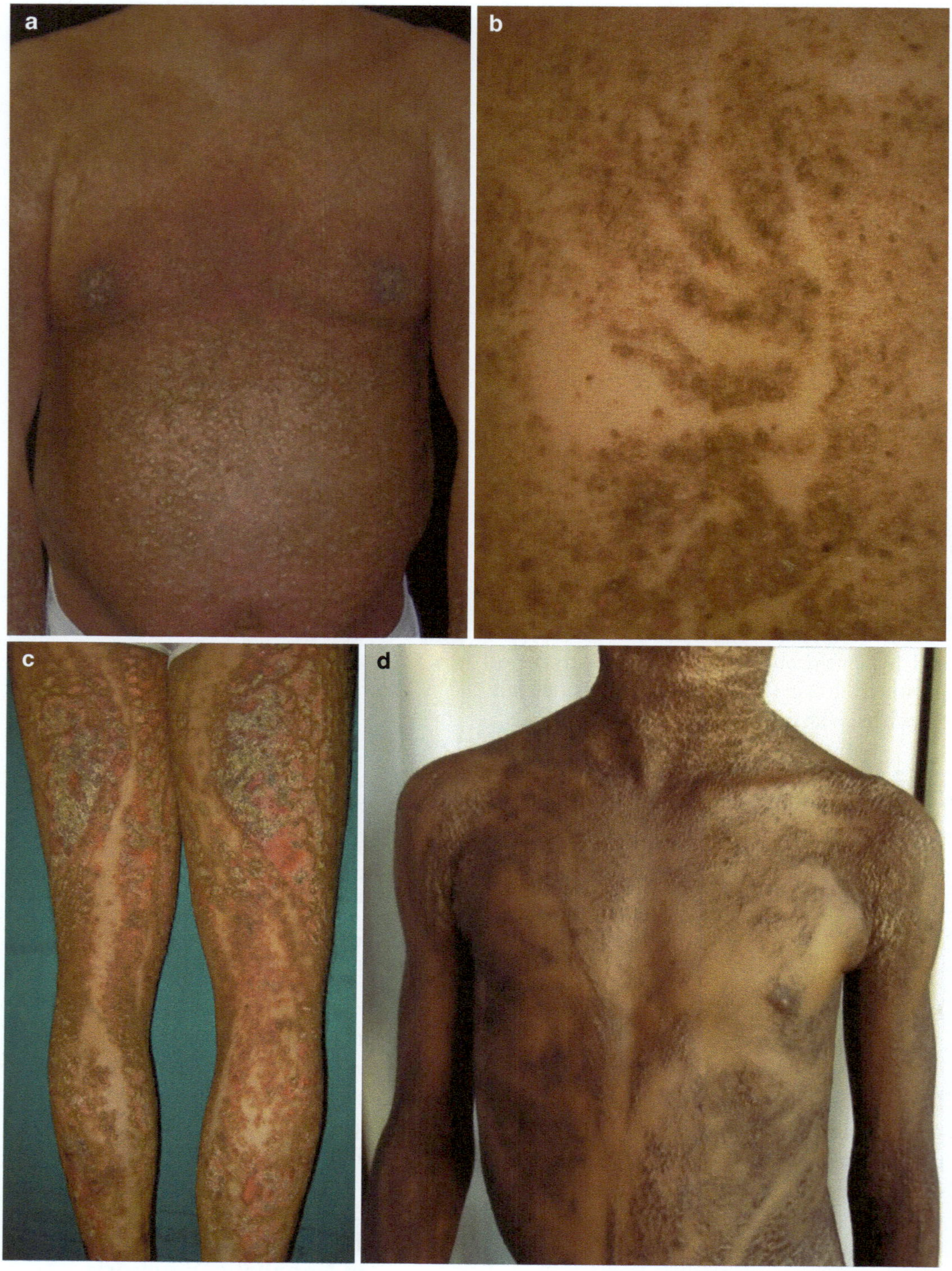

Fig. 8.5 Didymosis in Darier disease. (**a**) A 50-year-old man showing diffuse involvement of the trunk; (**b**) linear areas of either pronounced or absent involvement on the back; (**c**) didymotic involvement of legs [19]; (**d**) 12-year-old patient showing a similar type of twin spotting [26] (**a–c**: Reprinted with permission from John Wiley & Sons, USA; **d**: Courtesy of Dr. Shehu M. Yusuf, Kano, Nigeria)

pigmentokeratotica [9] or the different types of phacomatosis pigmentovascularis [12, 13]. Today, however, this view can no longer be upheld because evidence has recently been provided that the two components of phacomatosis pigmentokeratotica originate from one single pleiotropic *HRAS* mutation present in a heterozygous state [6]. This finding excludes the possibility of loss of heterozygosity. The same holds for phacomatosis cesioflammea because nevus flammeus was found to be caused by a heterozygous *GNAQ* mutation [23]. As a consequence, it has become rather unlikely that the remaining types of phacomatosis pigmentovascularis [12] and some other binary genodermatoses involving two different tissues (see Chap. 9) represent examples of nonallelic twin spotting.

On the other hand, various forms of nonallelic didymosis have experimentally been induced in *Drosophila melanogaster* [5, 24, 25]. Therefore, it seems reasonable to predict that nonallelic twin spotting may likewise occur in human skin.

References

1. Boente Mdel C, Bazan C, Montanari D (2011) Cutis tricolor parvimaculata in two patients with ring chromosome 15 syndrome. Pediatr Dermatol 28:670–673
2. Buoni S, Zannolli R, de Santi M, Macucci F, Hayek J, Orsi A, Scarinci R, Buscalferri A, Cuccia A, Zappella M, Miracco C (2006) Neurocutaneous syndrome with mental delay, autism, blockage in intracellular vescicular trafficking and melanosome defects. Eur J Neurol 13:842–851
3. Du Chekroun-Le L, Delaporte E, Catteau B, Destée A, Piette F (1998) Phacomatosis pigmentovascularis type II. Eur J Dermatol 8:569–572
4. Eng AM, Brody P, Rhee HL, Bronson DM (1991) Congenital ichthyosiform erythroderma and epidermal nevus. Int J Dermatol 30:284–287
5. Graf U, Würgler FE, Katz AJ, Frei H, Juon H, Hall CB, Kale PG (1984) Somatic mutation and recombination test in *Drosophila melanogaster*. Environ Mutagen 6:153–188
6. Groesser L, Herschberger E, Sagrera A, Shwayder T, Flux K, Ehmann L, Wollenberg A, Torrelo A, Bagazgoitia L, Diaz-Ley B, Tinschert S, Oschlies I, Siner S, Mickler M, Toll A, Landthaler M, Real FX, Hafner C (2013) Phacomatosis pigmentokeratotica is caused by a postzygotic HRAS mutation in a multipotent progenitor cell. J Invest Dermatol 133: 1998–2003
7. Hamm H, Happle R (1986) Mixed vascular nevus: report of 4 cases. Hautarzt 37:388–392
8. Happle R (1991) Allelic somatic mutations may explain vascular twin nevi. Hum Genet 86: 321–322
9. Happle R, Hoffmann R, Restano L, Caputo R, Tadini G (1996) Phacomatosis pigmentokeratotica: a melanocytic-epidermal twin nevus syndrome. Am J Med Genet 65:363–365
10. Happle R, Barbi G, Eckert D, Kennerknecht I (1997) "Cutis tricolor": congenital hyper- and hypopigmented macules associated with a sporadic multisystem birth defect: an unusual example of twin spotting? J Med Genet 34:676–678
11. Happle R, König A (1999) Dominant traits may give rise to paired patches of either excessive or absent involvement. Am J Med Genet 84:176–177
12. Happle R (2005) Phacomatosis pigmentovascularis revisited and reclassified. Arch Dermatol 141: 385–388
13. Happle R (2011) Didymosis cesioanemica: an unusual counterpart of phacomatosis cesioflammea. Eur J Dermatol 21:471
14. Itin PH, Happle R (2002) Darier disease with paired segmental manifestation of either excessive or absent involvement: a further step in the concept of twin spotting. Dermatology 205:344–347
15. Larralde M, Happle R (2005) Cutis tricolor parvimaculata: a distinct neurocutaneous syndrome? Dermatology 211:149–151
16. Leech SN, Taylor AE, Ramesh V, Birchall D, Ann Lynch S (2004) Widespread capillary malformation associated with global developmental delay and megalencephaly. Clin Dysmorphol 13:169–172
17. Lionetti E, Pavone P, Kennerknecht I, Failla G, Schepis C, De Pasquale R, Pavone L, Ruggieri M (2010) Neurological manifestations in individuals with pure cutaneous or syndromic (Ruggieri-Happle syndrome) phenotypes with "cutis tricolor": a study of 14 cases. Neuropediatrics 41:60–65
18. Niessen RC, Jonkman MF, Muis N, Hordijk R, van Essen AJ (2005) Pigmentary mosaicism following the lines of Blaschko in a girl with a double aneuploidy mosaicism: (47, XX,+7/45, X). Am J Med Genet A 137A:313–322
19. Rodríguez-Pazos L, Gomez-Bernal S, Loureiro M, Toribio J (2011) Type 2 segmental Darier disease with twin spot phenomenon. J Eur Acad Dermatol Venereol 25:496–497
20. Ruggieri M (2000) Cutis tricolor: congenital hyper- and hypopigmented lesions in a background of normal skin with and without associated systemic features: further expansion of the phenotype. Eur J Pediatr 159:745–749
21. Ruggieri M, Roggini M, Kennerknecht I, Polizzi A, Distefano A, Pavone V (2011) Spectrum of skeletal abnormalities in a complex malformation syndrome

with "cutis tricolor" (Ruggieri-Happle syndrome). Acta Paediatr 100:121–127

22. Ruggieri M, Milone P, Pavone P, Falsaperla R, Polizzi A, Caltabiano R, Fichera M, Gabriele AL, Distefano A, De Pasquale R, Salpietro V, Micali G, Pavone L (2012) Nevus vascularis mixtus (cutaneous vascular twin nevi) associated with intracranial vascular malformation of the Dyke-Davidoff-Masson type in two patients. Am J Med Genet A 158A:2870–2880
23. Shirley MD, Tang H, Gallione CJ, Baugher JD, Frelin LP, Cohen B, North PE, Marchuk DA, Comi AM, Pevsner J (2013) Sturge-Weber syndrome and port-wine stains caused by somatic mutation in GNAQ. N Engl J Med 368:1971–1979
24. Spano MA, Frei H, Wurgler FE, Graf U (2001) Recombinogenic activity of four compounds in the standard and high bioactivation crosses of *Drosophila melanogaster* in the wing spot test. Mutagenesis 16:385–394
25. Stern C (1936) Somatic crossing over and segregation in *Drosophila melanogaster*. Genetics 21:625–730
26. Yusuf SM, Mohammed AZ, Uloko AE (2009) Type 2 segmental Darier's disease in a twelve-year-old Nigerian male – a case report. Niger J Med 18:413–415

Other Binary Genodermatoses, Including "Pseudodidymosis"

9

Contents

Molecular research has provided evidence that not all genodermatoses occurring in a binary form reflect postzygotic recombination (see Sect. 8.2). Hence, in several binary disorders the concept of didymosis cannot be upheld.

In the following group of genodermatoses that tend to occur in a binary form, the concept of didymosis is, according to present knowledge, either wrong or rather unlikely.

9.1 Phacomatosis Pigmentokeratotica: An Example of "Pseudodidymosis"

This phenotype is characterized be a papular nevus spilus coexisting with nevus sebaceus (Fig. 9.1) [14, 30, 34, 37, 41]. The disorder can be taken as a variant of Schimmelpenning syndrome [12]. Sometimes the paired nevi may occur without involvement of internal organs [8, 24, 31], but numerous cases complicated by extracutaneous defects have been reported [4, 5, 25]. The associated sebaceous nevus was reported to be particularly prone to develop multiple basal cell carcinomas [6, 27, 35], but this may reflect a bias of ascertainment. Some patients develop other malignancies [13, 21].

Apparently, the two cutaneous components of the phenotype originate from one single pluripotent progenitor cell [12]. Because the mechanism of didymosis has been excluded, the term "pseudodidymosis" may be appropriate [18].

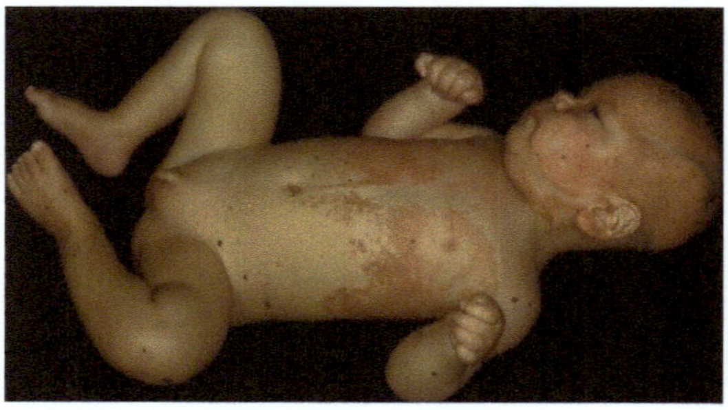

Fig. 9.1 Phacomatosis pigmentokeratotica (Courtesy of Dr. Ching-Ying Wu, Kaohsiung, Taiwan)

9.2 Paired Occurrence of Nevus Sebaceus and Melorheostosis

For heuristic reasons, an unusual case of unilateral melorheostosis coexistent with ipsilateral nevus sebaceus may be mentioned [36], although a nosological relationship is uncertain. Similar cases have so far not been reported, which is why this may be an arbitrary coincidence.

9.3 Paired Occurrence of Nevus Sebaceus and Aplasia Cutis Congenita

Aplasia cutis congenita is sometimes noted in close proximity to a sebaceous nevus (Fig. 9.2) [20], which renders a nosological relationship rather likely. A didymosis, however, as previously postulated [9, 15], can today be taken as excluded.

9.4 Paired Occurrence of Nevus Psiloliparus and Aplasia Cutis Congenita

Another disorder that appears to be sometimes paired with aplasia cutis congenita is nevus psiloliparus (Fig. 9.3). Such binary lesions are often noted in encephalocraniocutaneous lipomatosis [26, 28, 33]. According to present knowledge, the previously proposed term "didymosis aplasticopsilolipara" [38] is inappropriate.

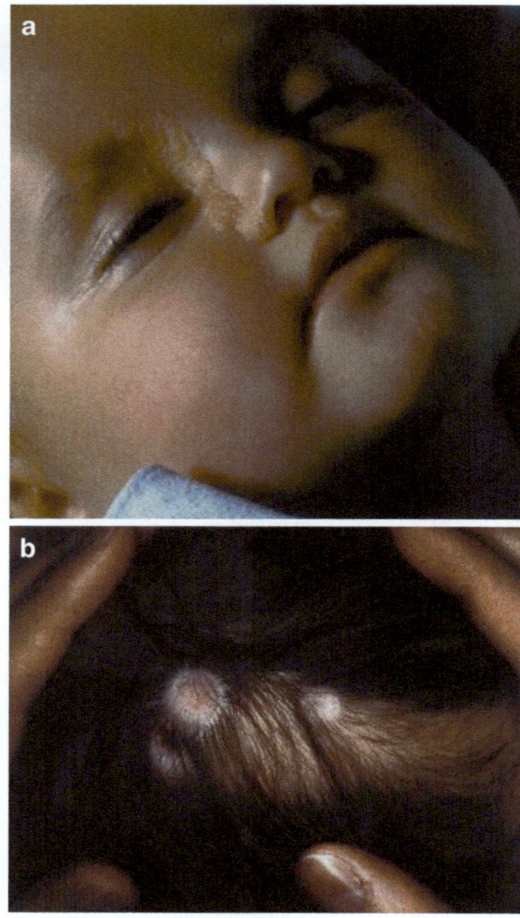

Fig. 9.2 Paired occurrence of nevus sebaceus and aplasia cutis congenita. (**a**) One-month-old girl with nevus sebaceus, intermingled with (**b**) aplasia cutis congenita on her scalp [20] (Reprinted with permission from John Wiley & Sons, USA)

9.5 The Group of Phacomatosis Pigmentovascularis

The different types of phacomatosis pigmentovascularis had so far been taken as possible examples of nonallelic twin spotting [16, 29]. Today this concept appears to be wrong (see Sect. 3.1.1.5).

9.5.1 Phacomatosis Cesioflammea

This phenotype is characterized by a coexistence of large blue macules that are asymmetrically arranged and an extensive nevus flammeus

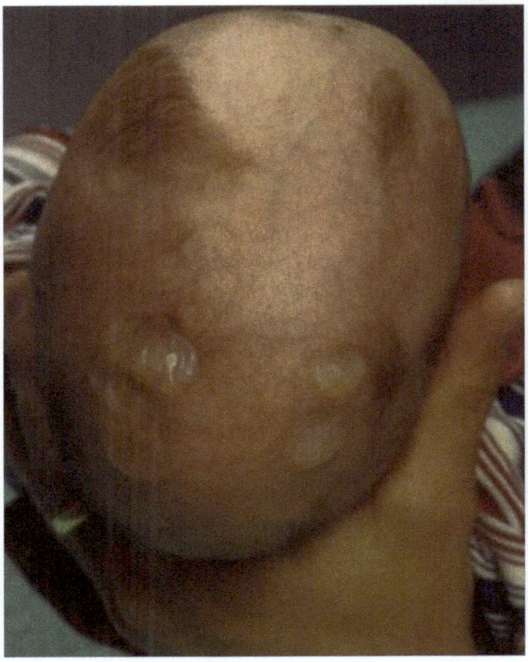

Fig. 9.3 Paired occurrence of nevus psiloliparus and aplasia cutis congenita [28] (Reprinted with permission from John Wiley & Sons, USA)

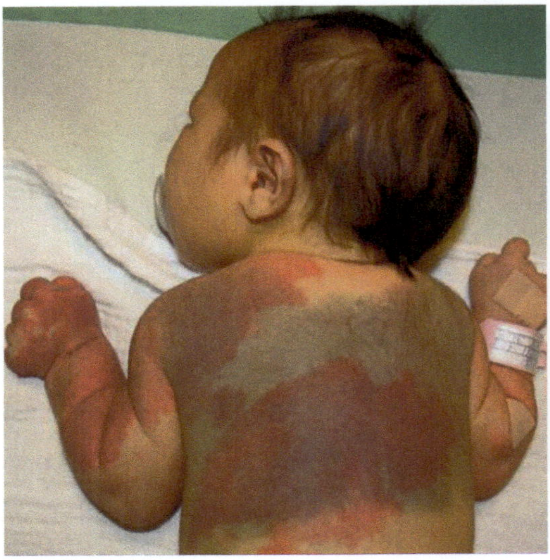

Fig. 9.4 Phacomatosis cesioflammea (Courtesy of Dr. Hansjörg Cremer, Heilbronn, Germany)

(Fig. 9.4). Extracutaneous features include brain defects, glaucoma, asymmetry of limbs, dysplastic veins or lymph vessels, and nevus anemicus [16]. Cases of monozygotic twins discordant for the disorder [7, 29] have supported the theory of

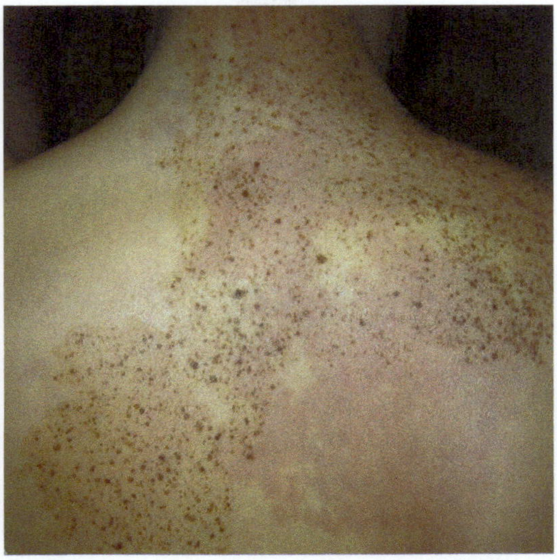

Fig. 9.5 Phacomatosis spilorosea [22] (Reprinted with permission from John Wiley & Sons, USA)

a postzygotic mutational event. The disorder was previously taken as a standard example of nonallelic didymosis [16, 32], but has recently turned out to be a "pseudodidymosis" (see Sect. 8.2).

9.5.2 Phacomatosis Spilorosea

This type of phacomatosis pigmentovascularis consists of a macular nevus spilus coexisting with nevus roseus that is of a much lighter hue than nevus flammeus (Fig. 9.5) (see Sect. 7.4.1.2). Associated extracutaneous anomalies include unilateral lymphedema, cerebral defects with hemiparesis or seizures, cataracts, glaucoma, oligodontia, and asymmetry of legs giving rise to scoliosis [10, 22, 23, 38, 40].

9.5.3 Phacomatosis Melanorosea

This type is characterized by one or more large, lateralized café-au-lait macules coexistent with nevus roseus (Fig. 9.6). So far, three cases have been reported [1–3]. Aguayo et al. [1] have argued that in this type of phacomatosis pigmentovascularis, the nevus roseus may be accompanied by cutis marmorata-like lesions. Further case reports are needed to settle this question.

9.5.4 Phacomatosis Cesiomarmorata

Large aberrant Mongolian spots may also coexist with cutis marmorata telangiectatica congenita (Fig. 9.7) (see Sect. 7.4.1.4). These skin lesions may be associated with blue sclerae, hypoplastic cornea, asymmetry of hemispheres and ventricles, or hyperplasia of a limb [11, 39]. It is so far not clear, however, whether all of these cases represent a binary skin disorder rather than an arbitrary coincidence.

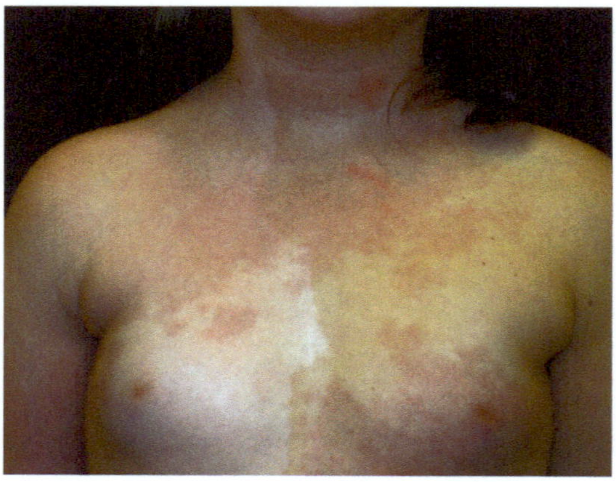

Fig. 9.6 Phacomatosis melanorosea [2] (Reprinted with permission from John Libbey Eurotext, Montrouge, France)

9.5.5 "Phacomatosis Cesioanemica"

A case of blue nevus coexistent with nevus anemicus [19] can be taken as a possible example of "phacomatosis cesioanemica" [17]. From a genetic point of view, this disorder may best be categorized as a particular variant of phacomatosis cesioflammea (see Sect. 9.5.1).

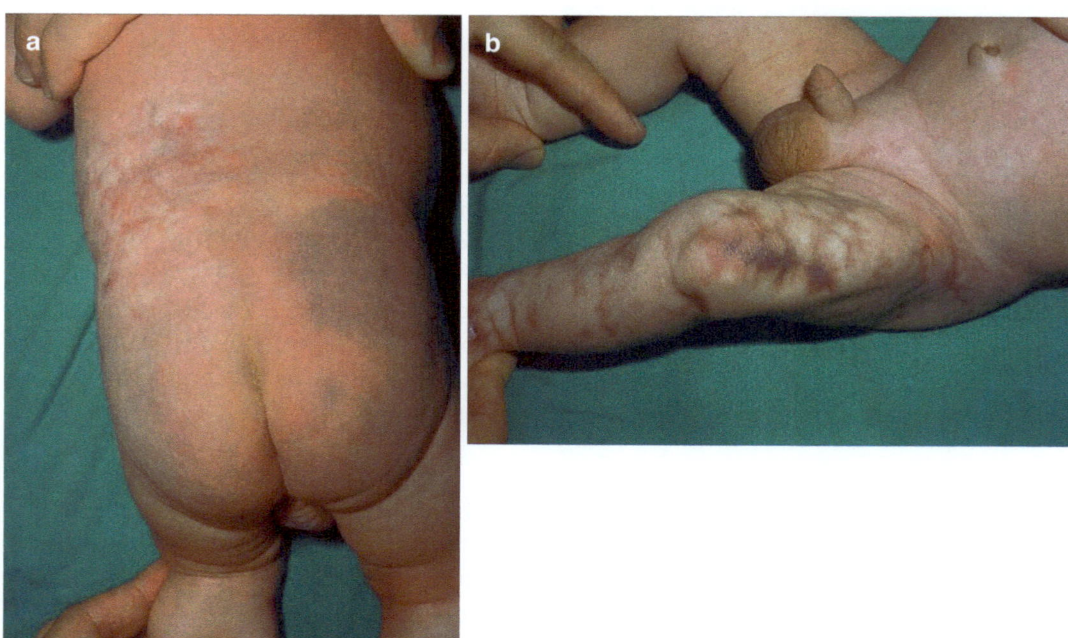

Fig. 9.7 Phacomatosis cesiomarmorata. (**a**) Dorsal aspect showing nevus cesius and cutis marmorata telangiectatica congenita; (**b**) vascular lesions involving the left leg [39] (Reprinted with permission from John Wiley & Sons, USA)

References

1. Aguayo-Leiva I, Alonso J, Echeverria B, Hernandez-Martin A, Torrelo A (2011) Phacomatosis melanovascularis: a new example of non-allelic twin spotting. Eur J Dermatol 21:487–489
2. Almeida H Jr, Happle R, Reginatto F, Basso F, Duquia R (2011) Phacomatosis melanorosea with heterochromia of scalp hair. Eur J Dermatol 21: 598–599
3. Arnold AW, Kleine MU, Happle R (2012) Phacomatosis melanorosea without extracutaneous features: an unusual type of phacomatosis pigmentovascularis. Eur J Dermatol 22:473–475
4. Baroni A, Staibano S, Russo T, Piccolo V, Satriano RA, Vozza A, Vozza G (2012) Verrucous epidermal naevus and naevus spilus associated with lower limb asymmetry and right bundle-branch block: a case of phacomatosis pigmentokeratotica? Clin Exp Dermatol 37:74–75
5. Boente MC, Pizzi de Parra N, Larralde de Luna M, Bonet HB, Santos Muñoz A, Parra V, Gramajo P, Moreno S, Asial RA (2000) Phacomatosis pigmentokeratotica: another epidermal nevus syndrome and a distinctive type of twin spotting. Eur J Dermatol 10:190–194
6. Bouthors J, Vantyghem MC, Manouvrier-Hanu S, Soudan B, Proust E, Happle R, Piette F (2006) Phacomatosis pigmentokeratotica associated with hypophosphataemic rickets, pheochromocytoma and multiple basal cell carcinomas. Br J Dermatol 155:225–226
7. Castori M, Sarazani S, Binni F, Pezzella FR, Cruciani G, Grammatico P (2011) Monozygotic twin discordance for phacomatosis cesioflammea further supports the post-zygotic mutation hypothesis. Am J Med Genet A 155A:2253–2256
8. Chantorn R, Shwayder T (2011) Phacomatosis pigmentokeratotica: a further case without extracutaneous anomalies and review of the condition. Pediatr Dermatol 28:715–719
9. Demerdjieva Z, Kavaklieva S, Tsankov N (2007) Epidermal nevus syndrome and didymosis aplasticosebacea. Pediatr Dermatol 24:514–516
10. Diociaiuti A, Guidi B, Aguilar Sanchez JA, Feliciani C, Capizzi R, Amerio P (2005) Phakomatosis pigmentovascularis type IIIb: a case associated with Sturge-Weber and Klippel-Trenaunay syndromes. J Am Acad Dermatol 53:536–539
11. Enjolras O, Boon LM, Mulliken JB (2006) Vascular malformations. In: Harper J, Oranje A, Prose N (eds) Textbook of pediatric dematology, vol 2, 2nd edn. Blackwell Publishing, Malden, pp 1153–1174
12. Groesser L, Herschberger E, Sagrera A, Shwayder T, Flux K, Ehmann L, Wollenberg A, Torrelo A, Bagazgoitia L, Diaz-Ley B, Tinschert S, Oschlies I, Siner S, Mickler M, Toll A, Landthaler M, Real FX, Hafner C (2013) Phacomatosis pigmentokeratotica is caused by a postzygotic *HRAS* mutation in a multipotent progenitor cell. J Invest Dermatol 133: 1998–2003
13. Gruson LM, Orlow SJ, Schaffer JV (2006) Phacomatosis pigmentokeratotica associated with hemihypertrophy and a rhabdomyosarcoma of the abdominal wall. J Am Acad Dermatol 55:S16–S20
14. Happle R, Hoffmann R, Restano L, Caputo R, Tadini G (1996) Phacomatosis pigmentokeratotica: a melanocytic-epidermal twin nevus syndrome. Am J Med Genet 65:363–365
15. Happle R, König A (2001) Didymosis aplasticosebacea: coexistence of aplasia cutis congenita and nevus sebaceus may be explained as a twin spot phenomenon. Dermatology 202:246–248
16. Happle R (2005) Phacomatosis pigmentovascularis revisited and reclassified. Arch Dermatol 141: 385–388
17. Happle R (2011) Didymosis cesioanemica: an unusual counterpart of phacomatosis cesioflammea. Eur J Dermatol 21:471
18. Happle R (2013) Phacomatosis pigmentokeratotica is a "pseudodidymosis". J Invest Dermatol 133: 1923–1925
19. Hernández-Núñez A, Borbujo J, Córdoba S, Martínez-Morán C, Martínez-Sánchez D, Romero-Maté A (2011) Nevus anemicus of the cheek with ipsilateral melanosis bulbi: an unusual example of didymosis. Eur J Dermatol 21:597–598
20. Högler W, Sidoroff A, Weber F, Baldissera I, Heinz-Erian P (1999) Aplasia cutis congenita, uvula bifida and bilateral retinal dystrophy in a girl with naevus sebaceus syndrome. Br J Dermatol 140:542–543
21. Jacobelli S, Leclerc-Mercier S, Salomon R, Hartmann O, Brunelle F, Happle R, Bodemer C, Hadj-Rabia S (2010) Phacomatosis pigmentokeratotica with nephroblastoma and juvenile hypertension. Acta Derm Venereol 90:279–282
22. Jordaan HF, Happle R (2008) Phacomatosis spilorosea associated with lymphoedema. Br J Dermatol 159:489–491
23. Karabudak O, Dogan B, Basekim C, Harmanyeri Y (2007) Phacomatosis spilorosea (phacomatosis pigmentovascularis type IIIb). Australas J Dermatol 48:256–258
24. Kinoshita K, Shinkai H, Utani A (2003) Phacomatosis pigmentokeratotica without extracutaneous abnormalities. Dermatology 207:415–416
25. Majmudar V, Loffeld A, Happle R, Salim A (2007) Phacomatosis pigmentokeratotica associated with a suprasellar dermoid cyst and leg hypertrophy. Clin Exp Dermatol 32:690–692
26. Martí N, Alonso V, Jordá E (2012) Encephalocraniocutaneous lipomatosis and didymosis aplasticopsilolipara. Actas Dermosifiliogr 103:341–342
27. Martínez-Menchón T, Mahiques Santos L, Vilata Corell JJ, Febrer Bosch I, Fortea Baixauli JM (2005) Phacomatosis pigmentokeratotica: a 20-year follow-up with malignant degeneration of both nevus components. Pediatr Dermatol 22:44–47

28. Moog U, Roelens F, Mortier GR, Sijstermans H, Kelly M, Cox GF, Robson CD, Kimonis VE (2007) Encephalocraniocutaneous lipomatosis accompanied by the formation of bone cysts: harboring clues to pathogenesis? Am J Med Genet A 143A: 2973–2980
29. Moutray T, Napier M, Shafiq A, Fryer A, Rankin S, Willoughby CE (2010) Monozygotic twins discordant for phacomatosis pigmentovascularis: evidence for the concept of twin spotting. Am J Med Genet A 152A:718–720
30. Park HY, Kim JH, Ji JH, Ahn SK, Hong SP (2011) Variant of phacomatosis pigmentokeratotica. J Dermatol 38:719–722
31. Polat M, Yalcin B, Ustun H, Caliskan D, Alli N (2008) Phacomatosis pigmentokeratotica without extracutaneous abnormalities. Eur J Dermatol 18:363–364
32. Shields CL, Kligman BE, Suriano M, Viloria V, Iturralde JC, Shields MV, Say EA, Shields JA (2011) Phacomatosis pigmentovascularis of cesioflammea type in 7 patients: combination of ocular pigmentation (melanocytosis or melanosis) and nevus flammeus with risk for melanoma. Arch Ophthalmol 129: 746–750
33. Stieler KM, Astner S, Bohner G, Bartels NG, Proquitté H, Sterry W, Haas N, Blume-Peytavi U (2008) Encephalocraniocutaneous lipomatosis with didymosis aplasticopsilolipara. Arch Dermatol 144:266–268
34. Tadini G, Restano L, Gonzáles-Pérez R, Gonzáles-Enseñat A, Vincente-Villa MA, Cambiaghi S, Marchettini P, Mastrangelo M, Happle R (1998) Phacomatosis pigmentokeratotica: report of new cases and further delineation of the syndrome. Arch Dermatol 134:333–337
35. Tévar E, Torrelo A, Contreras F, Colmenero I, Zambrano A (2006) Multiple basal cell carcinomas on phacomatosis pigmentokeratotica. Actas Dermosifiliogr 97:518–521
36. Tinschert S, Stein A, Goldner B, Dietel M, Happle R (2003) Melorheostosis with ipsilateral nevus sebaceus (didymosis melorheosebacea). Eur J Dermatol 13: 21–24
37. Torrelo A, Zambrano A (1998) What syndrome is this. Phakomatosis pigmentokeratotica (Happle). Pediatr Dermatol 15:321–323
38. Torrelo A, Boente M, Nieto O, Asial R, Colmenero I, Winik B, Zambrano A, Happle R (2005) Nevus psiloliparus and aplasia cutis: a further possible example of didymosis. Pediatr Dermatol 22: 206–209
39. Torrelo A, Zambrano A, Happle R (2006) Large aberrant Mongolian spots coexisting with cutis marmorata telangiectatica congenita (phacomatosis pigmentovascularis type V or phacomatosis cesiomarmorata). J Eur Acad Dermatol Venereol 20:308–310
40. Valdivielso-Ramos M, Mauleón C, Hernanz JM (2012) Phacomatosis spilorosea with oligodontia, scoliosis and fibrous cortical defects. J Eur Acad Dermatol Venereol 26:260–262
41. Wollenberg A, Butnaru C, Oppel T (2002) Phacomatosis pigmentokeratotica (Happle) in a 23-year-old man. Acta Derm Venereol 82:55–57

Mosaic Manifestation of Autosomal Dominant Skin Disorders

10

Contents

R. Happle, *Mosaicism in Human Skin*,
DOI 10.1007/978-3-642-38765-4_10, © Springer-Verlag Berlin Heidelberg 2014

In autosomal dominant cutaneous traits, a type 1 mosaic involvement that originates from a postzygotic new mutation should be distinguished from a type 2 segmental manifestation being superimposed on the ordinary, nonsegmental phenotype.

10.1 Hereditary Multiple Skin Tumors

Three forms of mosaicism can be distinguished in hereditary benign cutaneous neoplasias. Firstly, all of them can be categorized as mosaic lesions that tend to originate from LOH [177]. Secondly, a type 1 segmental manifestation of these tumors is sometimes found, giving rise to a linear or otherwise mosaic pattern. Thirdly, a type 2 segmental involvement may be superimposed on the ordinary nonsegmental trait. In the following paragraphs we shall consider the two types of segmental arrangement.

10.1.1 Trichoepithelioma

Multiple trichoepitheliomas showing a bona fide type 1 segmental involvement could so far not be retrieved in the available literature. On the other hand, a convincing example of type 2 segmental manifestation has been documented by Geffner et al. [143] in a girl who later developed bilateral facial lesions. Reportedly, her mother and one of her brothers were likewise affected with trichoepitheliomas. Another case fulfilling all of the criteria of superimposed linear trichoepitheliomatosis was reported by Schirren et al. [389].

Some other cases of pronounced linear trichoepitheliomas that were present at birth (Fig. 10.1) or developed during infancy [185, 331] are likewise suggestive of a type 2 segmental manifestation.

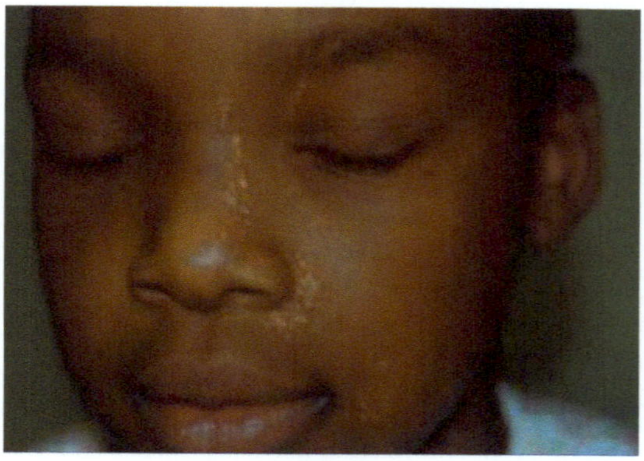

Fig. 10.1 This linear arrangement of trichoepitheliomas was present since birth [72], which is suggestive of a type 2 segmental involvement (Reproduced with permission from John Wiley & Sons, USA)

10.1.2 Trichodiscoma

Multiple trichodiscomas are inherited as an autosomal dominant trait [419]. A 39-year-old woman had a linear group of seven trichodiscomas involving her scalp [158]. This case can be categorized as a type 1 segmental manifestation of the disorder.

10.1.3 Pilomatricoma

Multiple pilomatricomas are a feature of several hereditary syndromes, but a segmental arrangement of these tumors has not been reported so far.

10.1.4 Basaloid Follicular Hamartoma

Multiple basaloid follicular hamartoma (BFH) can be inherited as an autosomal dominant trait. Some authors prefer the name "infundibulocystic basal cell carcinoma" to categorize these benign skin tumors [78, 364]. I strongly recommend, however, to avoid this misleading term because otherwise an essentially benign tumor can easily be mistaken as a malignant neoplasia, resulting in overtreatment.

Two cases of unilateral distribution of multiple non-syndromic BFH [202, 229] can be categorized as a type 1 segmental manifestation. Two cases of superimposed arrangement suggesting a type 2 segmental involvement have also been reported (Fig. 10.2) [78, 422].

On the other hand, segmentally arranged basaloid follicular hamartomas are a hallmark of Happle-Tinschert syndrome (see Sect. 12.2.1.3).

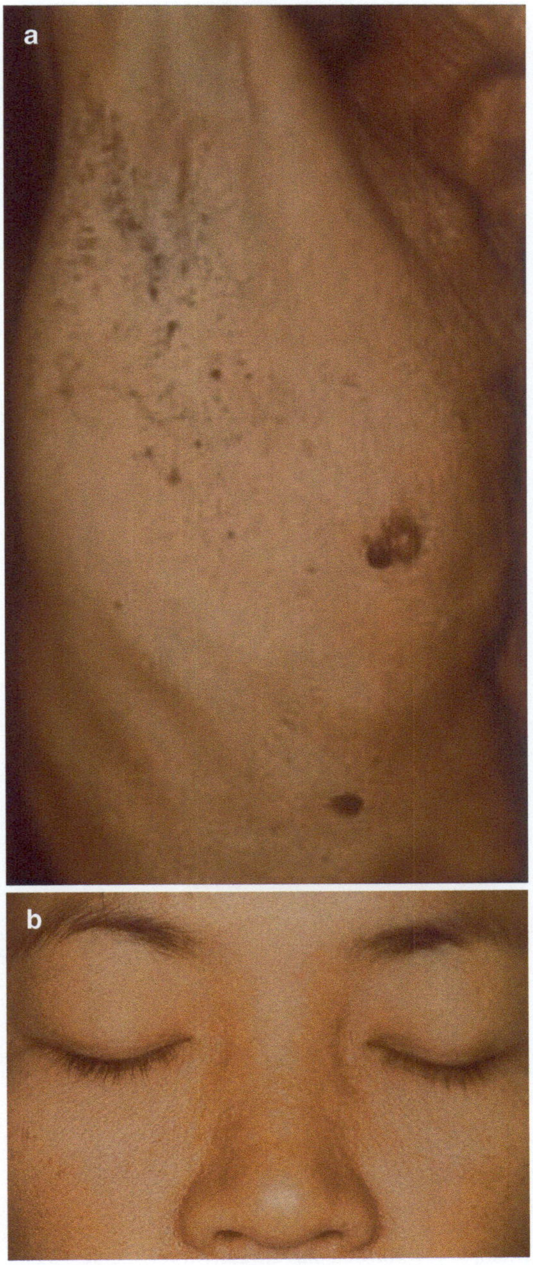

Fig. 10.2 (**a**) This 66-year-old man had multiple linear basaloid follicular hamartomas. Two brothers had nonsegmental lesions, and (**b**) his 39-year-old daughter showed likewise a diffuse involvement [78]. Because of such family constellation, a type 2 segmental manifestation is likely (Reprinted with permission from John Wiley & Sons, USA)

10.1.5 Fibrofolliculoma: A Hallmark of Hornstein-Knickenberg Syndrome (alias Birt-Hogg-Dubé Syndrome)

Multiple fibrofolliculomas (Fig. 10.3a) are a hereditary cutaneous trait that was previously described under the name "multiple perifollicular fibromas" [81]. They are a cutaneous hallmark of a syndrome that is caused by folliculin

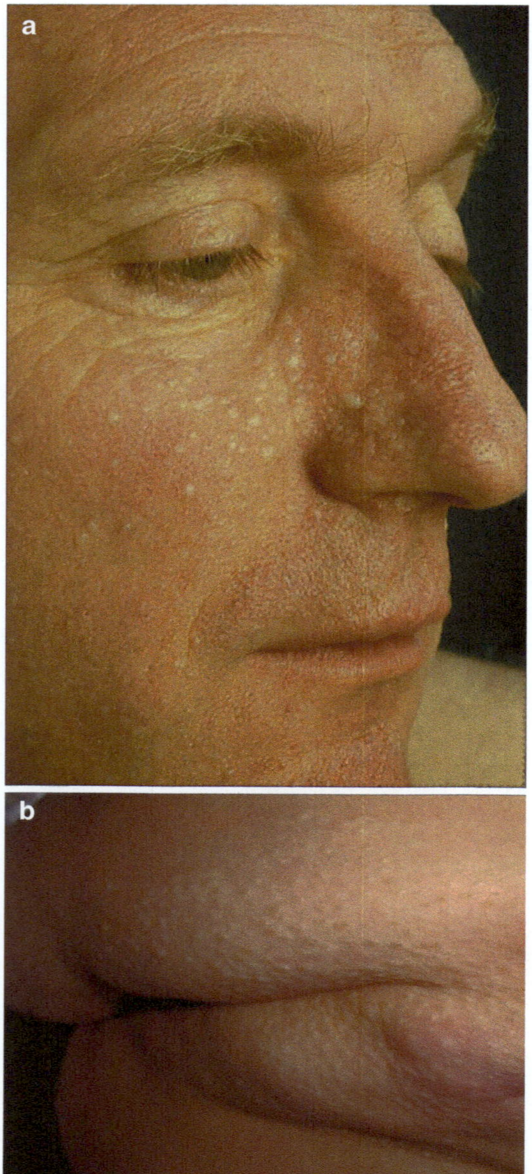

Fig. 10.3 Hornstein-Knickenberg syndrome. (**a**) Multiple fibrofolliculomas diffusely involving the face in 56-year-old man who died from colon carcinoma; (**b**) type 2 segmental involvement of the left chest wall in a 75-year-old woman. Note characteristic gooseskin-like surface of the huge lipomatous lump (**b**: Courtesy of Dr. Derek Lim and Dr. Eamonn R. Maher, Birmingham, UK)

mutations and includes lung cysts, pneumothorax, renal cysts, proneness to renal cell carcinoma, and colorectal polyps showing proclivity to malignant degeneration [301, 318]. The term "Birt-Hogg-Dubé syndrome" is an improper eponymic designation because these authors have not added any new data to what had completely and scholarly been described by Hornstein and coworkers [208–210] as a distinct autosomal dominant trait being associated with proneness to extracutaneous cancer. In fact, Birt et al. [34] referred to the pioneering work of Hornstein and Knickenberg [209]. They reiterated Hornstein's meticulous description of perifollicular fibromatosis cutis and, by giving it another name, incorrectly maintained that they had found "a previously unrecognized hereditary pilar hamartoma." The renaming "Hornstein-Knickenberg syndrome" gives credit to the original authors [193, 334].

10.1.5.1 Type 1 Segmental Manifestation of Hornstein-Knickenberg Syndrome

An unusual case of multiple unilateral facial fibrofolliculomas was published by Casalá et al. [66]. Whether this report can be taken as an example of type 1 segmental manifestation of Hornstein-Knickenberg syndrome cannot be said with certainty as long as molecular data are lacking.

10.1.5.2 Type 2 Segmental Manifestation of Hornstein-Knickenberg Syndrome

Possible examples of a type 2 segmental manifestation of the syndrome (Fig. 10.3b) have been documented by several authors. In a 49-year-old man with this disorder, Weintraub and Pinkus [472] described "a large connective tissue nevus" involving a unilateral segment of the thorax. The firm plaque was present "as long as the patient could remember." Its surface showed a "pigskin-like graining," being studded with umbilicated papules that contained central keratin plugs or hairs. Histopathologically these papules corresponded to fibrofolliculomas. Similar plaques have been noted by Toro et al. [439] in 3 out of 125 patients. Kluger et al. [246] found such plaques in 2 out of 22 patients and assumed that they may reflect a type 2 segmental manifestation of the trait.

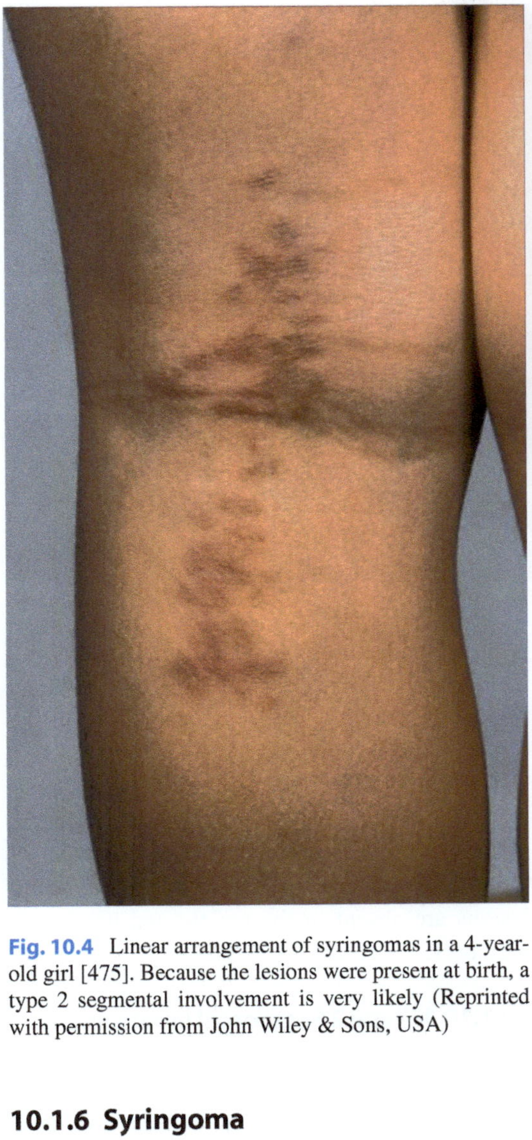

Fig. 10.4 Linear arrangement of syringomas in a 4-year-old girl [475]. Because the lesions were present at birth, a type 2 segmental involvement is very likely (Reprinted with permission from John Wiley & Sons, USA)

10.1.6 Syringoma

Multiple syringomas represent an autosomal dominant trait. A type 1 segmental involvement has been documented in many reports [70, 89, 175, 382, 489]. Cases of type 2 involvement superimposed on the ordinary nonsegmental phenotype have likewise been described [241, 288, 478]. A case of a 4-year-old girl with pronounced lesions of a congenital "linear syringomatous hamartoma" [475] can likewise be categorized as an example of type 2 segmental involvement (Fig. 10.4). In an additional case, type 2 mosaicism is likely but less certain [371].

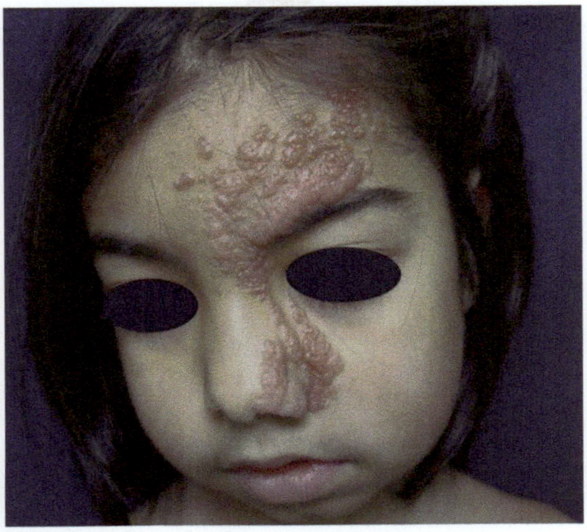

Fig. 10.5 An 8-year-old girl with congenital linear eccrine spiradenomas suggesting a type 2 segmental manifestation [117] (Reprinted with permission from John Libbey Eurotext, Montrouge, France)

10.1.7 Spiradenoma

Multiple spiradenomas are inherited as an autosomal dominant trait [431]. Several cases suggesting a type 1 involvement have been reported [8, 48, 321, 487].

An unquestionable case of type 2 segmental spiradenomatosis was described in a family with multiple eccrine spiradenoma and cylindroma present in three consecutive generations [479]. In three additional cases a type 2 segmental involvement is almost certain because a pronounced segmental involvement was already present at birth (Fig. 10.5) [117, 328, 370]. Some other cases of segmentally arranged eccrine spiradenomas [28, 162, 163, 267, 405, 450, 494] are difficult to categorize according to the dichotomous types of mosaicism.

10.1.8 Eccrine Poroma

A case of multiple eccrine poroma arranged in a linear pattern was described by Ogino [329].

10.1.9 Cylindromatosis

Cutaneous cylindromatosis is caused by *CYLD1* mutations. A case of multiple linear cylindromas

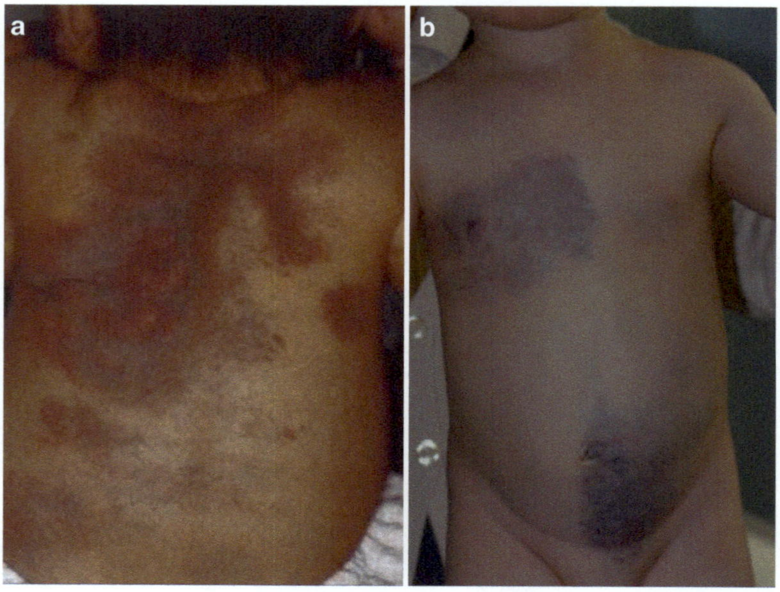

Fig. 10.6 Type 2 segmental glomangiomatosis in (**a**) a 2-month-old girl [486] and (**b**) a 1-year-old boy [154] (**a**: Reprinted with permission from Elsevier Limited, Oxford, UK; **b**: Reprinted with permission from John Wiley & Sons, USA)

as reported by Martinez et al. [291] probably represents a type 1 segmental manifestion.

10.1.10 Glomangiomatosis

The vascular lesions of this autosomal dominant trait are true angiomas, which is why the presently prevailing term "glomuvenous malformation" [45, 52] appears to be inappropriate. Glomangiomatosis is caused by mutations in the glomulin gene [52]. It is a rather rarely occurring phenotype, but in affected families a type 2 segmental involvement is very common. Apparently, the glomulin locus represents a hotspot for somatic recombination [189]. The lesions of segmental glomangiomatosis do not follow Blaschko's lines but tend to be arranged in a flag-like or checkerboard pattern [248, 286].

Cases suggesting a type 1 segmental involvement have rarely been reported [263, 438, 485], whereas a type 2 involvement (Fig. 10.6) appears to occur far more frequently [11, 24, 43, 53, 102, 107, 195, 205, 223, 286, 287, 303, 312, 315]. This may in part reflect a bias of ascertainment because a type 2 involvement is more conspicuous. Notwithstanding, the prevalence of such pronounced mosaic manifestation is extremely high as compared to most other autosomal dominant skin disorders. The superimposed segmental lesions tend to be much more painful than the nonsegmental glomangiomas [11, 77].

Cases of "congenital plaque-like" glomangioma [65, 232, 486] or "giant glomangioma" [411] can today be categorized as examples of type 2 mosaicism even when nonsegmental lesions are absent in the patient and his family members.

10.1.10.1 Type 2 Segmental Involvement of Internal Organs

The lesions of glomangiomatosis are usually limited to the skin. Remarkably, however, a type 2 segmental manifestation may also involve internal organs. Goujon et al. [154] described two newborn children with extensive mosaic plaque-type glomangiomatosis. They had been diagnosed *in utero* with pleural effusion and ascites, suggesting "a pathogenic link" between the conditions. In one of these cases, a glomulin mutation was found in the boy and his mother who had nonsegmental glomagiomatosis. No molecular analysis was perfomed in the other newborn. Similarly, Tejedor et al. [430] reported that in a newborn girl with plaque-type glomagiomatosis, isolated fetal ascites had been documented, 1 week prior to birth,

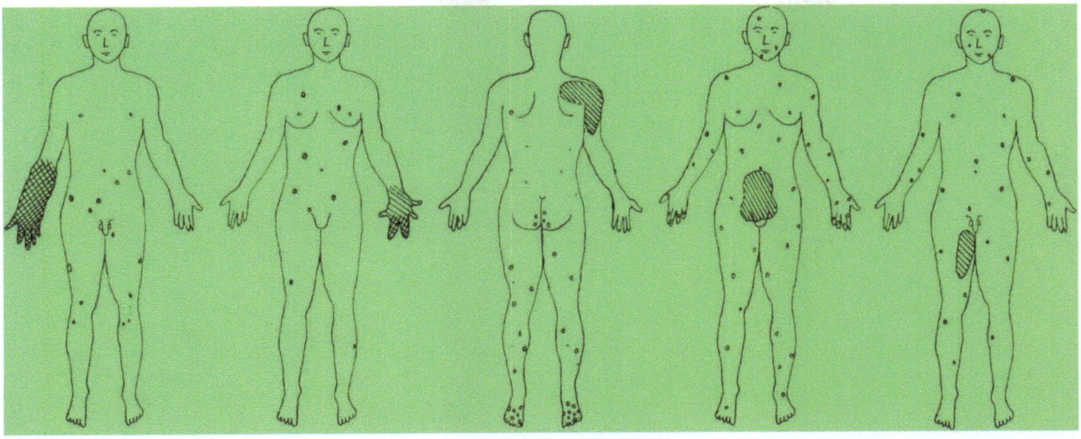

Fig. 10.7 Schematic representation of case reports suggesting a type 2 segmental manifestation of blue rubber bleb angiomatosis. Crossed hachures indicate areas of both cutaneous and subcutaneous involvement, whereas simple hachures symbolize either subcutaneous or extracutaneous lesions [190] (Reprinted with permission from John Libbey Eurotext, Montrouge, France)

by sonographic examination. After birth, the ascites was found to be of chylous origin. The patient's father had multiple glomangiomas being grouped on his left thigh. According to present knowledge, these cases cannot be understood without accepting the theory that at the underlying gene locus, the corresponding wild-type allele has been lost at an early developmental stage.

10.1.10.2 Practical Aspects

The lesions of type 2 segmental glomangiomatosis are far more difficult to treat than the nonsegmental glomangiomas [23].

10.1.11 Blue Rubber Bleb Angiomatosis ("Blue Rubber Bleb Nevus Syndrome")

In 1958, William Bean [27] coined the misnomer "blue rubber bleb nevus" to describe the peculiar color and consistence of cutaneous vascular tumors that also involve extracutaneous organs in the form of an autosomal dominant "blue rubber bleb nevus syndrome." These vascular lesions, however, are true angiomas and can, therefore, not be categorized as nevi [133, 190]. The molecular cause of blue rubber bleb angiomatosis is not known as yet.

10.1.11.1 Type 1 Segmental Involvement

Two cases of "unilateral dermatomal cavernous hemangiomatosis" have been reported [468, 477]. Most likely they represent examples of a type 1 segmental manifestation of blue rubber bleb angiomatosis.

10.1.11.2 Type 2 Segmental Involvement

In 2010, five cases suggesting a type 2 segmental manifestation have been reviewed (Fig. 10.7) [190]. In one of these patients, the right forearm was amputated in childhood because a giant congenital blue rubber bleb angioma had rendered the limb functionless [133]. A type 2 segmental involvement can render the disorder detectable even before birth (Fig. 10.8). In such cases, prenatal sonographic surveillance revealed very large cutaneous vascular tumors that were complicated by life threatening consumptive coagulopathy soon after birth [307].

10.1.11.3 Extracutaneous Type 2 Segmental Lesions

Type 2 segmental blue rubber bleb angiomatosis may also give rise to extensive extracutaneous lesions. For example, Atten et al. [19] described

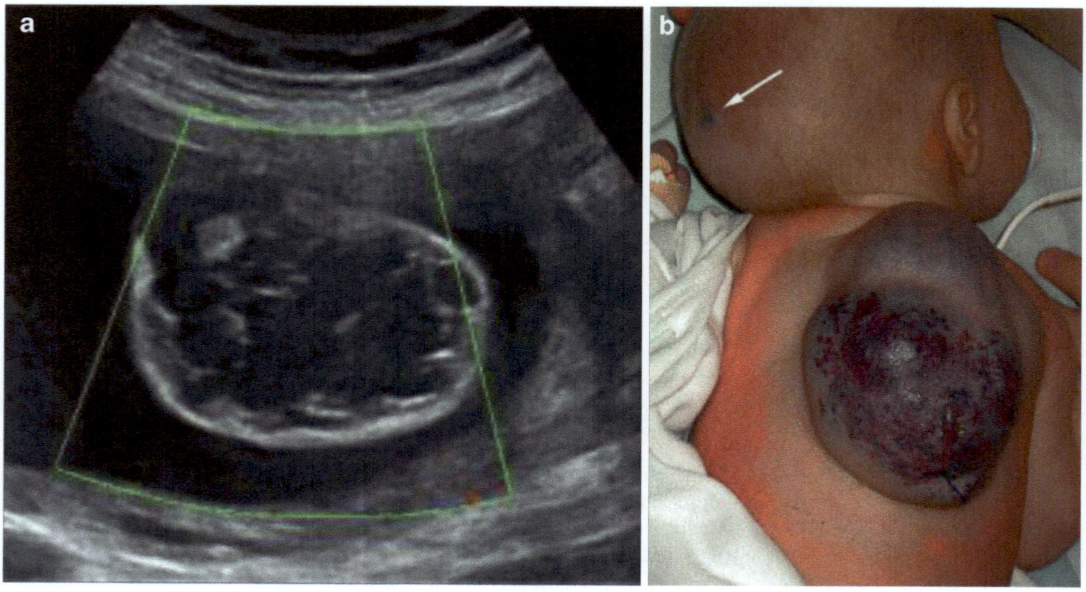

Fig. 10.8 Type 2 segmental blue rubber bleb angiomatosis [307]. (**a**) Prenatal ultrasound image at 25 weeks showing a "cystic" dorsal lesion with many septa and thromboses; (**b**) the same angiomatous tumor in the newborn. Note the additional small lesion (*arrow*) heralding nonsegmental involvement (Reprinted with permission from John Wiley and Sons, USA)

a woman with numerous cutaneous vascular lesions and episodes of rectal bleeding. During childhood she had an exploratory laparotomy and was told she had a large pelvic vascular malformation for which no treatment was proposed. At 20 years of age, magnetic resonance imaging showed a large pelvic hemangioma, measuring 60 cm and extending into the abdomen. It had eroded the rectal wall to produce rectal bleeding. Because of the enormous size of the pelvic tumor, bleeding was controlled by conservative measures and an expectant attitude was chosen. In retrospect, this appears to be a classic example of extracutaneous type 2 segmental manifestation of the disorder (see case 4 in Fig. 10.7). An excessive intraperitoneal and retroperitoneal involvement was also reported by Patel et al. [345]. In another case of blue rubber bleb angiomatosis, a giant mass of angiomatous lesions had to be resected in the tranverse colon [319].

10.1.12 Lipomatosis

Lipoma is a very common tumor. Its molecular basis is unknown. Lipomatosis is inherited as an autosomal dominant trait (OMIM 151900). A mosaic arrangement has so far not been reported.

However, a pronounced unilateral form of lipomatosis was documented as part of the hemihyperplasia-multiple lipomata syndrome [32, 47, 272, 393] (see Sect. 12.2.2). This sporadic phenotype is apparently not related to hereditary lipomatosis.

10.1.13 Neurofibromatosis 1

Neurofibromatosis 1 (NF1) is caused by mutations in the neurofibromin gene. Today the expression "segmental NF1" should be regarded as an umbrella term. The type 1 segmental NF1 is a well known phenomenon that has so far been reported in more than 150 cases. In my view, however, the type 2 segmental manifestation is even more common but widely neglected until today, although the existence of this superimposed form of mosaic NF1 has been proven at the molecular level.

10.1.13.1 Type 1 Segmental NF1

In the past century this type was simply called "segmental NF1" because it was generally

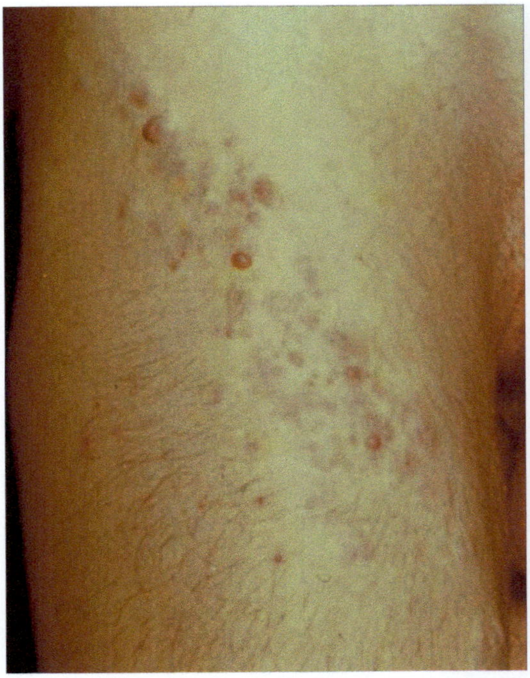

Fig. 10.9 Type 1 segmental neurofibromatosis 1 showing neurofibromas alone

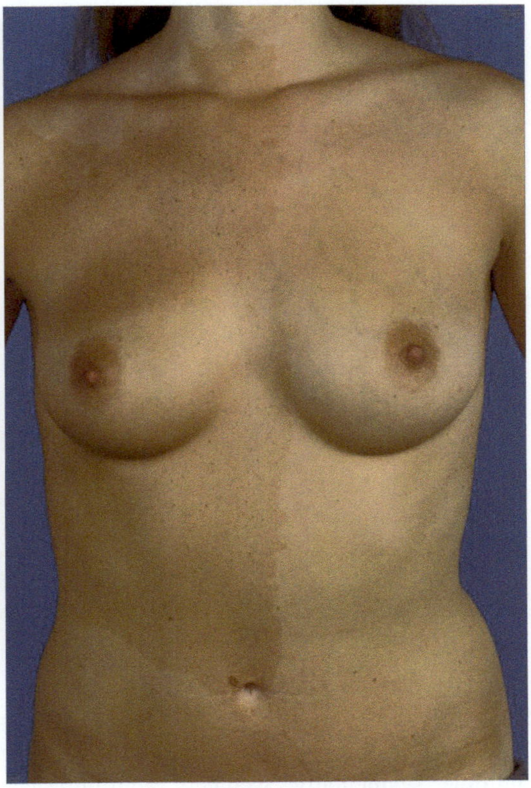

Fig. 10.10 Type 1 segmental neurofibromatosis 1 showing hyperpigmentation alone

believed that this was the only form of mosaic NF1. Around 1990 many experts erroneously considered segmental neurofibromatosis to represent a distinct entity in the form of "NF5," to be separated from NF1 [360, 367, 368]. Similarly, a classification of segmental neurofibromatosis into four subtypes as proposed by Roth et al. [373, 443] should today be taken as a historical error. This outdated concept implied that in patients with bilateral or otherwise systematized segmental neurofibromatosis, one had to assume multiple postzygotic mutations [233, 373, 425, 443], although one single postzygotic mutational event is sufficient to plausibly explain such cases.

Cases that can today be categorized as legitimate examples of type 1 segmental NF1 have been described by many authors [60, 61, 90, 97, 233, 306, 335, 375, 385]. In the involved area, either neurofibromas alone (Fig. 10.9) or pigmentary changes alone (Fig. 10.10) or both lesions are noted. Ipsilateral Lisch nodules may be an additional feature of this mosaic disorder [40, 376, 473].

Molecular proof that postzygotic mosaicism causes type 1 segmental NF1 has been provided by Tinschert et al. [434] and Vandenbroucke et al. [456]. Interestingly, in one of these cases [456] almost the entire surface of the body was affected by NF1 manifestations, leaving a few segments unaffected (Fig. 3.1). The authors were able to exclude revertant mosaicism, which clearly demonstrates that type 1 segmental NF1 may affect, by way of exception, far more than half of the body tissue.

10.1.13.2 Type 2 Segmental NF1

The lesions of type 2 segmental NF1 tend to be rather pronounced, and they are usually noted at birth or during early infancy. They may adopt various clinical appearances. In part this superimposed form of mosaicism may consist of a large band-like café-au-lait hyperpigmentation with intralesional cutaneous or subcutaneous neurofibromas [14, 242] (Fig. 10.11), or of

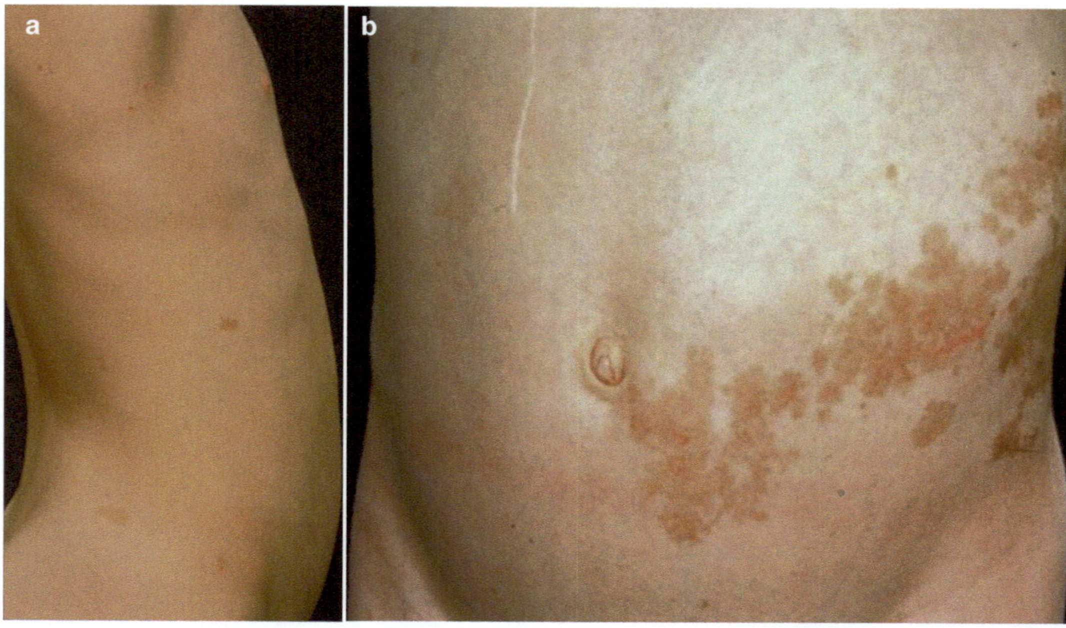

Fig. 10.11 Type 2 segmental neurofibromatosis 1 in an 11-year-old girl. (**a**) Right side showing nonsegmental café-au-lait macules; (**b**) left side showing pronounced segmental involvement [14] (**a**: Courtesy of Dr. David Atherton, London, UK; **b**: Reprinted with permission from John Wiley & Sons, USA)

multiple segmentally arranged neurofibromas [221], or of a giant café-au-lait macule without any hint of underlying neurofibromas [482]. The most frequently noted manifestation, however, is a large plexiform neurofibroma. Indeed, all plexiform neurofibromas of appreciable size represent a type 2 segmental NF1 [179]. It is in line with this view that virtually all of these tumors are present at birth [368]. In particular, all cases of "giant neurofibroma" [359] appear to result from LOH occurring at an early developmental stage.

According to Pivnick and Riccardi [352], 5 % of NF1 patients have large plexiform neurofibromas. In a population-based study, Huson et al. [215] found that even 26.7 % of individuals with NF1 had a plexiform neurofibroma evident by physical examination. These tumors may be inconspicuous during infancy but later tend to show aggressive growth. Involvement of the face may result in severe compromise and death [13], whereas overgrowth of a limb likewise constitutes a ponderous handicap [253]. The overlying skin is usually thickened and hyperpigmented. The acanthotic epidermis has sometimes been mistaken as an epidermal nevus [88]. Hypertrichosis

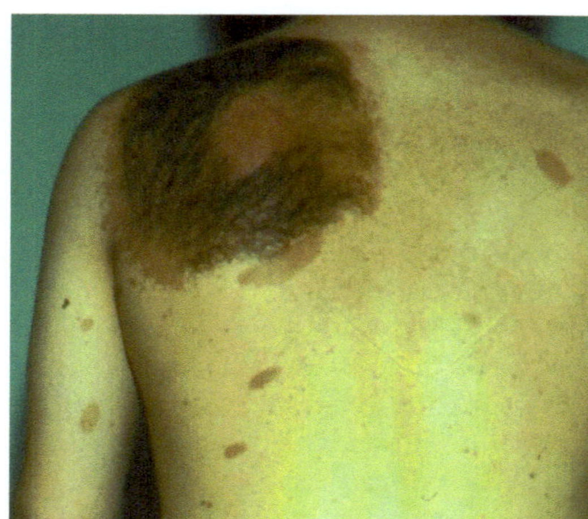

Fig. 10.12 Plexiform neurofibroma showing pronounced hypertrichosis (Courtesy of Dr. Howard Pride, Danville, Pennsylvania, USA)

in the form of dark, coarse terminal hair is usually noted (Fig. 10.12). On the other hand, plexiform neurofibroma may involve all of the underlying structures such as nerves, muscles, and bones, resulting in giant overgrowth of a limb (Fig. 10.13) or half of the face [253]. Large plexiform

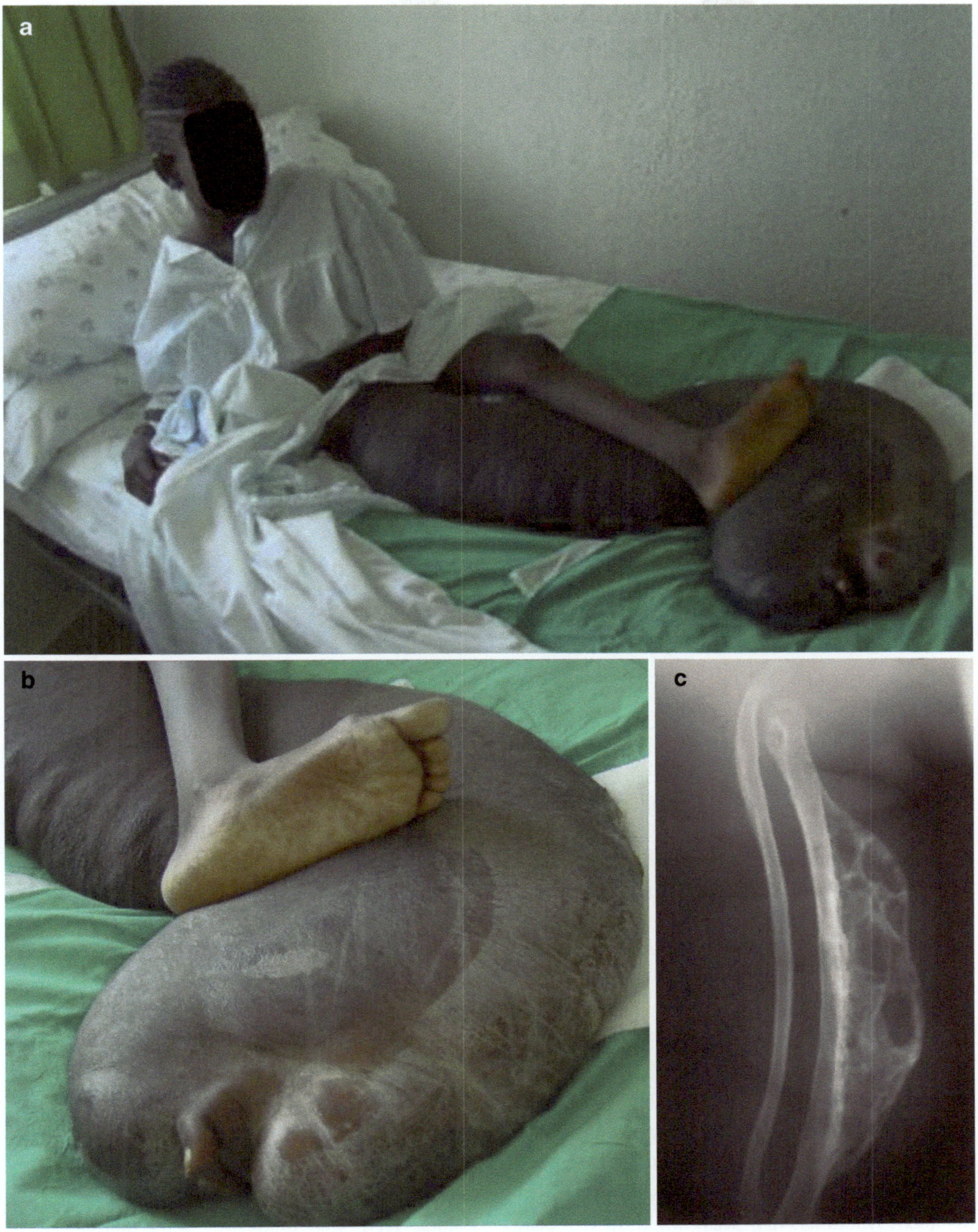

Fig. 10.13 Type 2 segmental neurofibromatosis 1 giving rise to huge overgrowth of a leg. (**a**) General view; (**b**) the right foot being hidden within the tumor mass; (**c**) cystic periosteous overgrowth of the tibia [292] (**a** and **c**: Reprinted with permission from John Wiley and Sons, USA; **b**: Courtesy of Dr. Silvestre Martínez-García, Málaga, Spain)

neurofibromas may involve the brain [41] or the liver [74], or show intrathoracic growth [449], or even originate from the colon [483]. By way of exception, such segmental overgrowth may also be caused by diffuse proliferation of the fusiform cells of classical neurofibromatous tissue [292].

Table 10.1 Some case reports that can today be reclassified as examples of type 2 segmental NF1

Reference	Age at onset of segmental lesions	Coexistence of nonsegmental lesions	Family members with nonsegmental lesions
Butterworth [58]	Birth	+	
Diekmann et al. [106]	Infancy	+	+
Bingham and Burrows [33]	Birth	+	+
Archer et al. [14]	Childhood	+	
Boltshauser et al. [41]	[17 patients]	+	
Chen et al. [74]	Early childhood	+	
Ettl et al. [122]	Birth	+	
Pivnick et al. [351]	Birth	+	
Kim et al. [242]	20 years	+	
Oguzkan et al. [330]	Childhood	+	+
Yang et al. [482]	Birth	+	
Pascual-Castroviejo [343]	[35 children]	+	
Rallis [359]	Infancy	+	
Menon and Kumar [302]	Birth	+	
Bano et al. [22]	Childhood	+	
Ji et al. [228]	Birth	+	
Thammaiah et al. [432]	<18 months	+	+
Kantaputra et al. [236]	Birth	+	
Sharma et al. [402]	Birth	+	
Nasir et al. [320]	<7 years	+	+
Salvitti et al. [383]	<3 months	+	+

In short, it is the type 2 segmental involvement that, to a large degree, contributes to morbidity in patients with NF1.

It should be noted that until today the concept of type 2 segmental NF1 is far from being generally accepted by the scientific community. For example, in 2008 Pascual-Castroviejo et al. [343] presented a series of 43 cases of segmental NF1 occurring in children or adolescents. The least frequently noted form was an exclusively cutaneous manifestation, whereas in 81 % of cases underlying structures such as the bones were additionally affected. In most patients the segmental involvement was characterized by a plexiform neurofibroma. Moreover, a family history of nonsegmental NF1 was noted in 15 cases (35 %). From this well-documented study we can conclude that most of the patients had in fact a type 2 segmental involvement in the form of plexiform neurofibroma. Similarly, numerous photographs as published in books or chapters on NF1 [271, 368, 377] inadvertently depict classical cases of type 2 segmental NF1. Many other examples of type 2 segmental NF1 can be found in the literature (Table 10.1)

Molecular evidence for the concept of type 2 segmental NF1 has been provided by several groups who documented absence of the corresponding wild-type allele in plexiform neurofibromas [96, 230, 247]. Moreover, Steinmann et al. [421] found postzygotic recombination to be a frequent cause of LOH in plexiform neurofibromas.

10.1.13.3 The Issue of "Genetic Transmission of Segmental NF1"

Because the concept of dichotomous types of segmental NF1 was unknown, some authors have developed, by conflating the two forms, the erroneous idea of "genetic transmission of segmental NF" [330, 373, 374, 395, 452]. In fact, authentic cases of familial occurrence of type 1 segmental NF1 have not been reported until today. Two asserted cases observed by Huson and Ruggieri [216] have not been documented in detail and can, therefore, not be accepted as legitimate examples of vertical transmission of type 1 segmental NF1. Most cases of alleged transmission of segmental NF1 can be explained in the following way. The parent had type 1 segmental NF1 heralding gonadal mosaicism. The child had nonsegmental NF1 being superimposed by a type 2 segmental involvement that may constitute, during childhood, the only recognizable feature of the disorder. On the other hand, the possibility that both parent and child had type 2 segmental lesions should also be considered.

10.1.13.4 Genetic Counseling in Cases of Segmental NF1

Because a type 1 segmental NF1 implies the possibility of gonadal mosaicism, patients run an increased, albeit rather small, risk to give birth to a child affected with nonsegmental NF1 [40, 87, 174, 216]. The risk cannot be expressed in terms of a given percentage, but apparently the danger of gonadal mosaicism increases with the size of the body area involved [377]. On the other hand, it should be borne in mind that germline mosaicism may even be present in individuals showing no clinical feature of NF1 [264].

Patients with type 2 segmental NF1 have a 50 % risk to give birth to a child with nonsegmental NF1. The possibility that an affected child will show, in addition, a type 2 segmental involvement is likewise present because all children born with nonsegmental NF1 run this risk. Hence, it is of utmost practical importance that clinicians are able to discriminate between a type 1 and type 2 segmental manifestation of NF1.

10.1.13.5 Other Practical Aspects

In type 2 segmental NF1, the involvement of extracutaneous structures can be recognized by application of the presently available imaging techniques [343].

Dermatologists in practice should be warned that attempts to remove plexiform neurofibromas by operative procedures may cause diffuse bleeding. Such therapeutic interventions should only be performed by a fully equipped and well-trained surgical team, including anesthesiological support.

Appropriate knowledge of type 2 segmental NF1 is important because the vast majority of malignant peripheral nerve sheath tumors develop from large plexiform neurofibromas [297, 368]. Apparently, the inherent early loss of the corresponding wild-type allele represents a first step in the multistage process of carcinogenesis giving rise to such malignant tumors.

10.1.14 Neurofibromatosis 2

In this disorder, a type 1 mosaicism appears to occur rather frequently [123], although cutaneous manifestations have so far not been reported. In such cases, a unilateral or otherwise segmental involvement of the brain has been confirmed by MRI examination [376]. A type 2 segmental involvement of the skin has been documented by Susan Huson from Manchester (Fig. 10.14).

10.1.15 Schwannomatosis

The disorder is characterized by multiple cutaneous and spinal schwannomas. The tumors that are also called neurilemmomas originate from mutations in *SMARCB1* [213]. Remarkably, however, evidence has been provided that these schwannomas may be caused by a four-hit mechanism [397] (see Sect. 3.1.1.1).

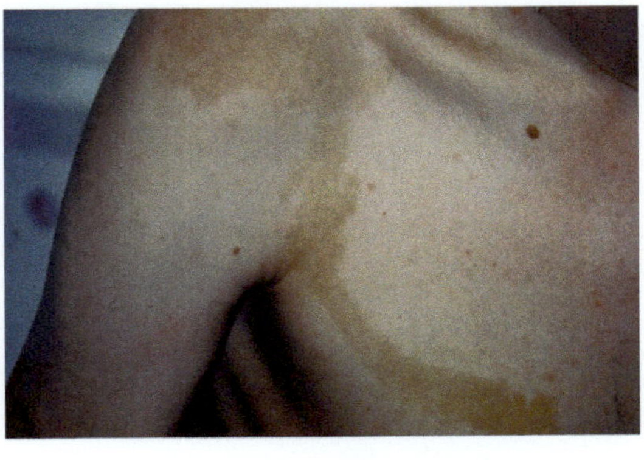

Fig. 10.14 Type 2 segmental manifestation of neurofibromatosis 2 (Courtesy of Dr. Susan M. Huson, Manchester, UK)

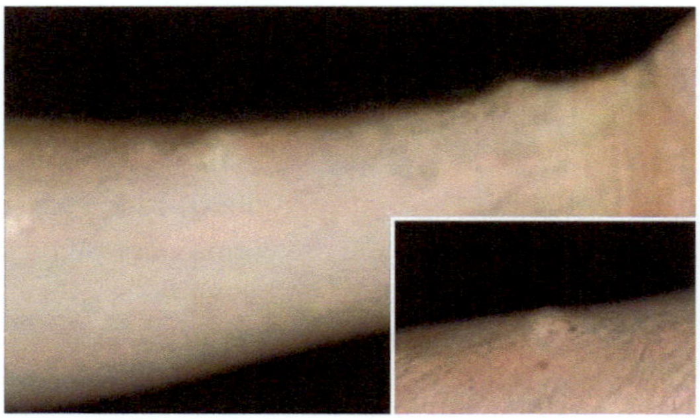

Fig. 10.15 Type 1 segmental schwannomatosis [266] (Reprinted with permission from John Wiley & Sons, USA)

Several patients with mosaic subcutaneous lesions have been reported [266, 355, 427, 442]. Most, if not all, of these cases appear to represent a type 1 segmental involvement (Fig. 10.15). Moreover, germline mosaicism in a mother of two affected children has been documented [212].

10.1.16 Legius Syndrome

The features of Legius syndrome include multiple NF1-like café-au-lait macules, axillary freckling, and learning problems, but contrasting with NF1 the patients have multiple lipomas, a Noonan-like facial appearance, and macrocephaly, whereas neurofibromas are absent [50]. The disorder is caused by mutations in *SPRED1*, a gene involved in MAPK (mitogen-activated protein kinase) signalling. The fact that neurofibromin is connected with the same pathway offers an explanation for the overlapping clinical features. In their first report, Brems et al. [50] have inadvertently documented a type 2 segmental manifestation in the form of a large, blocklike café-au-lait hyperpigmentation being sharply demarcated in the midline (Fig. 10.16).

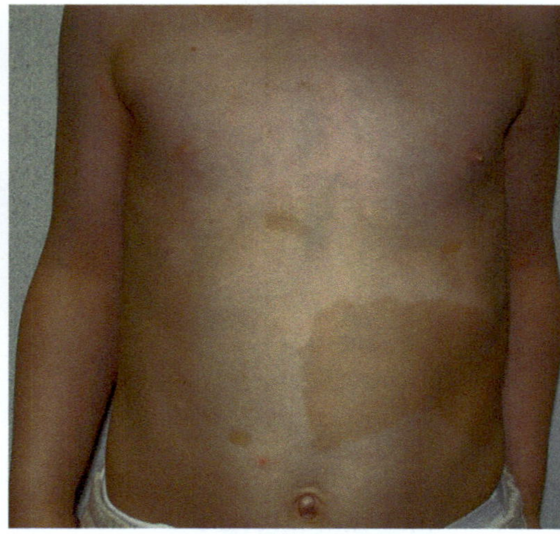

Fig. 10.16 Type 2 segmental manifestation of Legius syndrome [50] (Reprinted with permission from Nature Publishing Group)

10.1.17 Leiomyomatosis

Cutaneous leiomyomatosis is caused by mutations of the fumarate hydratase gene [437]. In some families the trait is associated with an increased risk to develop renal cancer [262, 390]. A type 1 segmental involvement has very rarely been reported [153, 255].

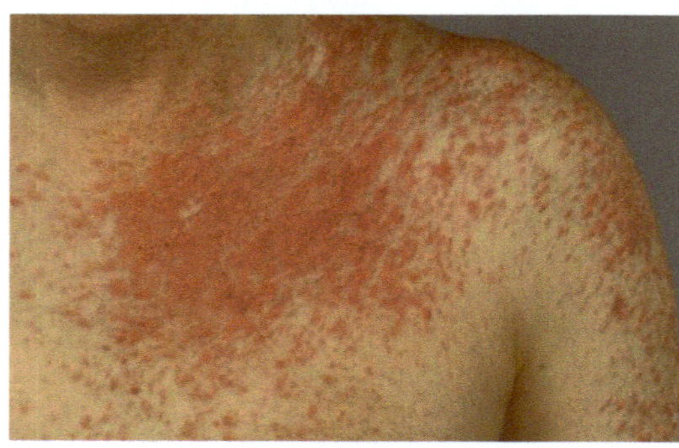

Fig. 10.17 Type 2 segmental leiomyomatosis [21] (Reprinted with permission from Elsevier Limited, Oxford, UK)

By contrast, type 2 segmental leiomyomatosis (Fig. 10.17) occurs with a strikingly high frequency [5, 21, 80, 177, 182, 322, 333, 369, 448, 474]. Alam et al. [4] found a combination of segmental and disseminated leiomyomas in 19/59 patients (32 %). This may reflect an increased proneness of the fumarate hydratase gene to postzygotic recombination [189]. As a characteristic feature, the superimposed segmental lesions are extremely painful after pressure, exposure to cold, or emotional stress [1, 5, 57, 62, 159, 224, 254, 270, 278, 281, 333, 409, 463].

Presumably, most cases of segmental leiomyoma without presence of nonsegmental skin lesions in the patient and his family [200, 217, 240, 243, 342, 428] represent examples of type 2 mosaicism. This view is supported by the description of symptom-free gene carriers in families of such patients [293]. On the other hand, female gene carriers may have uterine leiomyoma alone [293, 342, 390, 413].

10.1.17.1 Familial Occurrence of Type 2 Segmental Involvement

Type 2 segmental leiomyomatosis has been described in a mother and her daughter [251] and even in three relatives [197, 363], thus providing additional support in favor of the assumption that the fumarate hydratase gene is a hotspot for postzygotic recombination.

10.1.18 Gorlin Syndrome

Major features of this trait are multiple basal cell carcinomas, jaw cysts, palmoplantar pitting, calcification of the falx cerebri, and various skeletal abnormalities. The disorder is caused by *PTCH1* mutations.

10.1.18.1 Type 1 Segmental Involvement

Several cases suggesting a type 1 mosaicism have been reported [63, 285, 404]. This particular manifestation has often been equated with, but should today be distinguished from, segmentally arranged multiple basaloid follicular hamartomas (see Sects. 10.1.4 and 12.2.1.3).

10.1.18.2 Type 2 Segmental Involvement

Molecular proof of type 2 mosaicism was provided in a 12-year-old girl with pronounced unilateral lesions of Gorlin syndrome including congenital basal cell carcinomas, an odontogenic cyst, and rather large palmoplantar pits arranged along Blaschko's lines (Fig. 10.18) [441]. She had inherited a *PTCH1* mutation

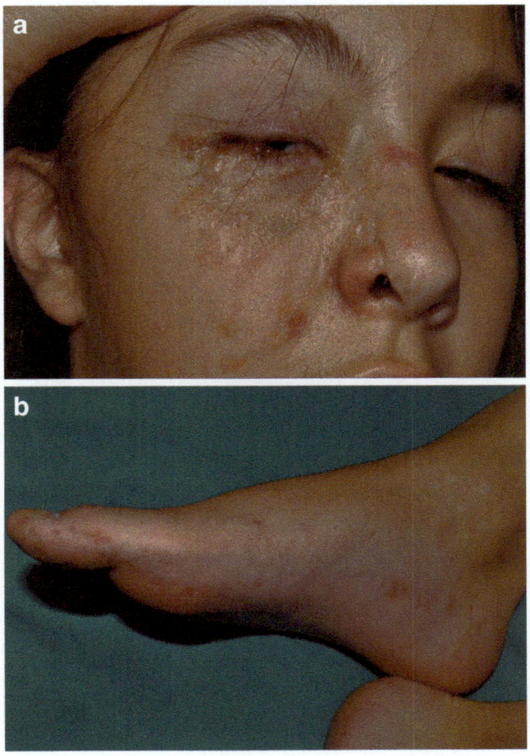

Fig. 10.18 Type 2 segmental Gorlin syndrome. (**a**) A 12-year-old girl with congenital basal cell carcinomas and linear atrophic lesions involving the right side of her face; (**b**) linear arrangement of rather large pits on the right foot [441] (**a**: Reprinted with permission from John Wiley & Sons, USA; **b**: Courtesy of Dr. Antonio Torrelo, Madrid, Spain)

from her father, and biopsies obtained from her unilateral skin lesions revealed an additional postzygotic mutation involving the corresponding *PTCH1* allele, thus giving rise to compound heterozygosity. Another impressive case suggesting a type 2 segmental involvement was documented by Gutierrez and Mora [165]. A 46-year-old man had numerous typical features including frontal bossing, maxillary prognathism, multiple basal cell carcinomas, scoliosis, bilateral thumb deformities, dense calcification of the falx cerebri, as well as coloboma and blindness of the right eye and contralateral cataract. The left side of his face was mildly affected with some basal cell carcinomas, whereas on the right side numerous, aggressively growing tumors formed tightly packed groups "with normal appearing skin interspersed." The pronounced lesions extended from the ear to the posterior midline of the neck and to the anterior aspect of the chest. Moreover, the right hand and foot showed a linear arrangement of atrophic depressions measuring up to 1 cm. The authors described these lesions as being "atypical and unlike pits in other patients with the syndrome." Hearing loss was likewise localized on the right side. Computed axial tomography showed "moderate atrophy, along with an enlarged right lateral ventricle with a probable area of porencephaly communicating with the frontal horns of the right ventricle." Hence, a rather severe unilateral involvement was undoubtedly superimposed on mild nonsegmental lesions of Gorlin syndrome.

10.1.19 Hereditary Nonsyndromic Multiple Basal Cell Carcinoma

This autosomal dominant trait (OMIM 605462) should be distinguished from Gorlin syndrome [178]. It is characterized by rather superficial basal cell carcinomas.

Type 1 mosaicism in the form of strictly unilateral involvement (Fig. 10.19) has been documented in several reports [46, 239, 261, 310]. Sometimes it was mistaken for a unilateral manifestation of Gorlin syndrome [403]. A case suggesting type 2 mosaicism was so far described only once [160].

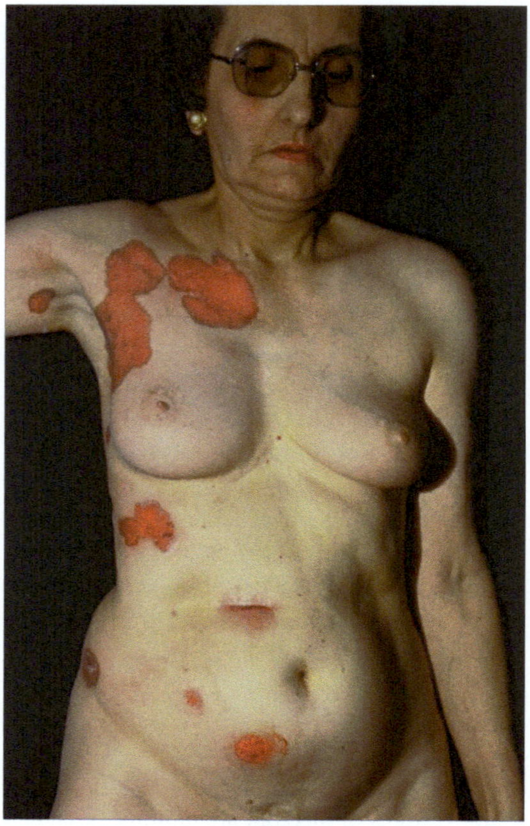

Fig. 10.19 Type 1 segmental manifestation of hereditary nonsyndromic multiple basal cell carcinoma (Courtesy of Dr. Georges Moulin, Lyon, France)

10.1.20 PTEN Hamartoma Syndrome (Cowden Disease Included)

PTEN hamartoma syndrome comprises two major clinical variants in the form of Cowden syndrome (OMIM 158350) and Bannayan-Riley-Ruvalcaba syndrome (OMIM 153480) [334]. Molecular analysis has shown that both variants can be caused by the same *PTEN* mutation and that both phenotypes may occur within the same family [201, 436], which is why Marsh et al. [289] proposed the new unifying term "PTEN hamartoma tumor syndrome." This designation appears to be redundant, because the name "PTEN hamartoma syndrome" contains all of the distinguishing information. On the other hand, the bold theory that cases of Proteus syndrome may be categorized as a particular form of PTEN hamartoma syndrome [349, 492] has turned out to be wrong (see below).

10.1.20.1 Cowden Variant of PTEN Hamartoma Syndrome

The Cowden variant is characterized by macrocephaly, multiple facial trichilemmomas, acral keratotic papules, and proneness to develop malignancies of breast, thyroid, and colon. The phenotype is predominantly reported in women [201].

10.1.20.2 Bannayan-Riley-Ruvalcaba Variant of PTEN Hamartoma Syndrome

The phenotype includes macrocephaly, mental deficiciency, pigmented macules of the penis, lipomatosis, hemangiomas, and intestinal polyps. Cancer proneness of various organs is present but appears to be less pronounced when compared to the Cowden variant [201]. This clinical variant is preponderantly described in males.

10.1.20.3 Lhermitte-Duclos Variant of PTEN Hamartoma Syndrome

In this variant (OMIM 158350 [334]), a dysplastic gangliocytoma of the cerebellum is associated with the features of Cowden disease [36].

10.1.20.4 Type 2 Segmental PTEN Hamartoma Syndrome

During the first decade of this century, type 2 segmental manifestations of PTEN hamartoma syndrome were often misdiagnosed as Proteus syndrome. The investigators erroneously asserted that some cases of Proteus syndrome were caused by a *PTEN* germline mutation [7, 119, 274, 290, 336, 414, 465, 492, 493], whereas other authors argued that this could not be true [84, 183]. Today this controversy has come to an end because Proteus syndrome was shown to originate from an *AKT1* mutation [269], whereas all cases of alleged "Proteus syndrome" caused by *PTEN* germline mutations can be categorized as examples of type 2 segmental PTEN hamartoma syndrome [186].

In 2007, the term "type 2 segmental Cowden disease" was proposed to denote a syndrome reflecting early postzygotic loss of the corresponding wild-type allele at the *PTEN* locus [186] (see also Table 3.4). A cutaneous hallmark is linear PTEN nevus that tends to be thicker and more papillomatous than common keratinocytic nevi (Fig. 10.20) [187] (see Sect. 7.3.1.4). Large connective tissue or vascular nevi, lipoblastomatosis, polyps of jejunum or colon, and focal segmental glomerulosclerosis may be associated, whereas the macrocephaly apparently reflects heterozygosity for the *PTEN* germline mutation [191]. As a synonym, the term "SOLAMEN syndrome" (segmental overgrowth, lipomatosis, arteriovenous malformation, epidermal nevus) was proposed by Caux et al. [69].

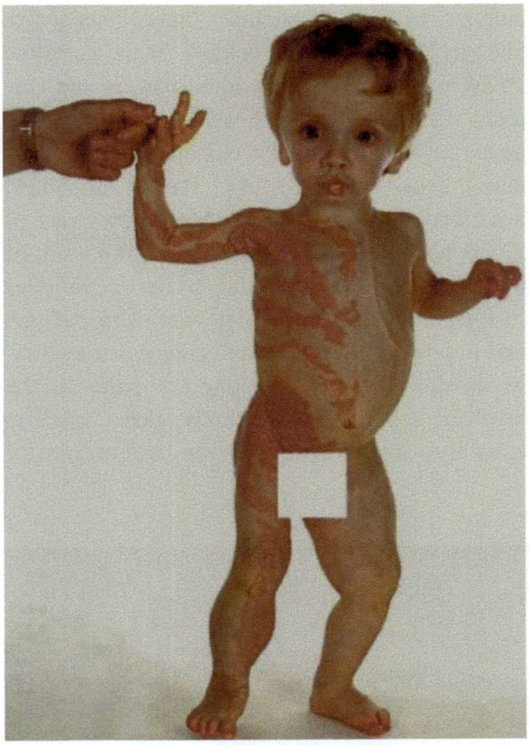

Fig. 10.20 Type 2 segmental PTEN hamartoma syndrome that was initially mistaken as "Proteus syndrome" [274]. Macrocephaly reflects heterozygosity for the *PTEN* mutation, whereas the unilateral linear PTEN nevus originated from early loss of heterozygosity (Reprinted with permission from John Wiley & Sons, USA)

An impressive case of type 2 segmental PTEN hamartoma syndrome has repeatedly been reported under various diagnostic terms such as "Proteus-like syndrome" [493], "arteriolovenular anomaly and epidermoid nevus involving right lower limb, scrotum, and penis" [426], or "PTEN hamartoma of soft tissue" [258]. This boy had giant overgrowth of his right leg that was covered by a linear epidermal nevus (Fig. 10.21). A R335X germline mutation was found in exon 5, whereas the lesional tissue of his right leg showed compound heterozygosity in the form of an additional R130X mutation involving exon 8 of the opposite allele [493]. Another case was reported under the name "hemigalencephaly as part of Jadassohn nevus sebaceus syndrome" [304]. The authors believed that this “may be another phenotypic finding associated with germline *PTEN* mutation.” For obvious reasons, the correct classification of the associated epidermal nevus is not nevus sebaceus but linear PTEN nevus (see Sect. 7.3.1.4). Tan et al. [426] described arteriovenous fistulas in 26 cases of PTEN hamartoma syndrome. Most, if not all, of these cases can today be categorized as examples of type 2 segmental involvement. The patients had lumps or “mass lesions” giving rise to cutaneous discoloration, swelling, or pain. In the involved segment of the body, these vascular anomalies tended to disrupt the architecture of neighboring muscles, and they were usually associated with variable amounts of ectopic fatty tissue. The high frequency of such complex vascular anomalies as noted by Tan et al. [426] may reflect a bias of ascertainment because all of the cases were documented in a specialized referral center for vascular anomalies. Iacobas et al. [219] described a boy affected with the Bannayan-Riley-Ruvalcaba variant who had a pronounced segmental involvement of his left arm with arteriovenous malformation, fatty infiltration of the muscles, and increased fibrous tissue, resulting in pain and loss of function of this arm.

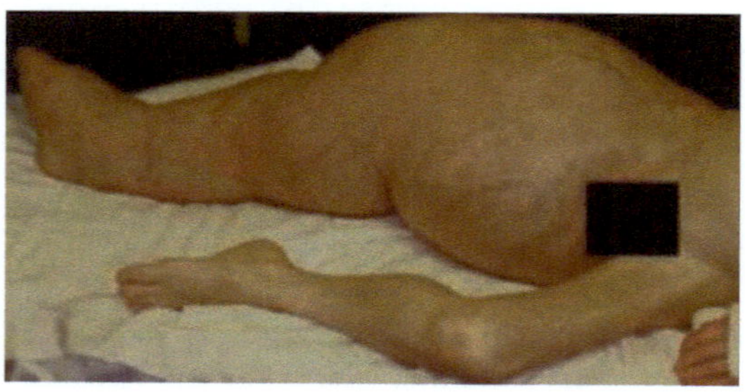

Fig. 10.21 Type 2 segmental manifestation of PTEN hamartoma syndrome [426], formerly diagnosed as “Proteus-like syndrome” [493] or “PTEN hamartoma of soft tissue (PHOST)” [258]. Note the linear epidermal nevus on the right lower leg (Reprinted with permission from BMJ Publishing Group Limited)

10.1.21 Cutaneous Mastocytosis

This disorder often affects several members of a family [10, 26, 82] and is presently considered to be inherited as an autosomal dominant trait (OMIM 154800 [334]). Familial childhood-onset mastocytosis can occur in both presence or absence of *c-KIT* mutations [37], whereas sporadic adult-onset mastocytosis is usually caused by gain-of-function mutations of *c-KIT* [276].

A type 1 mosaic involvement has been described in the form of linear or otherwise segmental arrangement of telangiectasia macularis eruptiva perstans (Fig. 10.22) [136, 152, 156, 249, 326, 344, 462].

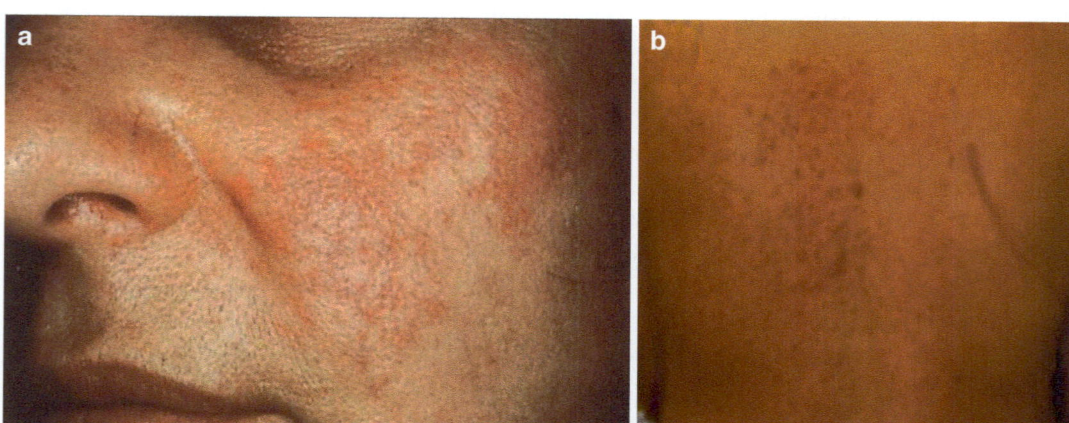

Fig. 10.22 Unilateral telangiectasia eruptiva perstans. (**a**) Facial lesions in a 36-year-old woman [152]; (**b**) dorsal lesions with strict midline separation in a 58-year-old man [326] (**a**: Reprinted with permission from John Wiley and Sons, USA; **b**: Reprinted with permission from John Libbey Eurotext, Montrouge, France)

10.2 Disorders of Keratinization

Examples of both type 1 and type 2 mosaicism were documented in various autosomal dominant disorders of keratinization.

10.2.1 Epidermolytic Ichthyosis of Brocq

A type 1 segmental involvement occurs rather frequently and is usually diagnosed as an "epidermal nevus of the epidermolytic type." Affected individuals run an increased risk to give birth to a child with diffuse epidermolytic ichthyosis of Brocq [31, 42, 98, 172, 207, 277, 323, 327, 362]. Molecular proof was provided by Paller et al. [341].

Cases of type 2 segmental manifestation in the form of pronounced linear lesions that are superimposed on the nonsegmental phenotype have also been reported [120, 166].

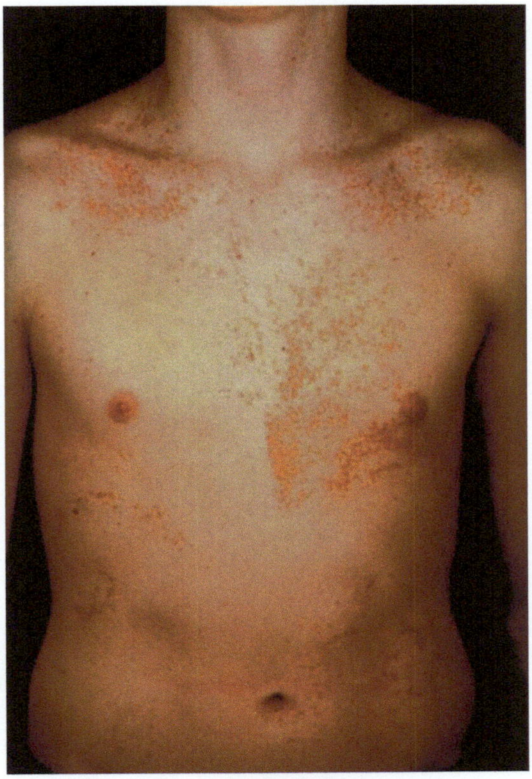

Fig. 10.23 Type 1 segmental Darier disease

10.2.2 Darier Disease

The lesions of mosaic Darier disease are always arranged along Blaschko's lines. Many examples of type 1 segmental involvement (Fig. 10.23) have been reported [49, 67, 99, 146, 378, 420, 488]. Molecular evidence of mosaicism was documented in some cases [381, 464]. Theoretically, one would expect that individuals with type 1 segmental Darier disease may sometimes give birth to a child with the nonsegmental form of the disorder (see Sect. 3.1.1.3). Surprisingly, however, no such case has so far been reported. Presumably it will take a long time until this riddle can be solved.

Several case reports can be categorized as examples of type 2 segmental Darier disease [76, 129, 194]. In one of these cases (Fig. 10.24), molecular proof of the postulated genetic mechanism has been provided [129]. Interestingly, cases of didymosis in the form of paired linear areas of either excessive or absent involvement were likewise reported (see Sect. 8.1.4). So far, such cases have not been investigated at the molecular level.

At this point in time, it seems justified to consider a diagnosis of type 2 segmental Darier disease in all cases of pronounced linear lesions of acantholytic dyskeratosis noted during infancy, even when the parents appear to be unaffected [38, 104, 265].

10.2.3 Hailey-Hailey Disease

A type 1 segmental manifestation was described by Hwang et al. [218]. On the other hand, the theory of early loss of heterozygosity giving rise to type 2 segmental Hailey-Hailey disease [177] was proven by molecular analysis of the ATP2C1 gene in a patient showing both linear and nonlinear involvement (Fig. 10.25) [354]. Quite

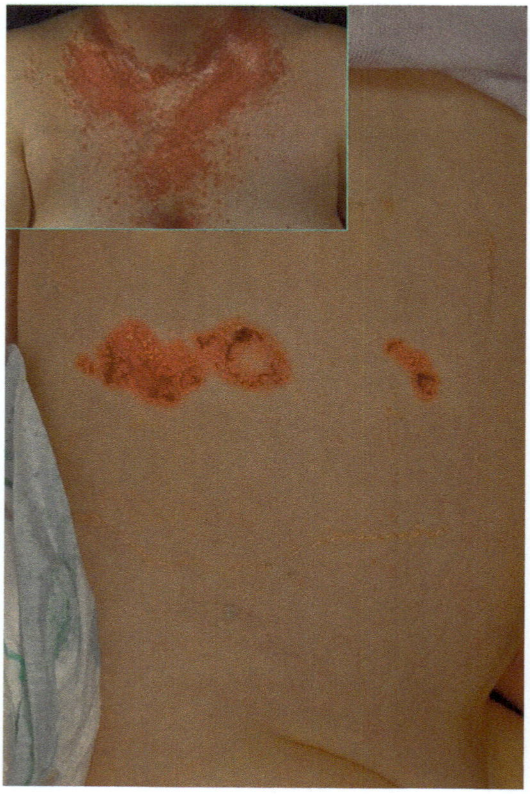

Fig. 10.24 Type 2 segmental Darier disease in a 1-month-old boy. *Inset*: Nonsegmental Darier disease in the boy's grandmother [129] (Reprinted with permission from John Wiley & Sons)

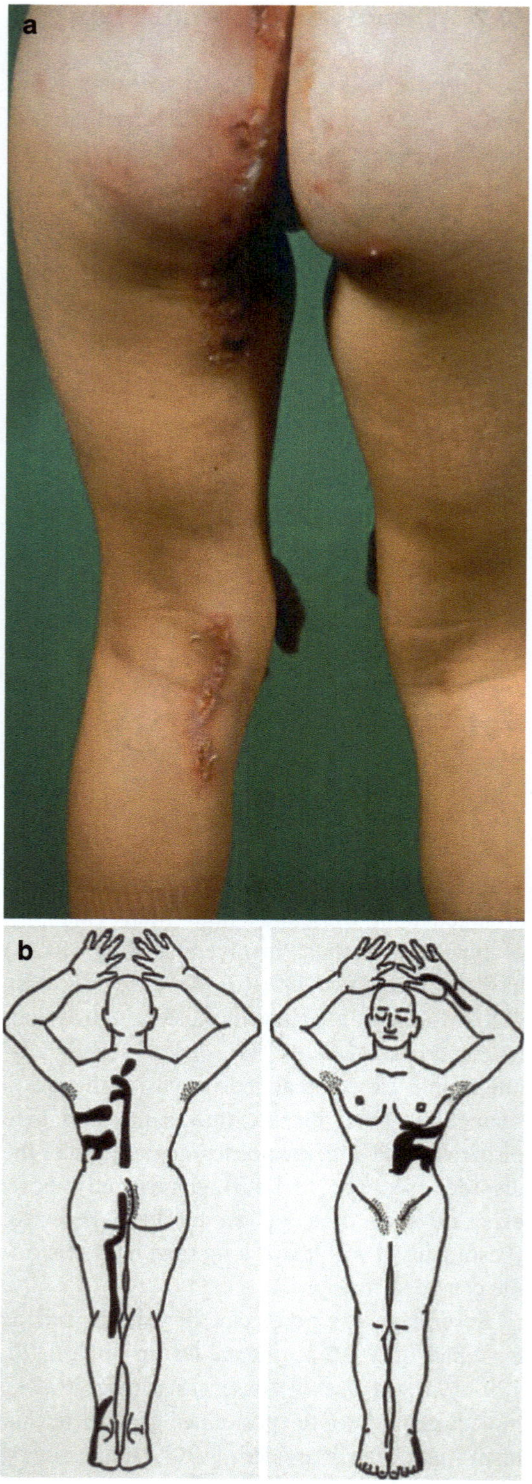

Fig. 10.25 Type 2 segmental Hailey-Hailey disease. (**a**) Isolated linear involvement at 5 years of age [454]; (**b**) additional nonsegmental involvement at 24 years of age [252] (**b**: Reprinted with permission from John Libbey Eurotext, Montrouge, France)

understandably, the pronounced linear lesions tend to show a rather poor therapeutic response to dermabrasion as compared to the nonsegmental lesions of the disorder [252].

Type 2 segmental Hailey-Hailey disease has sometimes been described as "relapsing linear acantholytic dermatosis," a term coined by Vakilzadeh and Kolde [454] who thought that this was a distinct entity. More recently, Arora et al. [16] proposed this diagnosis in a 4-year-old boy who had, since the age of 6 months, recurrent papulovesicles distributed along Blaschko's lines on one side of his body. Exacerbations occurred during summertime. Histopathological findings were typical of Hailey-Hailey disease. From these features we can conclude that later in life the patient will almost certainly develop, in addition, some nonlinear lesions of the disorder.

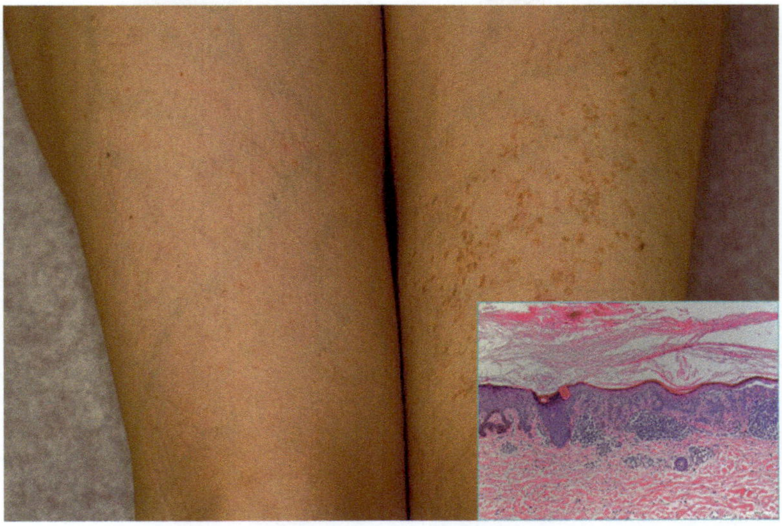

Fig. 10.26 Type 1 segmental Galli-Galli disease. *Inset*: hyperparakeratosis, acantholysis, hyperpigmentation of basal keratinocytes, and lymphocytic infiltrate in the upper dermis [15]

10.2.4 Dowling-Degos Disease, Including the Galli-Galli Variant

Dowling-Degos disease is characterized by reticulate or patchy hyperpigmentation with an affinity to the flexural areas. It is caused by *KRT5* mutations [30]. According to present knowledge, Galli-Galli disease is merely a variant of the same disorder, showing more pronounced features of acantholysis [418].

Arnold et al. [15] found a *KRT5* mutation in a type 1 segmental form of the Galli-Galli variant (Fig. 10.26), whereas in clinically unaffected skin and in the blood only the wild-type allele was present. No type 2 segmental involvement has so far been reported.

10.2.5 Acanthosis Nigricans

This disorder can be inherited as an autosomal dominant trait being caused by a K650T mutation in the FGFR3 gene [29]. There are several reports suggesting a type 1 mosaic manifestation of the disorder [86, 121, 226, 257, 348, 412]. Without molecular analysis, however, it seems difficult or even impossible to discriminate such lesions from other types of keratinocytic nevi.

During the years 1936–1976, Helen Ollendorff Curth [92–95] wrote many articles repeatedly describing a young man who had, since the age of 10 years, lesions of acanthosis nigricans showing a symmetrical distribution. In addition, a segmental and rather pronounced involvement was present since birth on one side of his body (Fig. 10.27a, b). During adulthood, the symmetric disorder improved to a large degree, whereas the nevus-like lesion remained unchanged. Ollendorff-Curth was fascinated by this case and wrote: "As the symmetric variety has been recognized as a genodermatosis, it seems justified to conclude that both the unilateral and the symmetric eruptions were caused by the same abnormal gene" [95]. Hence, she was nearby to sort out the problem that today has been solved by the theory of type 2 mosaicism [188]. A case of hystrix-like hyperkeratoses forming "wide stripes" and being superimposed on nonsegmental acanthosis nigricans, as reported by Babalian [20], probably represents another example of type 2 segmental acanthosis nigricans. In 2006, Ersoy-Evans et al. [121] described four cases of "acanthosis nigricans form of epidermal nevus." One of their patients was a 16-year-old boy who had the linear disorder since childhood and showed, in addition, typical bilateral lesions of acanthosis nigricans. The authors assumed that this represented type 2 mosaicism.

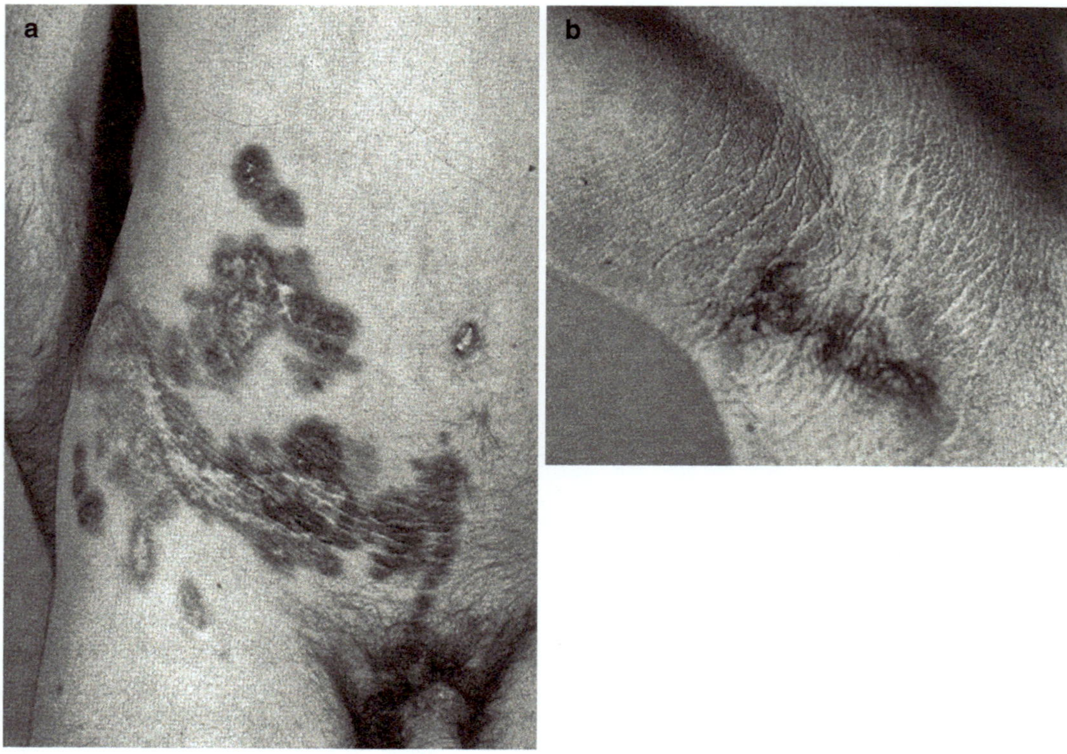

Fig. 10.27 A historical case of type 2 segmental acanthosis nigricans. (**a**) Pronounced unilateral involvement on the abdomen; (**b**) mild symmetrical skin changes in the axillary regions [93]

10.2.6 KID Syndrome

The keratitis-ichthyosis-deafness syndrome is an autosomal dominant trait being caused by *GJB2* mutations. The disorder is charaterized by corneal epithelial defects, erythrokeratoderma-like hyperkeratosis, and sensorineural hearing loss.

A systematized linear nevus indistuiguishable from porokeratotic eccrine nevus was documented in the mother of a child with KID syndrome [435]. Indeed, porokeratotic eccrine nevus (Fig. 7.38) appears to represent a type 1 segmental manifestation of KID syndrome [112] (see Sect. 7.3.2.6).

In a 13-year-old girl affected with KID syndrome, Restano et al. [365] noted a pronounced unilateral hyperkeratotic lesion arranged on her back in a Blaschko-linear pattern, being superimposed on the diffusely hyperkeratotic and thickened skin as usually noted in this trait. The linear lesion had been present since infancy. The authors inferred that this may represent a type 2 segmental manifestation of the disorder.

10.2.7 Autosomal Dominant Dyskeratosis Congenita

This type should be distinguished from the more common X-linked dyskeratosis congenita (see Sect. 12.1.7) and from an autosomal recessive form. All of these types are characterized by reticular hyperpigmentation, leukoplakia, nail dystrophy, and proneness to develop aplastic anemia. The autosomal dominant form appears to be milder than the other two types, with a lower incidence of aplastic anemia [111].

A type 2 segmental involvement was described in a 17-year-old boy with typical skin lesions of dyskeratosis congenita [25]. In this case, a V-shaped pattern of pronounced reticular hyperpigmentation, being superimposed on a diffuse,

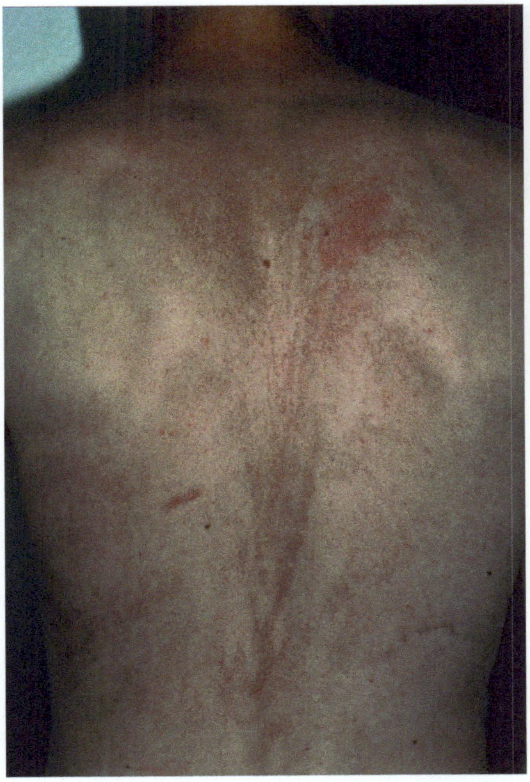

Fig. 10.28 Systematized type 2 segmental manifestation in a patient with autosomal dominant dyskeratosis congenita [25] (Reprinted with permission from John Wiley and Sons, USA)

less severe dyspigmentation, was noted on his back (Fig. 10.28). The patient was otherwise rather mildly affected and had a normal karyotype 46,XY. The authors considered an autosomal dominant trait with a pronounced linear manifestation, reflecting early loss of heterozygosity, to be the most likely diagnosis.

10.2.8 Pachyonychia Congenita of the Jadassohn-Lewandowsky Type

The disorder, also known as pachyonychia congenita type 1, is caused by keratin 16 mutations [300]. A type 1 segmental manifestation was reported under the diagnosis "unilateral palmoplantar verrucous nevus." A *KRT16* mutation was found to be present in DNA obtained from the affected palm and to be absent in the unaffected palm.

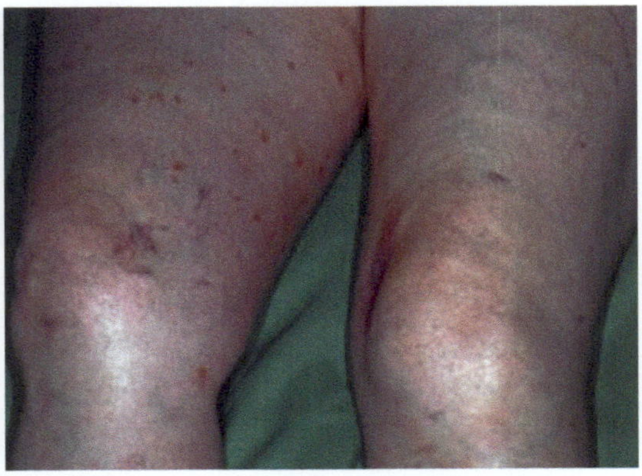

Fig. 10.29 Type 1 segmental manifestation of disseminated superficial actinic porokeratosis [54] (Reprinted with permission from Acta Dermato-Venereologica)

10.2.9 Disseminated Superficial Actinic Porokeratosis

Disseminated superficial actinic porokeratosis (DSAP) is the most frequently occurring form of porokeratosis [109]. Its genetic basis is so far unknown. The disorder is characterized by multiple, small, rather inconspicuous lesions. Their slightly elevated margin corresponds to the histopathological feature of "cornoid lamella." Both type 1 and type 2 mosaicism have been described.

Type 1 mosaicism was so far noticed only once, which may in part be explained by the fact the lesions of DSAP are rather tiny and easily overlooked. A 68-year-old woman who received immunosuppressive treatment for pemphigus vulgaris developed DSAP exclusively on her right leg (Fig. 10.29) [54]. Remarkably, the lesions were scattered within a large segmental area and did not show any linear arrangement.

By contrast, cases suggesting a type 2 segmental manifestation have rather frequently been reported (for reviews see [175, 177, 189]). Such reports are often published under the ambiguous diagnosis of "linear porokeratosis" [140]. The linear lesions often appear early in life and sometimes even during infancy [110, 115, 127, 149,

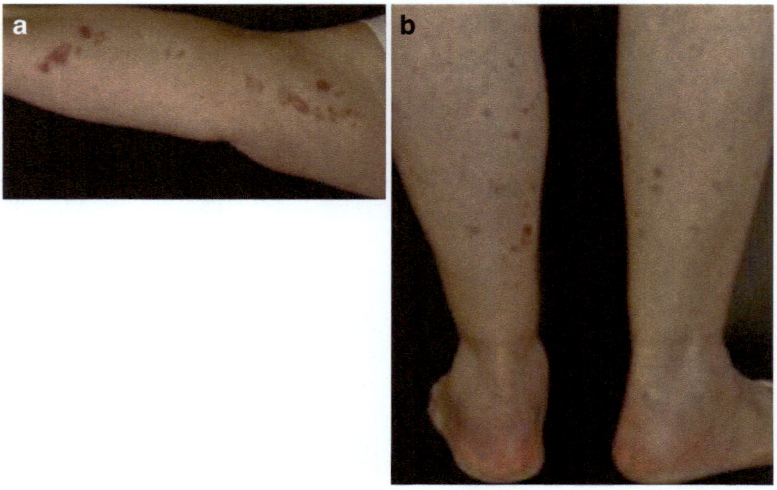

Fig. 10.30 Type 2 segmental manifestation of disseminated superficial actinic porokeratosis. (**a**) Congenital linear lesions on the right arm; (**b**) less severe bilateral involvement of legs since the age of 67 years [275] (Reprinted with permission from S. Karger AG, Basel, Switzerland)

238, 313, 325], or they may be noted at birth [309, 424], whereas the nonsegmental lesions tend to develop later in life (Fig. 10.30) [275, 346, 423]. Apparently, the underlying gene locus is a hotspot for postzygotic recombination [189]. Type 2 segmental DSAP may dramatically worsen when immunosuppressive drugs are given [284].

As a rule of thumb, if linear porokeratosis is noted at birth [75, 161, 250, 394] or during the first years of life, a type 2 segmental manifestation of either DSAP [151, 206, 295, 340, 396, 429, 460] or porokeratosis of Mibelli [317] (see Sect. 10.2.10) is very likely.

10.2.9.1 Practical Aspect

Superimposed linear forms of both DSAP and porokeratosis of Mibelli (see below) are especially prone to develop cancer [12, 161, 176, 316]. Apparently, the early allelic loss giving rise to such segmental lesions represents an initial step in a multistage mutational process of carcinogenesis.

10.2.10 Plaque-Type Porokeratosis of Mibelli

The designation "porokeratosis of Mibelli" is used by many authors as an umbrella term for all types of porokeratosis, whereas others use it to describe the classical plaque-type porokeratosis. Regarding this problem of ambiguity, I must confess that in several reviews published during the 1990s [175–177], I myself have erroneously subsumed the classical Mibelli type under the term DSAP. Autosomal dominant transmission of the Mibelli type has been documented beyond doubt [305, 384].

A case of type 1 mosaicism was described by Scholl [391]. The 13-year-old girl had scattered plaques involving the face, arm, hand, and foot exclusively on the left side of her body.

An impressive example of type 2 mosaicism was documented by Vittorio Mibelli in his original case report (Fig. 10.31) [305]. The patient had a pronounced linear porokeratosis present since the age of 2 years. Moreover, multiple nonsegmental plaques had first been noticed at the age of 7 years. Two siblings and the father were also affected with nonsegmental porokeratotic lesions. Additional nonambiguous cases of superimposed linear involvement occurring in families with nonsegmental porokeratosis of Mibelli have later been reported [35, 283]. In other reports the systematized linear disorder developed soon after birth [118, 361] or was accompanied in adulthood by "classic lesions of porokeratosis" [361], which is why a type 2 mosaicism is almost certain.

10.2.10.1 Familial Occurrence of Type 2 Segmental Involvement

This story of an unusual familial constellation is set in Pavia in northern Italy, where Mario Truffi [447] described in 1905 a 13-year-old boy who had since early infancy pronounced porokeratotic

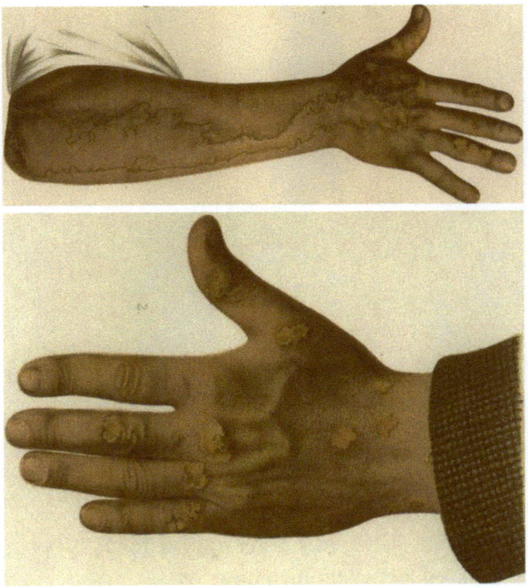

Fig. 10.31 Mibelli's original report on plaque-like porokeratosis can today be taken as an early documentation of type 2 segmental manifestation. A pronounced linear involvement of the right arm was present since the age of 2 years, whereas nonsegmental lesions as noted on the left hand developed since the age of 7 years [305]

lesions arranged in a systematized linear pattern on the *left* side of his body, including the oral mucosa. On the right side some nonsegmental lesions were noted. His mother had some scattered bilateral lesions of porokeratosis of Mibelli. Twenty-three years later Franco Flarer [128] presented a follow-up report with photographs of the conspicuous systematized linear involvement. He mentioned that the patient had developed some additional nonsegmental plaques on the right side of his body. Finally, Marco Gandola [137] presented in 1951 a revised family history. The patient had a younger brother who was less severely affected but, surprisingly, showed systematized lesions involving the *right* side of his body. This family observation can today be taken as a clinical hint that the underlying gene locus represents a hotspot for somatic recombination.

10.2.11 Porokeratosis Palmaris, Plantaris et Disseminata

This particular form of porokeratosis [164] is characterized by very small disseminated lesions that initially tend to appear on the palms and soles and later affect the entire integument. A type 2 segmental manifestation in the form of pronounced linear lesions involving the left arm of a 33-year-old woman has been documented [350]. Another possible example of superimposed linear porokeratosis palmaris, plantaris et disseminata was reported in a 31-year-old Afro-American woman with end-stage liver disease [214].

10.2.12 Type 2 Segmental Manifestation in Cases of Unclassifiable Porokeratosis

Some authors described linear porokeratosis superimposed on nonlinear lesions, but it is difficult to determine from their report which type of porokeratosis was present [150, 356, 357]. In some cases [142, 358], porokeratosis of Mibelli is probable but not certain, whereas in other reports [105] a diagnosis of DSAP is likely but not definite. A case of congenital linear and giant porokeratosis [280] represents almost certainly an example of type 2 mosaicism but can so far not be categorized further. The same holds true for a boy with systematized linear porokeratosis [387] who, as an adult, developed 14 squamous cell carcinomas on his limbs [132, 386].

10.2.13 Costello Syndrome

Costello syndrome includes a peculiar facial appearance, growth retardation, mental deficiency, musculoskeletal defects, and cardiomyopathy [334]. Cutaneous features comprise redundant skin (especially on the hands), periorificial papillomas, acanthosis nigricans-like hyperkeratosis, palmoplantar keratoderma with deep creases, sparse and curly hair, and dystrophic nails. The disorder is caused by heterozygous *HRAS* mutations [417].

A type 1 mosaic involvement was documented in a 15-year-old girl [157]. She had segmental acanthosis nigricans-like skin lesions on her

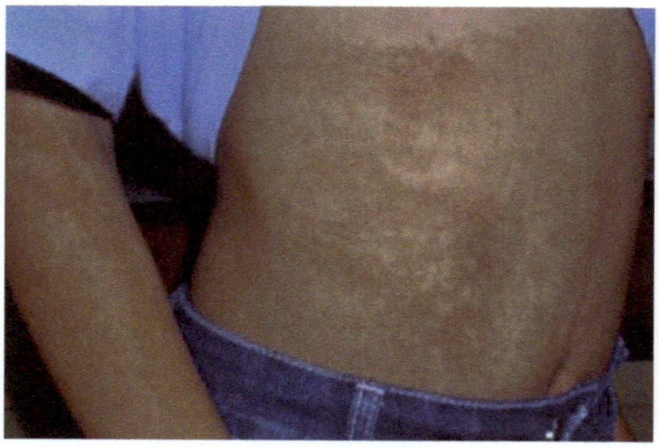

Fig. 10.32 Type 1 segmental manifestation of Costello syndrome. Mosaic arrangement of acanthosis nicgricans-like skin changes on the abdomen [157] (Reprinted with permission from John Wiley and Sons, USA)

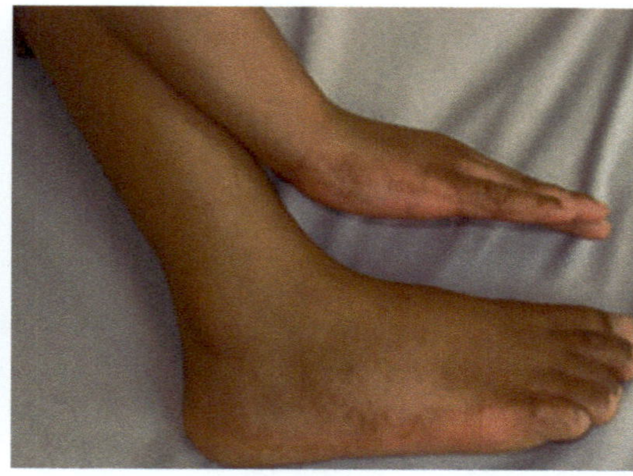

Fig. 10.33 Type 1 segmental acrokeratoelastoidosis [245] (Reprinted with permission from John Wiley and Sons, USA)

abdomen and right arm (Fig. 10.32), and a linear hyperkeratosis on her left sole. An *HRAS* mutation was found to be present in buccal DNA samples but absent in peripheral blood lymphocytes. Another patient with a mosaic *HRAS* mutation giving rise to type 1 segmental Costello syndrome, including patches of curly hair in his otherwise straight hair, had a son with nonsegmental Costello syndrome [417]. On the other hand, no clinical features of mosaicsm could be detected in a 1-year-old boy with Costello syndrome and a mosaic *HRAS* mutation present in blood lymphocytes [147].

10.2.14 Acrokeratoelastoidosis

In this trait multiple small, shiny, keratotic papules are arranged along the margins of hand and feet. They appear in childhood and progress during adolescence. Two cases of type 1 mosaicism in the form of unilateral involvement of both hand and foot have been reported (Fig. 10.33) [245, 311].

10.3 Disorders of Connective Tissue or Bones

In this group of disorders, mosaic manifestations are often noted. In particular, cases suggesting a type 2 segmental involvement have been documented in tuberous sclerosis, Buschke-Ollendorff syndrome, Albright's hereditary osteodystrophy, and hereditary osteoma cutis.

10.3.1 Tuberous Sclerosis

Tuberous sclerosis (TS) is characterized by multiple hamartomas of the connective tissue involving the brain and skin as well as various other organs such as the eyes, lungs, or kidneys. In most patients, progressive brain involvement gives rise to seizures, learning difficulties, or behavior abnormalities. The longer term "tuberous sclerosis complex (TSC)" has no additional meaning. It was created by geneticists for the simple reason that they needed, for the purpose of classification, a three-letter symbol. The disorder can be caused by two different genes. *TSC1* encodes for hamartin, whereas *TSC2* is producing tuberin.

Mosaicism is well documented in TS [259, 458]. So far, however, reviews of cases of mosaic TS do not discriminate between type 1 and type 2 mosaicism.

10.3.1.1 Type 1 Segmental TS

This type of mosaic TS has preponderantly been described in the form of unilateral facial angiofibromas (Fig. 10.34) [9, 103, 167, 298, 400, 410, 445]. Other forms of type 1 mosaic involvement may likewise occur. For example, a young man showed unilateral facial angiofibromas with ipsilateral presence of periungual fibromas of fingers and toes, small shagreen plaques, and hypopigmented macules involving the trunk, as well as retinal hamartomas [296]. A patient with postzygotic mosaicism confirmed at the molecular level [459] had multiple dental pits, and brain tomography showed one paraventricular calcification. Two cases of "solitary cortical tubers" [108] may likewise be best explained as type 1 mosaic TS. The nosological significance of a solitary periungual fibroma [491] is so far unclear.

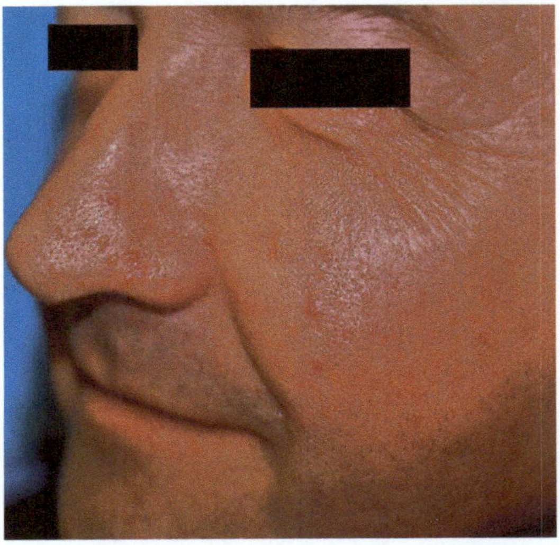

Fig. 10.34 Type 1 segmental tuberous sclerosis in the form of unilateral facial angiofibromas [445] (Reprinted with permission from Elsevier Limited, UK)

Reviews on unilateral facial angiofibromas should be read with great caution because the authors may have inadvertently included examples of type 2 segmental TS [44, 64].

10.3.1.2 Type 2 Segmental TS

This type of mosaic manifestation of TS occurs rather frequently but has so far been disregarded, which means that up to now we can find cases of classical type 2 segmental involvement erroneously categorized as a "forme fruste" [138, 139] or as a simple mosaic of TS [44, 64]. Large shagreen patches or cobblestone nevi most likely represent a type 2 segmental involvement [177]. Such lateralized lesions may sometimes show conspicuous growth during adolescence or adulthood (Fig. 10.35). A large linear lesion involving

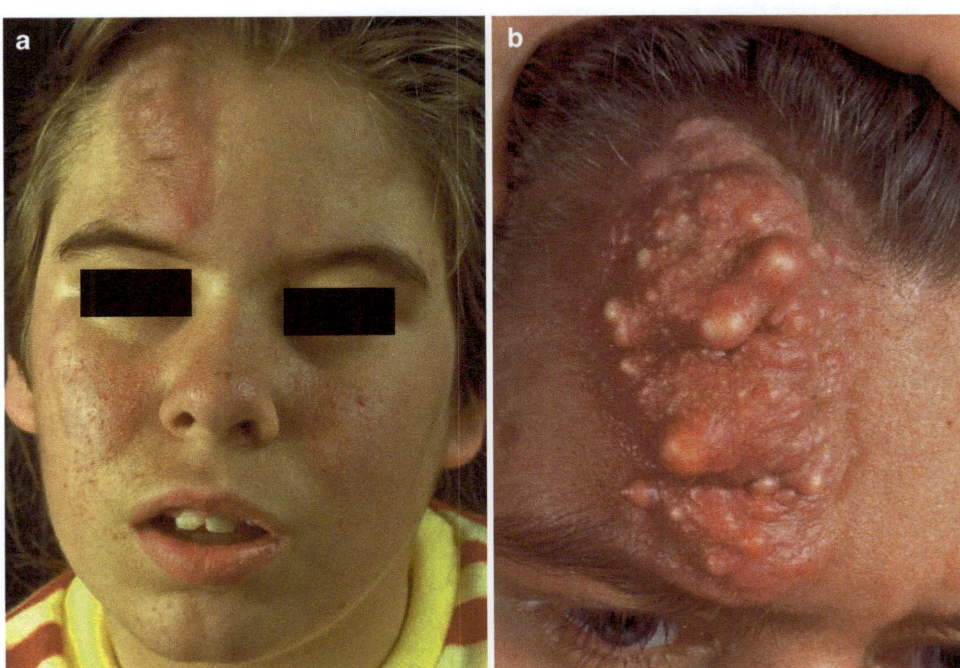

Fig. 10.35 (**a**) Type 2 segmental tuberous sclerosis involving the forehead of a young girl; (**b**) the same lesion some years later (Courtesy of Dr. Gerhard Kurlemann, Münster, Germany)

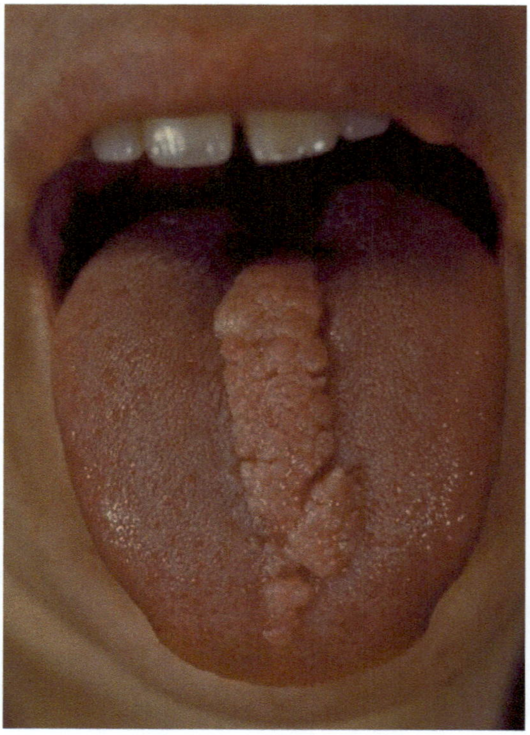

Fig. 10.36 Type 2 segmental involvement of the tongue in a patient with tuberous sclerosis [221] (Reprinted with permission from John Wiley & Sons, USA. Color photograph kindly provided by Dr. Peter H. Itin, Basel, Switzerland)

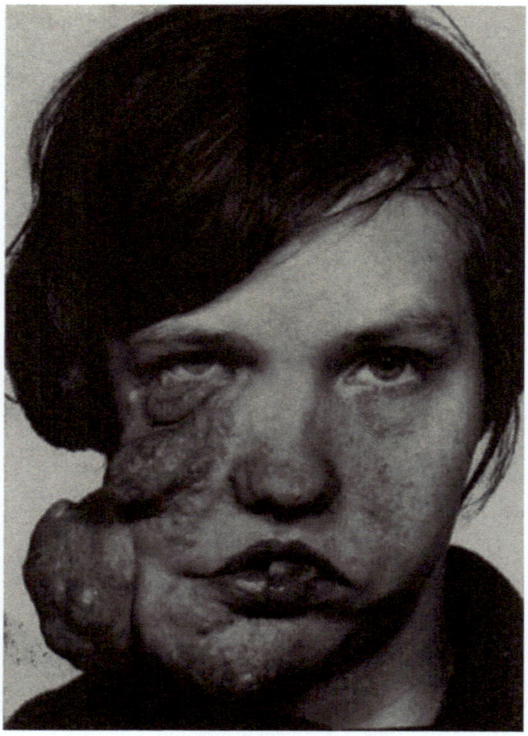

Fig. 10.37 Type 2 segmental involvement of maxilla and mandible in a patient with tuberous sclerosis [470] (Reprinted with permission from Elsevier Limited, UK)

the tongue was explained by the same mechanism (Fig. 10.36) [221]. Impressive examples of type 2 segmental TS with unilateral maxillary and mandibular involvement were inadvertently documented by oral surgeons (Fig. 10.37) [379, 470]. Moreover, all cases of "folliculocystic and collagen hamartoma" as observed in children with TS (Fig. 10.38) [440] may best be explained as a superimposed segmental manifestation of TS. The same holds for large "fibrous hamartomas of infancy" [169, 401, 451] or cases of macrodactyly reported in patients with TS [85, 282, 380, 401, 406, 451], or for overgrowth involving all tissues of a forearm [337, 469].

10.3.1.3 Cases of Unclassifiable Mosaic TS

Silvestre et al. [410] reported on a 12-year-old boy with multiple unilateral facial angiofibromas that had developed since the age of 5 years. In addition, he had a large hypopigmented macule involving the left side of his abdomen with a sharp midline separation. During the following years, the clinical course of the disorder may have shown whether this boy carried a germline mutation and thus was affected with a type 2 segmental TS.

10.3.1.4 Genetic Counseling

In cases of type 1 segmental TS, siblings and parents of the patient have a normal population risk of carrying a TS mutation, because such mosaicism excludes gonadal mosaicism of one of the parents [458]. For the next generation, the risk of nonsegmental TS is increased to some degree because gonadal mosaicism may be present. Children of patients with type 2 segmental TS have a 50 % risk to be affected with nonsegmental TS. Hence, it is of great practical importance to bear the dichotomy of mosaic TS in mind. For example, in a child with unilateral angiofibromas, it is mandatory to exclude a type 2 mosaic manifestation of TS.

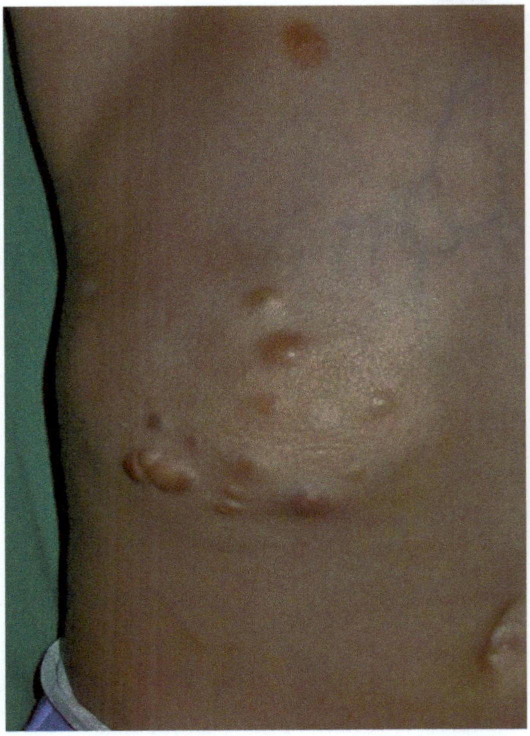

Fig. 10.38 Type 2 segmental tuberous sclerosis presenting as "folliculocystic and collagen hamartoma" in a 6-year-old boy [440] (Reprinted with permission from Elsevier Limited, Oxford, UK)

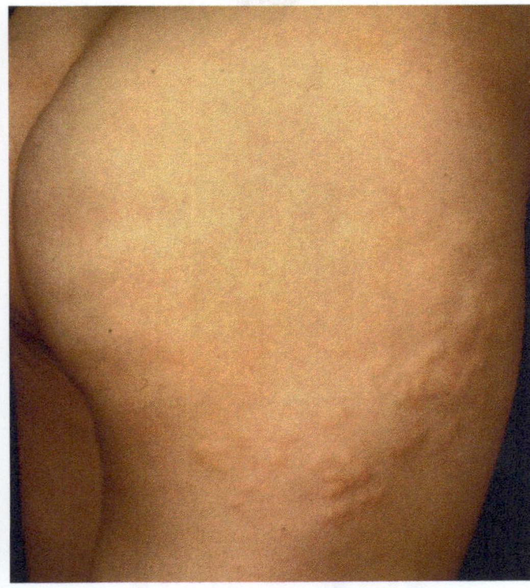

Fig. 10.39 Type 2 segmental Buschke-Ollendorff syndrome in a 5-year-old boy who had family members with nonsegmental lesions [17, 181] (Reprinted with permission from John Libbey Eurotext, Montrouge, France)

10.3.2 Buschke-Ollendorff Syndrome

The disorder is caused by loss-of-function mutations in *LEMD3* [199] and characterized by small disseminated connective tissue nevi of an elastin-rich type in combination with osteopoikilosis.

10.3.2.1 Type 2 Segmental Skin Lesions

In addition to the disseminated, skin-colored or yellowish, firm papules of dermatofibrosis, large plaques showing an asymmetrical arrangement have been called "juvenile elastoma" [170, 260, 471]. These plaques almost certainly represent a type 2 segmental manifestation of Buschke-Ollendorff syndrome [116, 181], especially when they are associated with nonsegmental lesions in the patient or his family members (Fig. 10.39) [17, 59, 180, 392, 481]. An exemplary type 2 segmental involvement has been described as "Buschke-Ollendorff syndrome of the scalp" [444]. Remarkably, such lesions have sometimes been mistaken for a "forme fruste" of Buschke-Ollendorff syndrome [130].

10.3.2.2 Familial Occurrence of Type 2 Segmental Skin Lesions

In Buschke-Ollendorff syndrome, asymmetrically arranged plaques or lumps are very often noted in several members of a family, for example, in a father and two of his children [415], in a father and his son or daughter [18, 59, 308], in two siblings [196, 211, 308], or even in three generations [457]. Because such lesions tend to appear in childhood or may even be noted at birth, they can best be explained as a type 2 mosaicism, which indicates that the underlying *LEMD3* locus represents a hotspot for postzygotic recombination.

10.3.2.3 Type 2 Segmental Involvement of Bones

Melorheostosis is a rare bone dysplasia characterized by skeletal sclerosis with periosteal hyperostosis reminiscent of dripping candle wax. The disorder usually shows an asymmetric arrangement [134] and is frequently associated

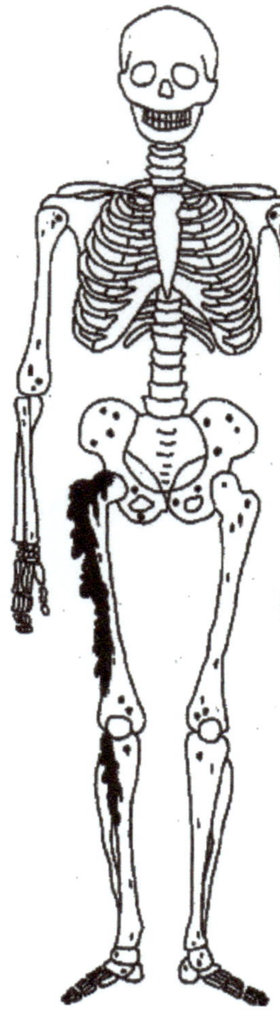

Fig. 10.40 Schematic drawing showing melorheostosis superimposed on mild disseminated osteopoikilosis, suggesting a type 2 segmental manifestation of Buschke-Ollendorff syndrome [184] (Reprinted with permission from John Wiley & Sons, USA)

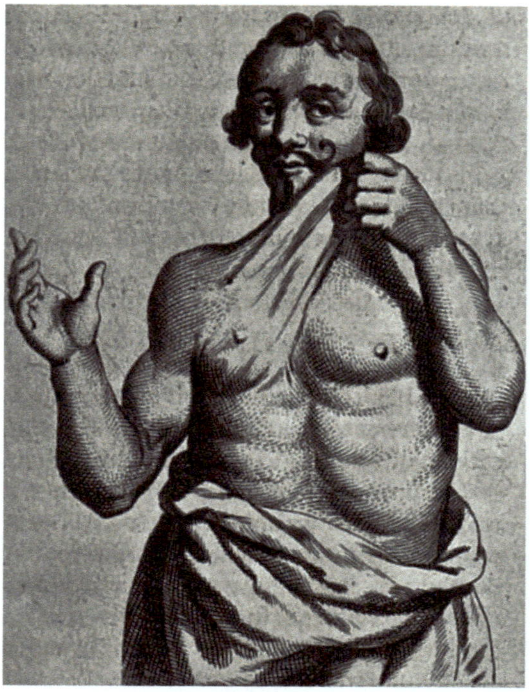

Fig. 10.41 A historical case of mosaic Ehlers-Danlos syndrome, reported in Latin in 1682: "De debilitate extraordinariâ cutis in viro quodam Hispano" (On an unusual weakness of skin in a Spanish man) [455]. Earlier descriptions had appeared in Dutch (1668) and German (1675) (Reproduced from Google e-books)

with hyperplasia and intensified density of the surrounding soft tissues, which may entail linear sclerodermatous skin changes [135, 416]. Cases of isolated melorheostosis are usually not caused by *LEMD3* mutations [198, 314]. On the other hand, this severe bone disease is sometimes found in patients with Buschke-Ollendorff syndrome [56, 198, 324, 338, 466, 476]. Such cases may be best explained as a type 2 segmental manifestation of the disorder (Fig. 10.40) [101, 184]. Limb reduction has likewise been described [399]. The terms "mixed sclerosing bone dystrophy" [338, 339, 466] or "mixed sclerosing bone dysplasia" [145, 235] appear to be less precise and should be avoided, because in such cases the melorheostosis is not simply mixed with, but superimposed on, the disseminated lesions of osteopoikilosis.

10.3.3 Ehlers-Danlos Syndromes

The various forms of Ehlers-Danlos syndromes are characterized by hyperelasticity of the skin. The disorder was first reported in 1668 by Job J. van Meekeren, a surgeon from Amsterdam, in a posthumously published book written in Dutch and subsequently translated into German and Latin [455]. The patient was able to stretch the skin of his right shoulder in a way that it covered his eyes (Fig. 10.41). Remarkably, however, the skin on the left side of his body was of a completely normal structure. Hence, this earliest

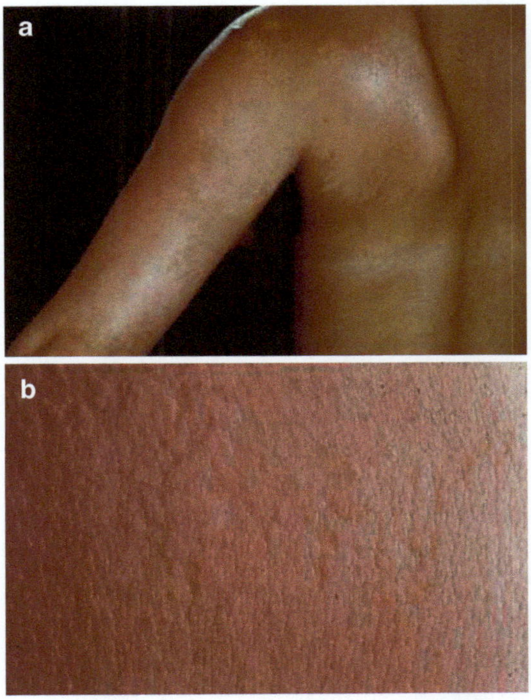

Fig. 10.42 Collagen nevus forming a large, pale plaque on the left shoulder and arm of a patient with Ehlers-Danlos syndrome III. (**a**) General view; (**b**) close-up [408]. This case can be taken as a possible example of type 2 mosaicism (Reprinted with permission from John Wiley and Sons, USA)

description of Ehlers-Danlos syndrome was an example of type 1 segmental manifestation. Some similar cases were reported later [91, 256].

A 25-year-old woman with bilateral, nonsegmental lesions of Ehlers-Danlos syndrome type III had a large connective tissue nevus of the collagen type on her left shoulder and arm [408]. The lesion was present since childhood and described as an "ill-defined, raised, pale, pink, warty, papular plaque" (Fig. 10.42). This case may represent an example of type 2 mosaicism [189].

10.3.4 Marfan Syndrome

This fibrillin disorder is characterized by ocular, cardiovascular, and skeletal defects including arachnodactyly [334]. Patients tend to be rather tall, thin, and loose jointed. As a typical cutaneous feature, extensisve atrophic striae distensae are often noted on the trunk and proximal parts of the limbs [83, 155]. Loveman et al. [279] described such skin lesions as a sign of Marfan syndrome in descendants of Abraham Lincoln.

A case of asymmetric manifestation of this disorder [55, 148] can now be reclassified as a superimposed form of mosaicism. The 3-year-old girl was thin and rather tall for her age. She showed pronounced overgrowth of the left side of her body (Fig. 10.43). Limb asymmetry had been noted at the age of 1 month. She had bilateral joint laxity, and arachnodactyly was present on both hands but more pronounced on the left. X-rays showed a bilaterally increased bone age, being more marked on the left. In both eyes megalocornea, shallow anterior chamber, iridic atrophy, and spherophakia were noted, whereas ectopia lentis was present only in the left eye. Ultrastructural examination showed that in the papillary dermis the diameter of fibrils was substantially larger than that of control individuals, but no significant difference between the two sides was noted. Biochemical abnormalities of collagen were found on both sides of the body, although more marked on the left. The authors, being unaware of the concept of type 2 mosaicism, speculated that "a mosaic of normal and abnormal cells" may have been "the consequence of an early somatic mutation" [55]. At the age of 6 ½ years, microfibrillar fiber staining in the papillary dermis was negative on the left side and "modestly diminished" on the right, as compared with healthy controls [148]. Today, all of these findings can best be explained by an early event of loss of heterozygosity giving rise to a type 2 segmental involvement.

10.3.5 Albright's Hereditary Osteodystrophy

Multiple cutaneous osteomas are a feature of Albright's hereditary osteodystrophy (AHO) [222, 227, 398]. Affected individuals have a round face, short and stout stature, and brachydactyly. The disorder is usually caused by *GNAS* mutations. Loss of function of the maternal allele results in pseudohypoparathyroidism (OMIM 103580), whereas loss of function of the paternal allele gives rise to pseudopseudohypoparathyroidism (OMIM 612463) [334].

Several cases can today be categorized as examples of type 2 segmental AHO [2, 3, 113]. Pronounced unilateral cutaneous lesions noted at birth, with subsequent bilateral appearance of scattered small osteomas, have been documented by Tijdens and Ruiter [433]. Another girl with AHO and congenital type 2 segmental skin lesions had a brother showing nonsegmental extracutaneous features of AHO and "a hard swelling on the lateral aspect of the left thigh" that was present since birth [51]. Initially the superimposed segmental lesions may show an inflammatory stage [433, 467]. Klaassens et al. [244] documented a patchy area of dermal and subcutaneous hypoplasia with palpable subcutaneous calcifications in a boy with hormone-resistant AHO (Fig. 10.44). His mother showed nonsegmental features of AHO. Hence, the boy's segmental lesion can be taken as another example of type 2 mosaicism.

10.3.6 Hereditary Osteomatosis Cutis

A nonsyndromic form of multiple osteoma cutis may occur as an autosomal dominant trait [126, 347]. It is caused by loss-of-function mutations of the GNAS gene on the paternal allele (OMIM 166350 [334]).

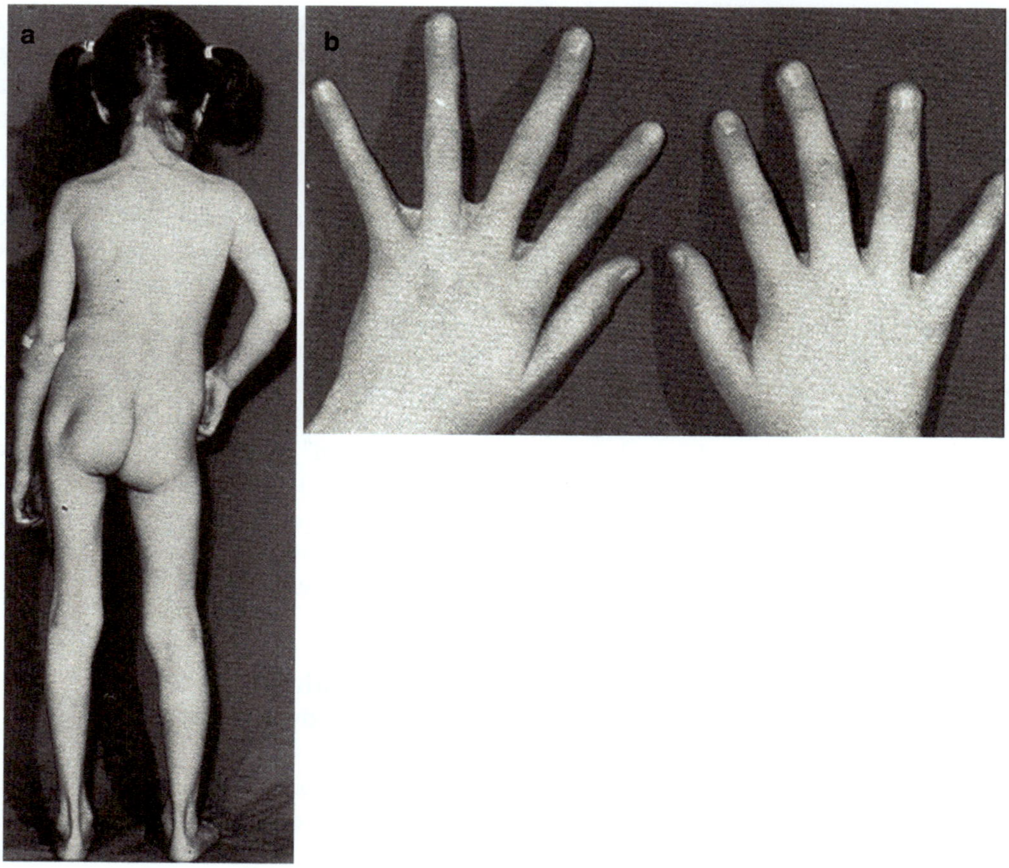

Fig. 10.43 "Asymmetric Marfan syndrome," now being reclassified as an example of type 2 segmental manifestation. (**a**) 3-year-old girl with marked overgrowth of the left side of her body; (**b**) bilateral arachnodactyly, being more pronounced on the left [55]; (**c**) increased length inequality of limbs at the age of 6½ years [148] (**a** and **b**: Reprinted with permission from John Wiley and Sons, USA; **c**: Reprinted with permission from Elsevier, Chennai, India)

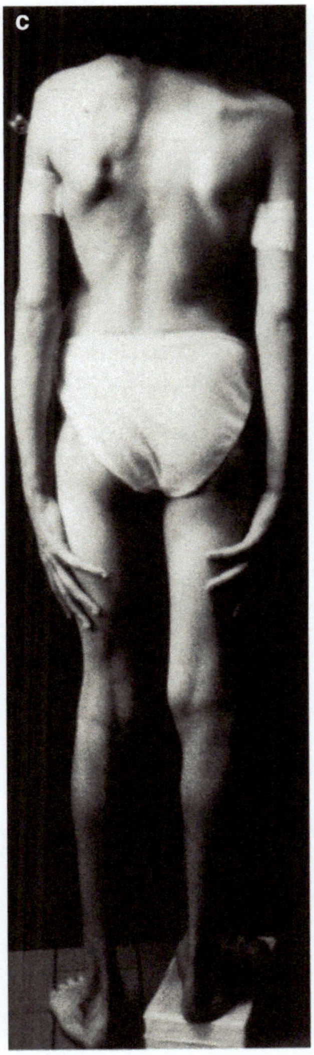

Fig. 10.43 (continued)

Several impressive cases suggesting a type 2 segmental osteomatosis cutis have been reported [141, 453, 461, 480]. A female infant described by Lim et al. [268] showed, at 1 h of age, an erythematous papular rash being strictly limited to several segmental areas preponderantly on her left side. Subsequently the involved regions became indurated and constituted a severe handicap (Fig. 10.45) [131]. Moreover, from the age of 6 years on, several isolated, nonsegmental lesions of intradermal ossification began to appear [131, 237]. In a similar case, Vero [461] documented involvement of the underlying deep structures including the muscles. (I do not accept the diagnosis of myositis ossificans as proposed by some authors [126, 131]). Gardner et al. [141] speculated that mitotic crossing-over may have caused conversion to homozygosity in the severely involved region of their patient who had six family members affected with nonsegmental cutaneous osteomas. Urtizberea et al. [453] described "progressive osseous heteroplasia" (see below) noted at birth in a girl whose father, younger sister, and paternal aunt had multiple nonsegmental lesions of osteoma cutis (Fig. 10.46). Another case was described as "severe congenital platelike osteoma cutis" involving the forehead in a linear configuration (Fig. 10.47) [446, 484].

10.3.6.1 A Note on "Progressive Osseous Heteroplasia"

Curiously enough, a type 2 segmental involvement has got, under the name "progressive osseous heteroplasia," its own entry #166350 in McKusick's Catalog OMIM [334], thus being listed as a "rare autosomal dominant trait" and equated with osteoma cutis. It should be realized, however, that the term "progressive osseous heteroplasia" [237, 388] does not represent a nosological entity because it may occur in osteomatosis cutis (MIM 166350), in AHO with hormone resistance (MIM 103580) [3, 51, 113, 144], and in AHO without hormone resistance (MIM 612463) [467]. Moreover, several other separate types of "progressive osseous heteroplasia" in the form of sporadic or familial traits may likewise exist [372, 407].

Contrasting with this view, Shore et al. [407] presented six pedigrees suggesting autosomal dominant inheritance of progressive osseous heteroplasia. Their report evoked a vehement debate [2, 124]. I see three different ways to explain these unusual family constellations. Firstly, the concept of type 2 segmental involvement may not be valid in patients with *GNAS* mutations. Pronounced asymmetric lesions would reflect the action of modifier genes or other mechanisms such as epigenetic modification of *GNAS* mutations [234]. Secondly, the familial cases may reflect paradominant inheritance [182]. However, a manifestation of the asymmetric disorder in no less than six sisters

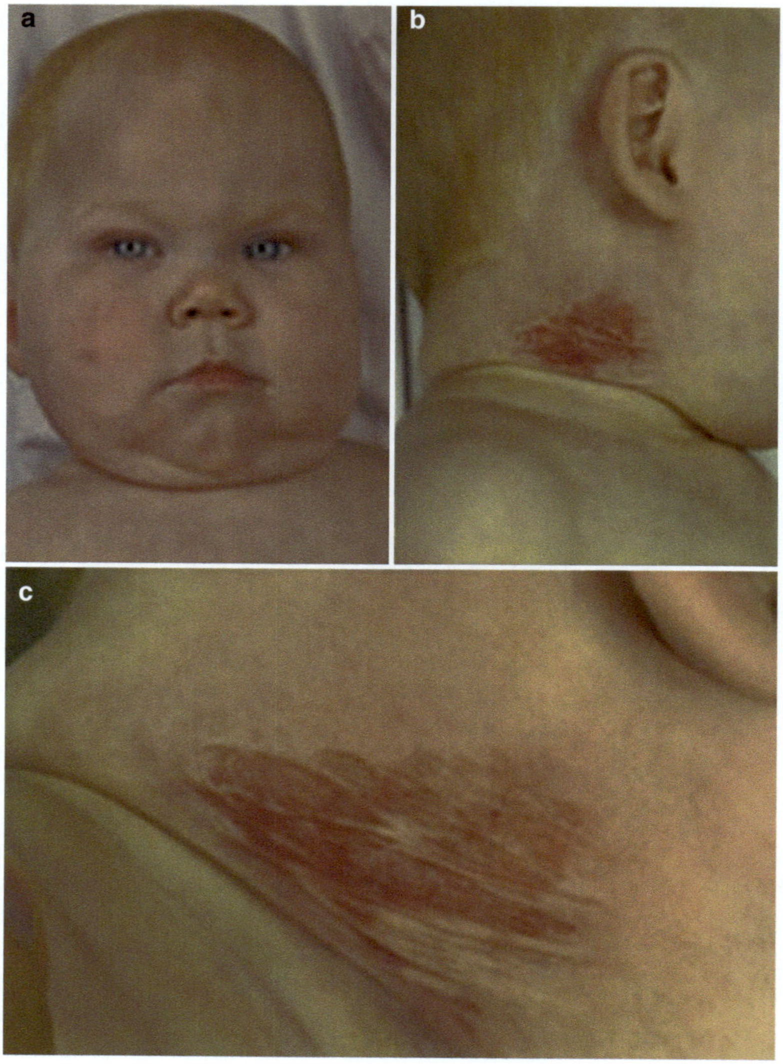

Fig. 10.44 A 2-year-old boy with Albright's hereditary osteodystrophy [244]. (**a**) Nonsegmental features reflecting haploinsufficiency include frontal bossing, round face, and depressed nasal bridge; (**b**, **c**) patchy area of atrophy with subcutaneous calcifications on the right side of his neck, suggesting a type 2 segmental manifestation (Reprinted with permission from John Wiley and Sons, USA)

[407] would not be compatible with this theory. Thirdly, the authors may have categorized nonsegmental cutaneous osteomas as mild forms of "progressive osseous heteroplasia." In my opinion the third explanation is the most likely one.

10.3.7 Zimmermann-Laband Syndrome

This autosomal dominant trait is characterized by coarse facial appearance with thick lips and bulbous nose, gingival fibromatosis, hypoplasia or aplasia of nails or terminal phalanges, mild hypertrichosis, increased joint mobility, hepatosplenomegaly, and mental deficiency [100]. A case of mosaic involvement was documented by Chacon-Camacho et al. [71]. The 8-year-old girl had hemihyperplasia with pronounced hypertrichosis preponderantly affecting the right side (Fig. 10.48a), ipsilateral hyperpigmentation and enlargement of her external genitalia (Fig. 10.48b), and a hemivertebra at T7. Moreover, postaxial polydactyly of the right hand and left foot were present. A type 2 segmental manifestation is possible but so far unproven.

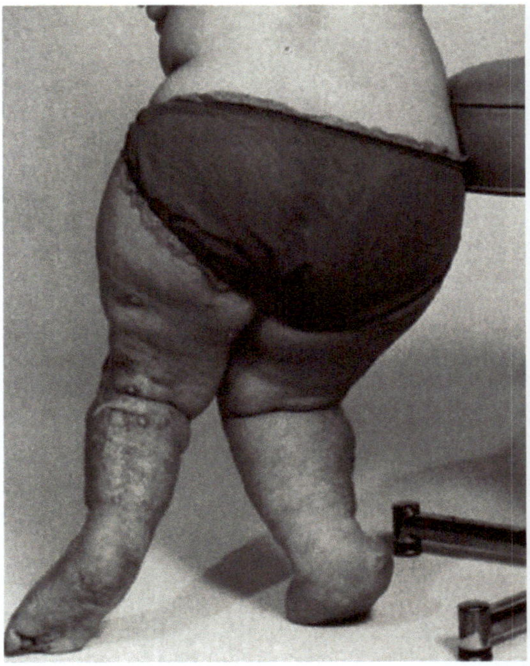

Fig. 10.45 Type 2 segmental hereditary osteomatosis cutis causing limb bowing and decreased joint mobility in a 9-year-old girl who had, in addition, some nonsegmental small lesions involving the upper part of her body [131] (Reprinted with permission from Elsevier Limited, UK)

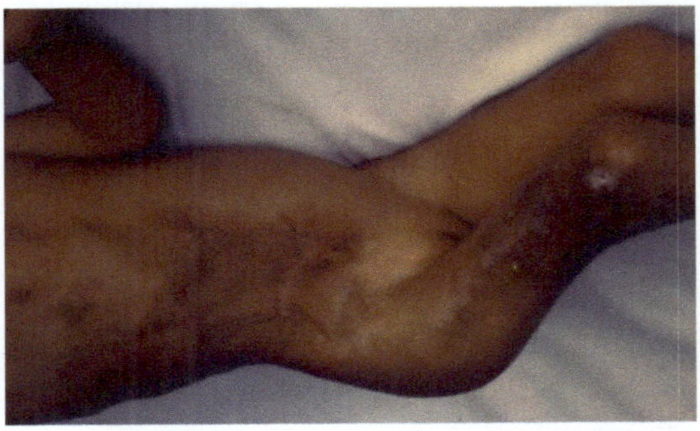

Fig. 10.46 Type 2 segmental hereditary osteomatosis cutis, reported under the diagnosis of "progressive osseous heteroplasia" [453] (Reproduced with permission and copyright © of the British Editorial Society of Bone and Joint Surgery. Color photograph was kindly provided by Dr. Adoni Urtizberea, Paris, France)

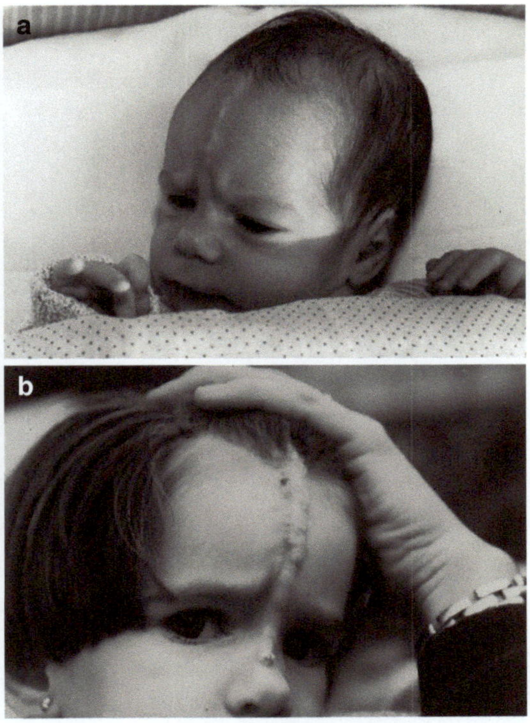

Fig. 10.47 Type 2 segmental hereditary osteomatosis cutis ("severe congenital platelike osteoma cutis"). (**a**) Appearance at birth; (**b**) progression during early childhood [484] (Reprinted with permission from John Wiley and Sons, USA)

10.3.8 Brachmann-de Lange Syndrome (Cornelia de Lange Syndrome)

This multisystem birth defect is included here because skeletal anomalies are rather typical features of the disorder [220]. Other major defects comprise mental deficiency, short stature, and a distinctive facial appearance with synophrys, anteverted nostrils, and a crescent-shaped mouth ("carp mouth"). Bone abnormalities mainly involve the arms. The fingers tend to be short. The disorder is genetically heterogeneous, but more than 60 % of cases are found to be caused by heterozygous *NIPBL* mutations [273]. The syndrome usually occurs sporadically, but almost all of these cases can be categorized as belonging to the various autosomal dominant traits (OMIM 122470).

Type 1 mosaicism was documented in an infant with a postzygotic *NIPBL* mutation [68]. The boy had numerous typical signs of Brachmann-de Lange syndrome. In addition, his right leg showed pigmentary disturbances following Blaschko's lines (Fig. 10.49). Another case suggesting type 1 mosaicism was reported in the premolecular era [490]. This girl showed

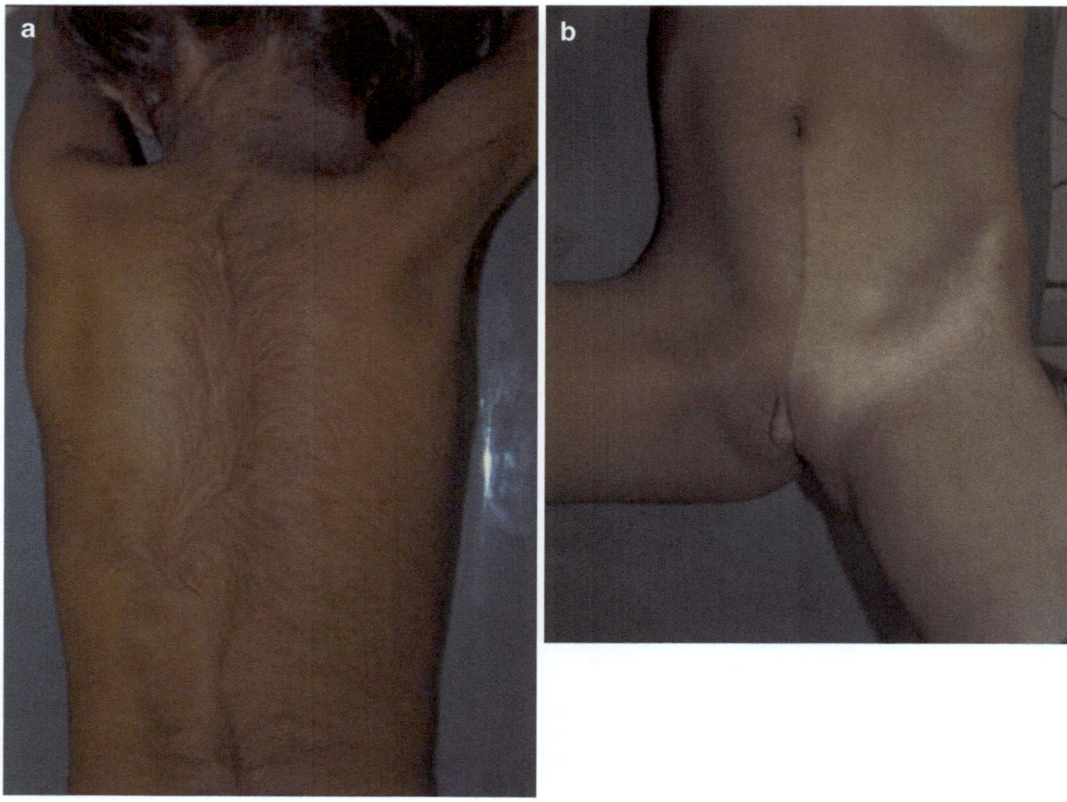

Fig. 10.48 Possible type 2 segmental manifestation of Zimmermann-Laband syndrome. (**a**) An 8-year-old girl with pronounced segmental hypertrichosis preponderantly involving her right side; (**b**) ipsilateral hyperpigmentation and hemihyperplasia with overgrowth of external genitalia [71] (Reprinted with permission from John Wiley & Sons, USA)

retarded growth of one half of her body and irregularly shaped areas of hypopigmentation involving her trunk and legs.

10.4 Vascular Disorders

Cutaneous mosaicism has been documented in both hereditary hemorrhagic telangiectasia and rhodoid nevus syndrome.

10.4.1 Hereditary Hemorrhagic Telangiectasia (Osler-Rendu-Weber Syndrome)

Hereditary hemorrhagic telangiectasia is characterized by multiple lesions of vascular dysplasia. Mucosal involvement and other extracutaneous lesions tend to cause bleeding that is often severe and may even be lethal. The disorder is genetically heterogeneous but usually caused by mutations in the *ENG* gene encoding endoglin (OMIM 187300) [334].

A case of type 1 segmental involvement of one side of the face has been reported [353]. In another patient, a pronounced segmental lesion of the forehead suggesting a type 2 mosaicism was documented (Fig. 10.50). So far it is unclear whether some cases of arteriovenous fistulas involving the lungs or other internal organs [6, 225] may likewise reflect type 2 mosaicism.

10.4.2 Rhodoid Nevus Syndrome ("Capillary Malformation-Arteriovenous Malformation")

The phenotype is characterized by multiple rhodoid nevi (Fig. 10.51) and caused by *RASA1*

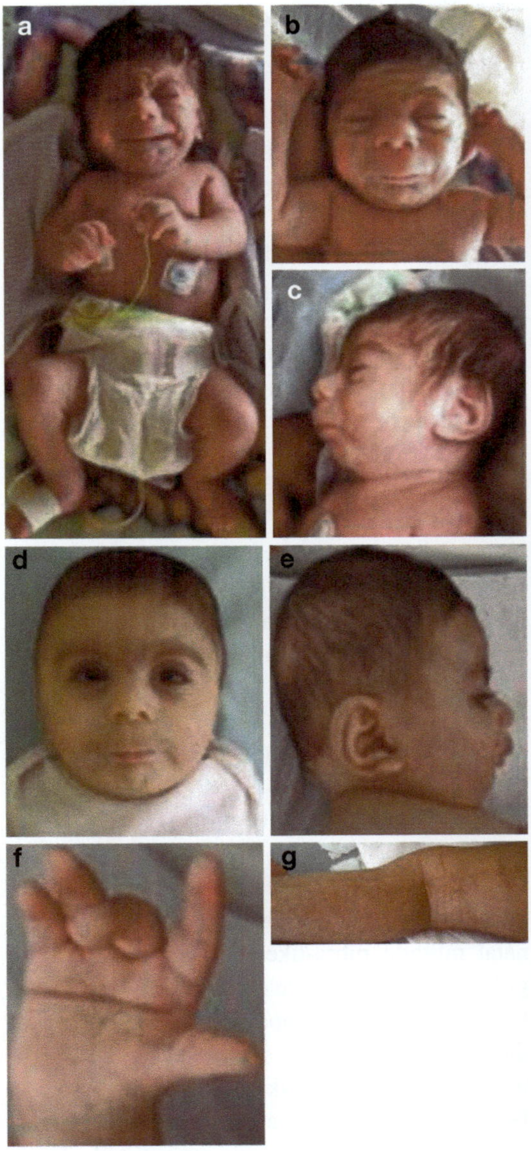

Fig. 10.49 Mosaicism in Brachmann-de Lange syndrome. (**a–c**) Frontal and lateral view of the boy at 1 week of life and (**d, e**) at 9 months; (**f**) transverse palmar crease; (**g**) linear pigmentary disturbance on the right leg [68] (Reprinted with permission from John Wiley and Sons, USA)

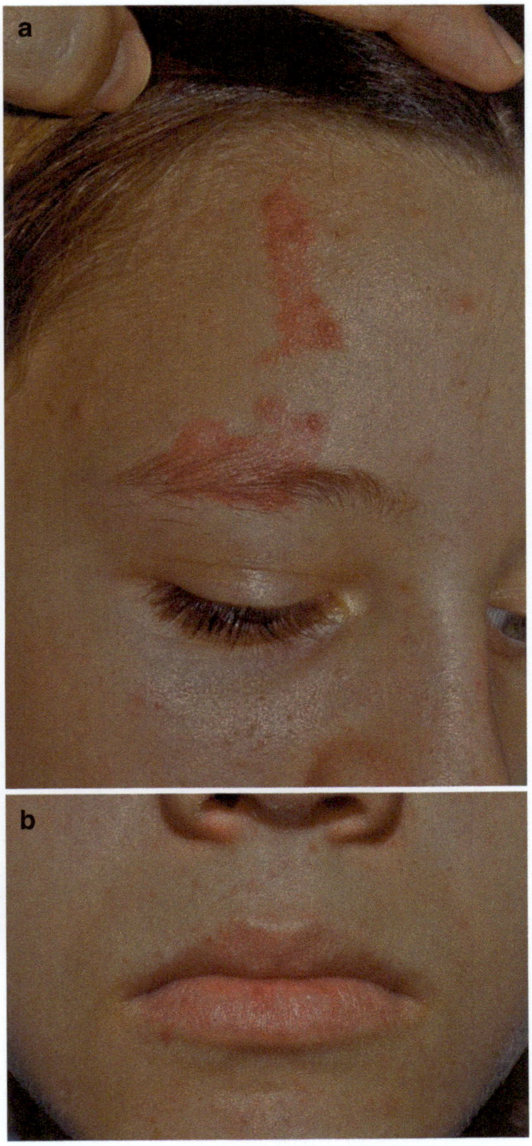

Fig. 10.50 Type 2 segmental manifestation of hereditary hemorrhagic teleangiectasia in a girl. (**a**) Pronounced linear lesion on the forehead; (**b**) nonsegmental involvement of the lips (Photographs and diagnosis kindly offered by Dr. Antonio Torrelo, Madrid, Spain)

mutations (see Sect. 7.4.1.3). An arteriovenous malformation is sometimes associated with this cutaneous trait and has prompted Eerola et al. [114] to create the ambiguous name "capillary malformation-arteriovenous malformation." Today, the occurrence of such arteriovenous fistulas (Fig. 10.52) [73, 204, 366] can best be categorized as a type 2 segmental manifestation of rhodoid nevus syndrome [192].

For obvious reasons, rhodoid nevus syndrome will often occur in a family without any type 2 segmental involvement [203]. According

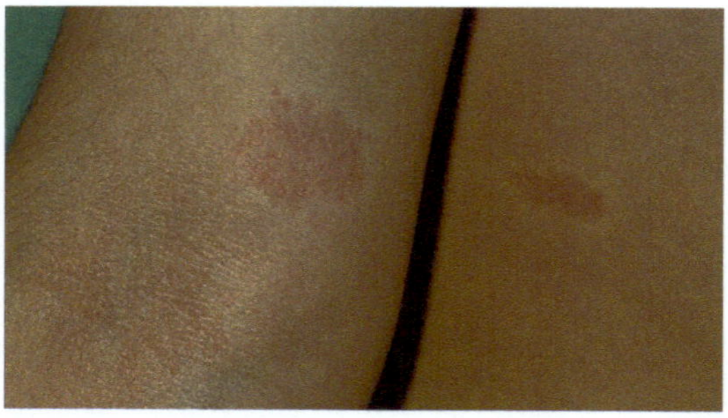

Fig. 10.51 Rhodoid nevi with characteristic anemic halo (Courtesy of Dr. Antonio Torrelo, Madrid, Spain)

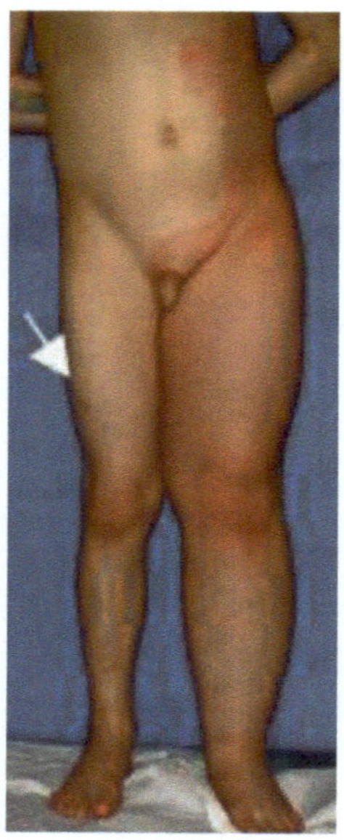

Fig. 10.52 Rhodoid nevus syndrome showing type 2 segmental involvement of the left leg. Note the contralateral nonsegmental lesion (*arrow*) [366] (Reprinted with permission from John Wiley and Sons, USA)

to OMIM [334], "the phenotypic spectrum of *RASA1*-related CM-AVM can include patients with only capillary malformations." Hence, the presence of an arteriovenous malformation cannot be taken as a prerequisite for the diagnosis of rhodoid nevus syndrome.

10.5 Blistering Skin Disorders

10.5.1 Self-Limited Dystrophic Epidermolysis Bullosa (Transient Bullous Dermolysis of the Newborn)

This peculiar form of dystrophic epidermolysis bullosa tends to heal spontaneously with scarring [171]. The disorder is caused by *COL7A1* mutations [79, 125]. It was first described as an autosomal dominant trait [79], but cases of compound heterozygosity have likewise been reported [168].

A newborn girl had linear lesions exclusively involving her right leg (Fig. 10.53) [332]. In the blood, compound heterozygosity for a splicing site mutation and an in-frame mutation was found. It is so far unclear which mechanism has caused this type of mosaicism.

10.5.2 Transient Superficial Acantholysis Arranged Along Blaschko's Lines in a Newborn

Jokiaho et al. [231] described a case of mild dysmorphism, parietal meningocele, and "neonatal miliaria rubra-like lesions" in a newborn girl. A nonmosaic deletion 3q27→3qter was found in blood lymphocytes. A photograph of the neonate showed a systematized, Blaschko-linear pattern of the skin lesions (Fig. 10.54). Histopathologically, acantholysis of the uppermost part of the epidermis was noted. These changes were related by the authors to the infundibula of sweat gland ducts. The inflammatory lesions resolved completely within 1 week.

This unusual mosaic phenotype is difficult to categorize because no molecular analysis could be done at the time of publication. It may or may not

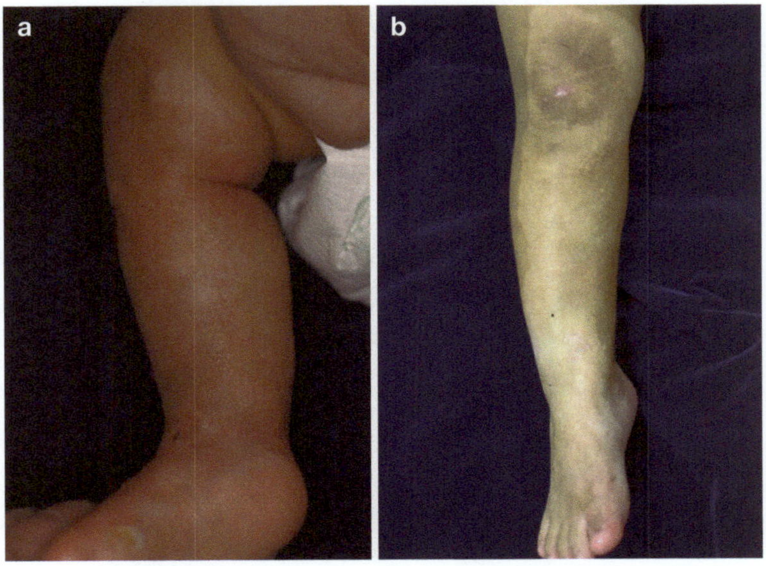

Fig. 10.53 Mosaic manifestation of self-limited dystrophic epidermolysis bullosa. Note segmentally arranged scars (**a**) at 2 months and (**b**) at 22 months after birth [332] (Reprinted with permission from John Wiley and Sons, USA)

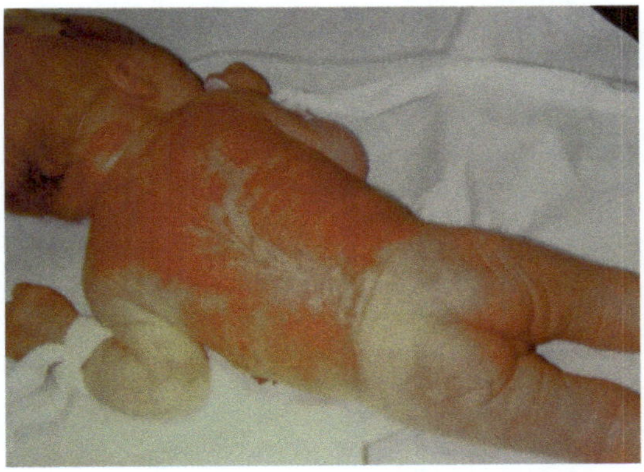

Fig. 10.54 A newborn girl with linear inflammatory skin lesions showing infundibular acantholysis of sweat glands [231] (Courtesy of Dr. Kirsti-Maria Niemi, Helsinki, Finland)

represent a new skin disorder [173]. A nosological relationship with the nonmosaic deletion 3q27→3qter is very unlikely. Today one would search for a desmoplakin mutation as noted in lethal acantholytic epidermolysis bullosa [39, 299] or for a *COL7A1* mutation as found in epidermolysis bullosa simplex superficialis [294].

References

1. Agarwalla A, Thakur A, Jacob M, Joshi A, Garg VK, Agrawal S (2000) Zosteriform and disseminated lesions in cutaneous leiomyoma. Acta Derm Venereol 80:446
2. Ahmed SF, Barr DG, Bonthron DT (2002) *GNAS1* mutations and progressive osseous heteroplasia. N Engl J Med 346:1669–1671
3. Ahmed SF, Dixon PH, Bonthron DT, Stirling HF, Barr DG, Kelnar CJ, Thakker RV (1998) *GNAS1* mutational analysis in pseudohypoparathyroidism. Clin Endocrinol (Oxf) 49:525–531
4. Alam NA, Barclay E, Rowan AJ, Tyrer JP, Calonje E, Manek S, Kelsell D, Leigh I, Olpin S, Tomlinson IP (2005) Clinical features of multiple cutaneous and uterine leiomyomatosis: an underdiagnosed tumor syndrome. Arch Dermatol 141:199–206
5. Alam M, Rabinowitz AD, Engler DE (2002) Gabapentin treatment of multiple piloleiomyoma-related pain. J Am Acad Dermatol 46: S27–S29
6. Al-Saleh S, Dragulescu A, Manson D, Golding F, Traubici J, Mei-Zahav M, Maclusky IB, Faughnan ME, Carpenter S, Ratjen F (2012) Utility of contrast echocardiography for pulmonary arteriovenous malformation screening in pediatric hereditary

hemorrhagic telangiectasia. J Pediatr 160:1039–1043, e1031

7. Alvarez-Breckenridge CA, Waite KA, Eng C (2007) *PTEN* regulates phospholipase D and phospholipase C. Hum Mol Genet 16:1157–1163
8. Angulo J, González-Guerra E, Vargas-Machuca I, Fariña MC, Martín L, Requena L (2006) [Multiple linear spiradenoma with a linear distribution]. Actas Dermosifiliogr 97:31–34
9. Anliker MD, Dummer R, Burg G (1997) Unilateral agminated angiofibromas: a segmental expression of tuberous sclerosis? Dermatology 195:176–178
10. Anstey A, Lowe DG, Kirby JD, Horton MA (1991) Familial mastocytosis: a clinical, immunophenotypic, light and electron microscopic study. Br J Dermatol 125:583–587
11. Antony FC, Cliff S, Cowley N (2003) Complete pain relief following treatment of a glomangiomyoma with the pulsed dye laser. Clin Exp Dermatol 28:617–619
12. Anzai S, Takeo N, Yamaguchi T, Sato T, Takasaki S, Terashi H, Takayasu S, Takahashi S (1999) Squamous cell carcinoma in a renal transplant recipient with linear porokeratosis. J Dermatol 26:244–247
13. Apter N, Chemke J, Hurwitz N, Levin S (1975) Neonatal neurofibromatosis: unusual manifestations with malignant clinical course. Clin Genet 7:388–393
14. Archer CB, Glover M, Atherton DJ (1988) Segmental neurofibromatosis with generalized café au lait spots. Br J Dermatol 119:96–97
15. Arnold AW, Kiritsi D, Happle R, Kohlhase J, Hausser I, Bruckner-Tuderman L, Has C, Itin PH (2012) Type 1 segmental Galli-Galli disease resulting from a previously unreported keratin 5 mutation. J Invest Dermatol 132:2100–2103
16. Arora S, Arora G, Ranjan P (2005) Relapsing linear acantholytic dermatosis in a four-year-old boy. Indian J Dermatol Venereol Leprol 71:351–353
17. Assmann A, Mandt N, Geilen CC, Blume-Peytavi U (2001) Buschke-Ollendorff syndrome–differential diagnosis of disseminated connective tissue lesions. Eur J Dermatol 11:576–579
18. Atherton DJ, Wells RS (1982) Juvenile elastoma and osteopoikilosis (the Buschke-Ollendorf syndrome). Clin Exp Dermatol 7:109–113
19. Atten MJ, Ahmed S, Attar BM, Richter H 3rd, Mehta B (2000) Massive pelvic hemangioma in a patient with blue rubber bleb nevus syndrome. South Med J 93:1122–1125
20. Babalian L (1947) Juvenile acanthosis nigricans and ichthyosis. Arch Dermatol 55:411–412
21. Badeloe S, Bladergroen RS, Jonkman MF, Burrows NP, Steijlen PM, Poblete-Gutiérrez P, van Steensel MA, van Geel M, Frank J (2008) Hereditary multiple cutaneous leiomyoma resulting from novel mutations in the fumarate hydratase gene. J Dermatol Sci 51:139–143
22. Bano S, Prasad A, Yadav SN, Chaudhary V, Sachdeva N (2010) Elephantiasis neuromatosa of the lower limb in a patient with neurofibromatosis type-1: a case report with imaging findings. J Pediatr Neurosci 5:59–63
23. Barnes L, Estes SA (1986) Laser treatment of hereditary multiple glomus tumors. J Dermatol Surg Oncol 12:912–915
24. Baselga E, Drolet BA, Fleming MS, Esterly NB (1997) Multiple acquired vascular nodules. Pediatr Dermatol 14:327–329
25. Baselga E, Drolet BA, van Tuinen P, Esterly NB, Happle R (1998) Dyskeratosis congenita with linear areas of severe cutaneous involvement. Am J Med Genet 75:492–496
26. Bazex A, Dupré A, Christol B, Andrieu H (1971) [Familial mastocytosis: presentation of 2 cases; general review; nosologic importance]. Ann Dermatol Syphiligr (Paris) 98:241–260
27. Bean WB (1958) Blue rubber bleb nevi of the skin and gastrintestinal mucosa. In: Bean WB (ed) Vascular spiders and related lesions of the skin. Charles C Thomas, Springfield, pp 175–178
28. Bedlow AJ, Cook MG, Kurwa A (1999) Extensive naevoid eccrine spiradenoma. Br J Dermatol 140:154–157
29. Berk DR, Spector EB, Bayliss SJ (2007) Familial acanthosis nigricans due to K650T *FGFR3* mutation. Arch Dermatol 143:1153–1156
30. Betz RC, Planko L, Eigelshoven S, Hanneken S, Pasternack SM, Bussow H, Van Den Bogaert K, Wenzel J, Braun-Falco M, Rütten A, Rogers MA, Ruzicka T, Nöthen MM, Magin TM, Kruse R (2006) Loss-of-function mutations in the keratin 5 gene lead to Dowling-Degos disease. Am J Hum Genet 78:510–519
31. Bhat YJ, Baba AN, Manzoor S, Qayoom S, Ahmed SM (2010) Bullous ichthyosiform erythroderma with rickets in child of a parent with naevus unius lateris. Indian J Dermatol Venereol Leprol 76:192–194
32. Biesecker LG, Peters KF, Darling TN, Choyke P, Hill S, Schimke N, Cunningham M, Meltzer P, Cohen MM Jr (1998) Clinical differentiation between Proteus syndrome and hemihyperplasia: description of a distinct form of hemihyperplasia. Am J Med Genet 79:311–318
33. Bingham EA, Burrows D (1984) Neurofibromatosis (elephantiasis neuromatosa). Br J Dermatol 111:65–66
34. Birt AR, Hogg GR, Dubé WJ (1977) Hereditary multiple fibrofolliculomas with trichodiscomas and acrochordons. Arch Dermatol 113: 1674–1677
35. Bloom D, Abramowitz EW (1943) Porokeratosis Mibelli. Arch Dermatol Syphilol 47:1–15
36. Blumenthal GM, Dennis PA (2008) PTEN hamartoma tumor syndromes. Eur J Hum Genet 16: 1289–1300

37. Bodemer C, Hermine O, Palmérini F, Yang Y, Grandpeix-Guyodo C, Leventhal PS, Hadj-Rabia S, Nasca L, Georgin-Lavialle S, Cohen-Akenine A, Launay JM, Barete S, Feger F, Arock M, Catteau B, Sans B, Stalder JF, Skowron F, Thomas L, Lorette G, Plantin P, Bordigoni P, Lortholary O, de Prost Y, Moussy A, Sobol H, Dubreuil P (2010) Pediatric mastocytosis is a clonal disease associated with D816V and other activating *c-KIT* mutations. J Invest Dermatol 130:804–815
38. Boente Mdel C, Frontini Mdel V, Primc NB, Asial RA (2004) [Linear Darier disease in two siblings: an example of loss of heterozygosity]. Ann Dermatol Venereol 131:805–809
39. Bolling MC, Veenstra MJ, Jonkman MF, Diercks GF, Curry CJ, Fisher J, Pas HH, Bruckner AL (2010) Lethal acantholytic epidermolysis bullosa due to a novel homozygous deletion in *DSP*: expanding the phenotype and implications for desmoplakin function in skin and heart. Br J Dermatol 162:1388–1394
40. Boltshauser E, Stocker H, Mächler M (1989) Neurofibromatosis type 1 in a child of a parent with segmental neurofibromatosis (NF-5). Neurofibromatosis 2:244–245
41. Boltshauser E, Stocker H, Sailer H, Valavanis A (1989) Intracranial abnormalities associated with facial plexiform neurofibromas in neurofibromatosis type 1. Neurofibromatosis 2:274–277
42. Bonafé JL, Blanchet-Bardon C, Christol B, Rolland M (1987) Naevus verruqueux systématisé épidermolytique et érythrodermie ichtyosiforme congénitale bulleuse. Ann Dermatol Venereol 114:916
43. Boon LM, Mulliken JB, Enjolras O, Vikkula M (2004) Glomuvenous malformation (glomangioma) and venous malformation: distinct clinicopathologic and genetic entities. Arch Dermatol 140:971–976
44. Bordel-Gomez MT, Monteagudo-Sanchez B, Alvarez-Fernandez JC (2008) [Multiple unilateral facial angiofibromas: description of a new case]. Actas Dermosifiliogr 99:824–827
45. Borroni RG, Narula N, Diegoli M, Grasso M, Concardi M, Rosso R, Cerica A, Brazzelli V, Arbustini E (2011) A novel mutation of the glomulin gene in an Italian family with autosomal dominant cutaneous glomuvenous malformations. Exp Dermatol 20:1032–1034
46. Bouscarat F, Avril MF, Prade M, Aurias A, Guillaume JC (1990) [Unilateral basal cell epitheliomatosis]. Ann Dermatol Venereol 117:864–866
47. Boybeyi O, Alanay Y, Kayikçioğlu A, Karnak I (2010) Hemihyperplasia-multiple lipomatosis syndrome: an underdiagnosed entity in children with asymmetric overgrowth. J Pediatr Surg 45:E19–E23
48. Braun-Falco M, Bonel H, Ring J, Hein R (2003) Linear spiradenoma with focal malignant transformation. J Eur Acad Dermatol Venereol 17: 308–312
49. Brazzelli V, Prestinari F, Barbagallo T, Vassallo C, Agozzino M, Borroni G (2006) Linear Darier's disease successfully treated with 0.1% tazarotene gel "short-contact" therapy. Eur J Dermatol 16:59–61
50. Brems H, Chmara M, Sahbatou M, Denayer E, Taniguchi K, Kato R, Somers R, Messiaen L, De Schepper S, Fryns JP, Cools J, Marynen P, Thomas G, Yoshimura A, Legius E (2007) Germline loss-of-function mutations in *SPRED1* cause a neurofibromatosis 1-like phenotype. Nat Genet 39: 1120–1126
51. Brook CG, Valman HB (1971) Osteoma cutis and Albright's hereditary osteodystrophy. Br J Dermatol 85:471–475
52. Brouillard P, Boon LM, Mulliken JB, Enjolras O, Ghassibe M, Warman ML, Tan OT, Olsen BR, Vikkula M (2002) Mutations in a novel factor, glomulin, are responsible for glomuvenous malformations ("glomangiomas"). Am J Hum Genet 70:866–874
53. Brouillard P, Ghassibe M, Penington A, Boon LM, Dompmartin A, Temple IK, Cordisco M, Adams D, Piette F, Harper JI, Syed S, Boralevi F, Taïeb A, Danda S, Baselga E, Enjolras O, Mulliken JB, Vikkula M (2005) Four common glomulin mutations cause two thirds of glomuvenous malformations ("familial glomangiomas"): evidence for a founder effect. J Med Genet 42:e13
54. Buhl T, Wienrich BG, Sieblist C, Schön MP, Seitz CS (2010) Development of segmental superficial actinic porokeratosis during immunosuppressive therapy for pemphigus vulgaris. Acta Derm Venereol 90:212–213
55. Burgio RG, Martini A, Cetta G, Zanaboni G, Vitellaro L, Danesino C (1988) Asymmetric Marfan syndrome. Am J Med Genet 30:905–909
56. Butkus CE, Michels VV, Lindor NM, Cooney WP 3rd (1997) Melorheostosis in a patient with familial osteopoikilosis. Am J Med Genet 72:43–46
57. Butler DF (1984) Pathology quiz case 1: cutaneous leiomyoma (familial). Arch Dermatol 120:1618–1620
58. Butterworth T (1956) Neurofibrosarcoma; report of a case in a seven-year-old girl. Arch Dermatol 74:457–458
59. Cairns RJ (1967) Familial juvenile elastoma: osteopoikilosis (2 cases). Proc R Soc Med 60:1267
60. Calzavara PG, Carlino A, Anzola GP, Pasolini MP (1988) Segmental neurofibromatosis. Case report and review of the literature. Neurofibromatosis 1:318–322
61. Calzavara-Pinton PG, Carlino A, Marini D (1989) [Segmental neurofibromatosis: description of our second clinical case and review of the literature]. G Ital Dermatol Venereol 124:231–234
62. Camacho F, Jorquera E, Vasquez FJ, Hevia A (1994) Giant zoniform leiomyoma: light and electron microscopy study. Eur J Dermatol 4:384–386
63. Camisa C, Rossana C, Little L (1985) Naevoid basal-cell carcinoma syndrome with unilateral neoplasms and pits. Br J Dermatol 113:365–367

64. Camprubí M, Balaguer A, Azon Masoliver A, Jimenez-Feijoo R, Escribano Subias J (2006) Unilateral facial angiofibromas; a review of the literature. Pediatr Dermatol 23:303–305
65. Carvalho VO, Taniguchi K, Giraldi S, Bertogna J, Marinoni LP, Fillus JN, Reis Filho JS (2001) Congenital plaquelike glomus tumor in a child. Pediatr Dermatol 18:223–226
66. Casalá AM, Kowalczuk AM, Abulafia J, Bianchi C, Bianchi O, Galimberti R (1983) Fibromas perifoliculares multiples unilaterales. Arch Argent Dermatol 33:337–342
67. Casals M, Campoy A, Aspiolea F, Carrasco MA, Camps A (2009) Successful treatment of linear Darier's disease with topical adapalene. J Eur Acad Dermatol Venereol 23:237–238
68. Castronovo P, Delahaye-Duriez A, Gervasini C, Azzollini J, Minier F, Russo S, Masciadri M, Selicorni A, Verloes A, Larizza L (2010) Somatic mosaicism in Cornelia de Lange syndrome: a further contributor to the wide clinical expressivity? Clin Genet 78:560–564
69. Caux F, Plauchu H, Chibon F, Faivre L, Fain O, Vabres P, Bonnet F, Selma ZB, Laroche L, Gérard M, Longy M (2007) Segmental overgrowth, lipomatosis, arteriovenous malformation and epidermal nevus (SOLAMEN) syndrome is related to mosaic *PTEN* nullizygosity. Eur J Hum Genet 15: 767–773
70. Ceulen RP, Van Marion AM, Steijlen PM, Frank J, Poblete-Gutiérrez P (2008) Multiple unilateral skin tumors suggest type 1 segmental manifestation of familial syringoma. Eur J Dermatol 18:285–288
71. Chacon-Camacho OF, Vazquez J, Zenteno JC (2011) Expanding the phenotype of gingival fibromatosis-mental retardation-hypertrichosis (Zimmermann-Laband) syndrome. Am J Med Genet A 155A:1716–1720
72. Chang YC, Colome-Grimmer M, Kelly E (2006) Multiple trichoepitheliomas in the lines of Blaschko. Pediatr Dermatol 23:149–151
73. Chee D, Phillips R, Maixner W, Southwell BR, Hutson JM (2010) The potential of capillary birthmarks as a significant marker for capillary malformation-arteriovenous malformation syndrome in children who had nontraumatic cerebral hemorrhage. J Pediatr Surg 45:2419–2422
74. Chen TC, Kuo WR, Chai CY, Juan KH (1991) Huge plexiform neurofibroma of the head and liver: case report. Gaoxiong Yi Xue Ke Xue Za Zhi 7: 650–656
75. Chen HH, Liao YH (2002) Onychodystrophy in congenital linear porokeratosis. Br J Dermatol 147:1272–1273
76. Chester BJ, Brown L (1959) Darier's disease resembling linear verrucous epidermal nevus. Arch Dermatol 80:625–626
77. Chia MW, Tay YK, Ng WL (2008) Multiple pebbly blue papules in a segmental distribution. Pediatr Dermatol 25:381–382
78. Choonhakarn C, Meesingha A (2006) Multiple hereditary infundibulocystic basal cell carcinomas: a distinctive genodermatosis. Br J Dermatol 154:1007–1010
79. Christiano AM, Fine JD, Uitto J (1997) Genetic basis of dominantly inherited transient bullous dermolysis of the newborn: a splice site mutation in the type VII collagen gene. J Invest Dermatol 109:811–814
80. Chuang GS, Martinez-Mir A, Geyer A, Engler DE, Glaser B, Cserhalmi-Friedman PB, Gordon D, Horev L, Lukash B, Herman E, Cid MP, Brenner S, Landau M, Sprecher E, Garcia Muret MP, Christiano AM, Zlotogorski A (2005) Germline fumarate hydratase mutations and evidence for a founder mutation underlying multiple cutaneous and uterine leiomyomata. J Am Acad Dermatol 52:410–416
81. Civatte J, Le Tréguilly JP (1971) [Multiple perifollicular fibromas of the face and neck]. Bull Soc Fr Dermatol Syphiligr 78:100–103
82. Clark DP, Buescher L, Havey A (1990) Familial urticaria pigmentosa. Arch Intern Med 150:1742–1744
83. Cohen PR, Schneiderman P (1989) Clinical manifestations of the Marfan syndrome. Int J Dermatol 28:291–299
84. Cohen MM Jr, Turner JT, Biesecker LG (2003) Proteus syndrome: misdiagnosis with *PTEN* mutations. Am J Med Genet A 122A:323–324
85. Colamaria V, Zambelli L, Tinazzi P, Dalla Bernardina B (1988) Tuberous sclerosis associated with partial gigantism in a child. Brain Dev 10:178–181
86. Colegio OR, McNiff JM, Antaya RJ (2008) Nevoid acanthosis nigricans with subtle melanocyte hyperplasia. J Am Acad Dermatol 58:S102–S103
87. Consoli C, Moss C, Green S, Balderson D, Cooper DN, Upadhyaya M (2005) Gonosomal mosaicism for a nonsense mutation (R1947X) in the NF1 gene in segmental neurofibromatosis type 1. J Invest Dermatol 125:463–466
88. Courville P, Thomine E, Surlemont Y, Hemet J, Métayer J, Lauret P (2000) [Epidermal nevus associated with a type I neurofibromatosis and a nephroblastoma: a new epidermal nevus syndrome?]. Ann Pathol 20:616–619
89. Creamer D, Macdonald A, Griffiths WA (1999) Unilateral linear syringomata: a case report. Clin Exp Dermatol 24:428–430
90. Crowe FW, Schull WJ, Neel JV (1956) A clinical, pathologial, and genetic study of multiple neurofibromatosis. Charles C. Thomas, Springfield
91. Cullen SI (1979) Localized Ehlers-Danlos syndrome. Arch Dermatol 115:332–333
92. Curth HO (1936) Benign type of acanthosis nigricans. Arch Dermatol 34:353–366
93. Curth HO (1964) [Problems of acanthosis nigricans]. Hautarzt 15:433–439
94. Curth HO (1976) Unilateral epidermal naevus resembling acanthosis nigricans. Br J Dermatol 95:433–436
95. Curth HO, Aschner BM (1959) Genetic studies on acanthosis nigricans. AMA Arch Derm 79:55–66

96. Däschner K, Assum G, Eisenbarth I, Krone W, Hoffmeyer S, Wortmann S, Heymer B, Kehrer-Sawatzki H (1997) Clonal origin of tumor cells in a plexiform neurofibroma with LOH in *NF1* intron 38 and in dermal neurofibromas without LOH of the *NF1* gene. Biochem Biophys Res Commun 234:346–350
97. Dawson TAJ (1984) Regional eruptive neurofibromatosis. Br J Dermatol 111:65
98. De la Brassinne M, Nazzaro V, Mimoz C, Blanchet-Bardon C (1986) Érythrodermie congénitale ichtyosiforme bulleuse chez une fille dont la mère est atteinte d'un naevus verruqueux épidermolytique. Ann Dermatol Venereol 113:619
99. de la Hera I, Chico R, Llamas R, Vanaclocha F (2011) [Linear Darier disease]. Actas Dermosifiliogr 102:299–301
100. de Pina Neto JM, Soares LR, Souza AH, Petean EB, Velludo MA, de Freitas AC, Ribas JP (1988) A new case of Zimmermann-Laband syndrome with mild mental retardation, asymmetry of limbs, and hypertrichosis. Am J Med Genet 31:691–695
101. Debeer P, Pykels E, Lammens J, Devriendt K, Fryns JP (2003) Melorheostosis in a family with autosomal dominant osteopoikilosis: report of a third family. Am J Med Genet A 119A:188–193
102. Degos R, Guilaine J, Civatte J (1964) [Multiple glomus tumors]. Ann Dermatol Syphiligr (Paris) 91:40–41
103. Del Pozo J, Martínez W, Calvo R, Almagro M, Fonseca E (2002) Unilateral angiofibromas: an oligosymptomatic and segmentary form of tuberous sclerosis. Eur J Dermatol 12:262
104. Demetree JW, Lang PG, St Clair JT (1979) Unilateral, linear, zosteriform epidermal nevus with acantholytic dyskeratosis. Arch Dermatol 115:875–877
105. Diederen PV, Lammers AM (1999) Porokeratosis linearis. Ned Tijdschr Dermatol Venereol 9:40–41
106. Diekmann L, Hüther W, Pfeiffer RA (1967) [Unusual manifestations of neurofibromatosis (von Recklinghausen's disease) in childhood. Abdominal tumor with clitoris hypertrophy. Renal hypertension caused by renal artery stenosis. Monozygotic twins–plexiform neuroma in the head region and glaucoma]. Z Kinderheilkd 101:191–222
107. Dijkstra HJ, Blanken R (1999) Multipele glomustumoren. Ned Tijdschr Dermatol Venereol 9:73–74
108. DiPaolo D, Zimmerman RA (1995) Solitary cortical tubers. AJNR Am J Neuroradiol 16:1360–1364
109. Donald GF, Hunter GA (1968) Disseminated superficial actinic porokeratosis: a report of eight cases. Australas J Dermatol 9:335–344
110. Dover JS, Phillips TJ, Burns DA, Krafchik BR (1986) Disseminated superficial actinic porokeratosis: coexistence with other porokeratotic variants. Arch Dermatol 122:887–889
111. Drachtman RA, Alter BP (1995) Dyskeratosis congenita. Dermatol Clin 13:33–39
112. Easton JA, Donnelly S, Kamps MA, Steijlen PM, Martin PE, Tadini G, Janssens R, Happle R, van Geel M, van Steensel MA (2012) Porokeratotic eccrine nevus may be caused by somatic connexin26 mutations. J Invest Dermatol 132:2184–2191
113. Eddy MC, Jan De Beur SM, Yandow SM, McAlister WH, Shore EM, Kaplan FS, Whyte MP, Levine MA (2000) Deficiency of the alpha-subunit of the stimulatory G protein and severe extraskeletal ossification. J Bone Miner Res 15:2074–2083
114. Eerola I, Boon LM, Mulliken JB, Burrows PE, Dompmartin A, Watanabe S, Vanwijck R, Vikkula M (2003) Capillary malformation-arteriovenous malformation, a new clinical and genetic disorder caused by *RASA1* mutations. Am J Hum Genet 73:1240–1249
115. Ehlers G, Rothe A (1971) [Porokeratosis Mibelli with multiple precancerous and epidermoid carcinomas]. Hautarzt 22:68–73
116. Ehrig T, Cockerell CJ (2003) Buschke-Ollendorff syndrome: report of a case and interpretation of the clinical phenotype as a type 2 segmental manifestation of an autosomal dominant skin disease. J Am Acad Dermatol 49:1163–1166
117. Ekmekci TR, Koslu A, Sakiz D (2005) Congenital blaschkoid eccrine spiradenoma on the face. Eur J Dermatol 15:73–74
118. Emmerson RW (1965) Porokeratosis (Mibelli). Br J Dermatol 77:462–463
119. Eng C (2003) *PTEN*: one gene, many syndromes. Hum Mutat 22:183–198
120. Eng AM, Brody P, Rhee HL, Bronson DM (1991) Congenital ichthyosiform erythroderma and epidermal nevus. Int J Dermatol 30:284–287
121. Ersoy-Evans S, Sahin S, Mancini AJ, Paller AS, Guitart J (2006) The acanthosis nigricans form of epidermal nevus. J Am Acad Dermatol 55:696–698
122. Ettl A, Marinkovic M, Koornneef L (1996) Localized hypertrichosis associated with periorbital neurofibroma: clinical findings and differential diagnosis. Ophthalmology 103:942–948
123. Evans DG, Wallace AJ, Wu CL, Trueman L, Ramsden RT, Strachan T (1998) Somatic mosaicism: a common cause of classic disease in tumor-prone syndromes? Lessons from type 2 neurofibromatosis. Am J Hum Genet 63:727–736
124. Farfel Z (2002) *GNAS1* mutations and progressive osseous heteroplasia. N Engl J Med 346:1669–1671
125. Fassihi H, Diba VC, Wessagowit V, Dopping-Hepenstal PJ, Jones CA, Burrows NP, McGrath JA (2005) Transient bullous dermolysis of the newborn in three generations. Br J Dermatol 153:1058–1063
126. Fawcett HA, Marsden RA (1983) Hereditary osteoma cutis. J R Soc Med 76:697–699
127. Feldman SR, Crosby DL, Tomsick RS (1991) Scaly atrophic lesions both scattered and in linear arrays: disseminated superficial actinic porokeratosis in a patient with linear porokeratosis. Arch Dermatol 127:1219, 1222
128. Flarer F (1928) Nota clinica ed istologica su alcune particolarità di evoluzione in un caso di porocheratosi. Arch Ital Dermatol Sifilogr Venereol 3:524–548

129. Fölster-Holst R, Nellen RG, Jensen JM, Poblete-Gutiérrez P, Steijlen PM, Schwarz T, Happle R, Van Geel M, Frank J (2012) Molecular genetic support for the rule of dichotomy in type 2 segmental Darier disease. Br J Dermatol 166:464–466
130. Foo CC, Kumarasinghe SP (2005) Juvenile elastoma: a forme fruste of the Buschke-Ollendorff syndrome? Australas J Dermatol 46:250–252
131. Foster CM, Levin S, Levine M, Mukherjee A, Costa JL, Eanes ED, Triche T, Zasloff M (1986) Limited dermal ossification: clinical features and natural history. J Pediatr 109:71–76
132. Foulds IS, Slater DN (1983) Porokeratosis of Mibelli and immune complex glomerulonephritis. Clin Exp Dermatol 8:69–75
133. Fretzin DF, Potter B (1965) Blue rubber bleb nevus. Arch Intern Med 116:924–929
134. Freyschmidt J (2001) Melorheostosis: a review of 23 cases. Eur Radiol 11:474–479
135. Freyschmidt J, Freyschmidt G (1999) SKIBO diseases. Disorders affecting the skin and bones: a clinical, dermatologic and radiologic synopsis. Springer, Berlin
136. Fried SZ, Lynfield YL (1987) Unilateral facial telangiectasia macularis eruptiva perstans. J Am Acad Dermatol 16:250–252
137. Gandola M (1951) La porocheratosi di Mibelli: aspetti evolutivi tardivi e associazioni morbose insolite. Boll Soc Medicochir 65:273–292
138. Garcia Muret MP, Pujol RM, de Moragas JM (1998) [Multiple and unilateral angiofibromas of the face: forme fruste of Bourneville tuberous sclerosis]. Ann Dermatol Venereol 125:325–327
139. Garcia-Muret M, Pujol R (2000) [Multiple unilateral angiofibromas of the face]. Ann Dermatol Venereol 127:211
140. García-Navarro X, Garcés JR, Baselga E, Alomar A (2009) Linear porokeratosis: excellent response to photodynamic therapy. Arch Dermatol 145:526–527
141. Gardner RJ, Yun K, Craw SM (1988) Familial ectopic ossification. J Med Genet 25:113–117
142. Garg T, Ramchander VB, Barara M, Nangia A (2011) Generalized linear porokeratosis: a rare entity with excellent response to acitretin. Dermatol Online J 17:3
143. Geffner RE, Goslen JB, Santa Cruz DJ (1986) Linear and dermatomal trichoepitheliomas. J Am Acad Dermatol 14:927–930
144. Gelfand IM, Hub RS, Shore EM, Kaplan FS, Dimeglio LA (2007) Progressive osseous heteroplasia-like heterotopic ossification in a male infant with pseudohypoparathyroidism type Ia: a case report. Bone 40:1425–1428
145. Ghai S, Sharma R (2003) Mixed sclerosing bone dysplasia–a case report with literature review. Clin Imaging 27:203–205
146. Gilaberte M, Puig L, Vidal D, Alomar A (2003) Acantholytic dyskeratotic naevi following Blaschko's lines: a mosaic form of Darier's disease. J Eur Acad Dermatol Venereol 17:196–199
147. Girisha KM, Lewis LE, Phadke SR, Kutsche K (2010) Costello syndrome with severe cutis laxa and mosaic *HRAS* G12S mutation. Am J Med Genet A 152A:2861–2864
148. Godfrey M, Olson S, Burgio RG, Martini A, Valli M, Cetta G, Hori H, Hollister DW (1990) Unilateral microfibrillar abnormalities in a case of asymmetric Marfan syndrome. Am J Hum Genet 46: 661–671
149. Goebeler M, Bröcker EB, Hamm H (2006) Keratotic plaques on the left trunk area and all extremities: porokeratosis. Arch Dermatol 142:1059–1064
150. Goldman GD, Milstone LM (1995) Generalized linear porokeratosis treated with etretinate. Arch Dermatol 131:496–497
151. Goldner R (1971) Zosteriform porokeratosis of Mibelli. Arch Dermatol 104:425–426
152. Gonzalez-Castro U, Luelmo-Aguilar J, Castells-Rodellas A (1993) Unilateral facial telangiectasia macularis eruptiva perstans. Int J Dermatol 32:123–124
153. Götz H, Klüken N (1966) Leiomyome (linker Unterarm). Zeitschr Haut Geschlechtskr 41:470
154. Goujon E, Cordoro KM, Barat M, Rousseau T, Brouillard P, Vikkula M, Frieden IJ, Vabres P (2011) Congenital plaque-type glomuvenous malformations associated with fetal pleural effusion and ascites. Pediatr Dermatol 28:528–531
155. Grahame R, Pyeritz RE (1995) The Marfan syndrome: joint and skin manifestations are prevalent and correlated. Br J Rheumatol 34:126–131
156. Grees SA, Valle LE (2004) Cutaneous mastocytosis: segmentary lesions of telangiectatic mastocytosis: an unusual clinical presentation. Rev Argent Dermatol 85:20–28
157. Gripp KW, Stabley DL, Nicholson L, Hoffman JD, Sol-Church K (2006) Somatic mosaicism for an *HRAS* mutation causes Costello syndrome. Am J Med Genet A 140:2163–2169
158. Grosshans E, Dungler T, Hanau D (1981) [Pinkus' trichodiscoma (author's transl)]. Ann Dermatol Venereol 108:837–846
159. Grzybowski M (1933) Contribution à l'étude de l'histologie des myomes cutanés. Ann Dermatol Syphiligr (Paris) 4:852–860
160. Guarneri B, Borgia F, Cannavò SP, Vaccaro M, Happle R (2000) Multiple familial basal cell carcinomas including a case of segmental manifestation. Dermatology 200:299–302
161. Guenova E, Hoetzenecker W, Metzler G, Rocken M, Schaller M (2007) Multicentric Bowen disease in linear porokeratosis. Eur J Dermatol 17: 439–440
162. Gupta S, Jain VK, Singh U (2000) Multiple eccrine spiradenomas in zosteriform distribution in a child. Pediatr Dermatol 17:384–386
163. Gupta S, Radotra BD, Kaur I, Handa S, Kumar B (2001) Multiple linear eccrine spiradenomas with eyelid involvement. J Eur Acad Dermatol Venereol 15:163–166

164. Guss SB, Osbourn RA, Lutzner MA (1971) Porokeratosis plantaris, palmaris, et disseminata: a third type of porokeratosis. Arch Dermatol 104: 366–373
165. Gutierrez MM, Mora RG (1986) Nevoid basal cell carcinoma syndrome: a review and case report of a patient with unilateral basal cell nevus syndrome. J Am Acad Dermatol 15:1023–1030
166. Hadlich J, Linse R (1989) Keratosen mit granulärer Degeneration und ihre Beziehungen zueinander. 2. Mitteilung: Heterophänie von epidermolytic Hyperkeratosis (Erythrodermia ichthyosiformis bullosa). Naevus verrucosus hystricoides und Keratosis palmoplantaris cum degeneratione granulosa Vörner. Dermatol Monatsschr 175:418–424
167. Hall MR, Kovach BT, Miller JL (2007) Unilateral facial angiofibromas without other evidence of tuberous sclerosis: case report and review of the literature. Cutis 80:284–288
168. Hammami-Hauasli N, Raghunath M, Küster W, Bruckner-Tuderman L (1998) Transient bullous dermolysis of the newborn associated with compound heterozygosity for recessive and dominant *COL7A1* mutations. J Invest Dermatol 111:1214–1219
169. Han HJ, Lim GY, You CY (2009) A large infiltrating fibrous hamartoma of infancy in the abdominal wall with rare associated tuberous sclerosis. Pediatr Radiol 39:743–746
170. Hanke P (2002) Buschke-Ollendorff syndrome. Zeitschr Hautkr 77:606–607
171. Hanson SG, Fine JD, Levy ML (1999) Three new cases of transient bullous dermolysis of the newborn. J Am Acad Dermatol 40:471–476
172. Happle R (1990) [Acanthokeratolytic epidermal nevus: acanthokeratolysis is hereditary, not the nevus]. Hautarzt 41:117–118
173. Happle R (1990) Deletion 3q27→3qter associated with a new skin disorder? Hum Genet 85:563–564
174. Happle R (1991) [Segmental neurofibromatosis and germ-line mosaicism]. Hautarzt 42:739–740
175. Happle R (1997) A rule concerning the segmental manifestation of autosomal dominant skin disorders: review of clinical examples providing evidence for dichotomous types of severity. Arch Dermatol 133:1505–1509
176. Happle R (1997) Cancer proneness of linear porokeratosis may be explained by allelic loss. Dermatology 195:20–25
177. Happle R (1999) Loss of heterozygosity in human skin. J Am Acad Dermatol 41:143–164
178. Happle R (2000) Nonsyndromic type of hereditary multiple basal cell carcinoma. Am J Med Genet 95:161–163
179. Happle R (2001) Large plexiform neurofibromas may be explained as a type 2 segmental manifestation of neurofibromatosis 1. Am J Med Genet 98:363–364
180. Happle R (2001) [Type 2 segmental manifestation of autosomal dominant skin diseases: development of a new formal genetic concept]. Hautarzt 52:283–287
181. Happle R (2001) Buschke-Ollendorff syndrome: early, unilateral and pronounced involvement may be explained as a type 2 segmental manifestation. Eur J Dermatol 11:505
182. Happle R (2002) Dohi memorial lecture. New aspects of cutaneous mosaicism. J Dermatol 29: 681–692
183. Happle R (2004) The manifold faces of Proteus syndrome. Arch Dermatol 140:1001–1002
184. Happle R (2004) Melorheostosis may originate as a type 2 segmental manifestation of osteopoikilosis. Am J Med Genet A 125A:221–223
185. Happle R (2007) Type 2 segmental trichoepitheliomatosis: a genetic concept that may explain congenital multiple trichoepitheliomas in the lines of Blaschko. Pediatr Dermatol 24:448–449
186. Happle R (2007) Type 2 segmental Cowden disease vs. Proteus syndrome. Br J Dermatol 156:1089–1090
187. Happle R (2007) Linear Cowden nevus: a new distinct epidermal nevus. Eur J Dermatol 17: 133–136
188. Happle R (2008) Type 2 segmental acanthosis nigricans: a historical case explained by a new concept. Arch Dermatol 144:1637
189. Happle R (2009) Superimposed segmental manifestation of both rare and common cutaneous disorders: a new paradigm. Actas Dermosifiliogr 100: 77–85
190. Happle R (2010) Type 2 segmental manifestation of "blue rubber bleb nevus syndrome": a reappraisal of 5 case reports. Eur J Dermatol 20:266–268
191. Happle R (2010) The group of epidermal nevus syndromes Part I. Well defined phenotypes. J Am Acad Dermatol 63:1–22; quiz 23–24
192. Happle R (2010) The rhodoid nevus: a proposed term for a so far unnamed capillary malformation. Dermatology 221:317–319
193. Happle R (2012) Hornstein-Birt-Hogg-Dubé syndrome: a renaming and reconsideration. Am J Med Genet A 158A:1247–1251
194. Happle R, Itin PH, Brun AM (1999) Type 2 segmental Darier disease. Eur J Dermatol 9:449–451
195. Happle R, König A (1999) Type 2 segmental manifestation of multiple glomus tumors: a review and reclassification of 5 case reports. Dermatology 198:270–272
196. Harman RRM (1969) Juvenile elastoma with osteopoikilosis. Br J Dermatol 81:165
197. Heinritz W, Paasch U, Sticherling M, Wittekind C, Simon JC, Froster UG, Renner R (2008) Evidence for a founder effect of the germline fumarate hydratase gene mutation R58P causing hereditary leiomyomatosis and renal cell cancer (HLRCC). Ann Hum Genet 72:35–40
198. Hellemans J, Debeer P, Wright M, Janecke A, Kjaer KW, Verdonk PC, Savarirayan R, Basel L, Moss C, Roth J, David A, De Paepe A, Coucke P, Mortier GR (2006) Germline *LEMD3* mutations are rare in sporadic patients with isolated melorheostosis. Hum Mutat 27:290

199. Hellemans J, Preobrazhenska O, Willaert A, Debeer P, Verdonk PC, Costa T, Janssens K, Menten B, Van Roy N, Vermeulen SJ, Savarirayan R, Van Hul W, Vanhoenacker F, Huylebroeck D, De Paepe A, Naeyaert JM, Vandesompele J, Speleman F, Verschueren K, Coucke PJ, Mortier GR (2004) Loss-of-function mutations in LEMD3 result in osteopoikilosis, Buschke-Ollendorff syndrome and melorheostosis. Nat Genet 36:1213–1218
200. Henderson CA, Ruban E, Porter DI (1997) Multiple leiomyomata presenting in a child. Pediatr Dermatol 14:287–289
201. Hendriks YM, Verhallen JT, van der Smagt JJ, Kant SG, Hilhorst Y, Hoefsloot L, Hansson KB, van der Straaten PJ, Boutkan H, Breuning MH, Vasen HF, Brocker-Vriends AH (2003) Bannayan-Riley-Ruvalcaba syndrome: further delineation of the phenotype and management of *PTEN* mutation-positive cases. Fam Cancer 2:79–85
202. Herman AR, Busam KJ, Greenberg RA, Nehal KS (2003) Multiple infundibulocystic basal cell carcinomas: case report of unique unilateral presentation. Dermatol Surg 29:436–439
203. Hershkovitz D, Bercovich D, Sprecher E, Lapidot M (2008) *RASA1* mutations may cause hereditary capillary malformations without arteriovenous malformations. Br J Dermatol 158:1035–1040
204. Hershkovitz D, Bergman R, Sprecher E (2008) A novel mutation in *RASA1* causes capillary malformation and limb enlargement. Arch Dermatol Res 300:385–388
205. Hoekzema R, Zonneveld IM, van der Wal AC (2010) Type 2 segmental glomangiomas. Dermatol Online J 16:8
206. Hong JB, Hsiao CH, Chu CY (2009) Systematized linear porokeratosis: a rare variant of diffuse porokeratosis with good response to systemic acitretin. J Am Acad Dermatol 60:713–715
207. Hornstein O (1965) [Congenital bullous ichthyosiform erythroderma (Brocq)]. Dermatol Wochenschr 151:1255–1265
208. Hornstein OP (1976) Generalized dermal perifollicular fibromas with polyps of the colon. Hum Genet 33:193–197
209. Hornstein OP, Knickenberg M (1975) Perifollicular fibromatosis cutis with polyps of the colon: a cutaneo-intestinal syndrome sui generis. Arch Dermatol Res 253:161–175
210. Hornstein OP, Knickenberg M, Mörl M (1976) Multiple dermal perifollicular fibromas with polyps of the colon – report of a peculiar clinical syndrome. Acta Hepatogastroenterol (Stuttg) 23:53–58
211. Huilgol SC, Griffiths WA, Black MM (1994) Familial juvenile elastoma. Australas J Dermatol 35:87–90
212. Hulsebos TJ, Kenter SB, Jakobs ME, Baas F, Chong B, Delatycki MB (2010) *SMARCB1/INI1* maternal germ line mosaicism in schwannomatosis. Clin Genet 77:86–91
213. Hulsebos TJ, Plomp AS, Wolterman RA, Robanus-Maandag EC, Baas F, Wesseling P (2007) Germline mutation of *INI1/SMARCB1* in familial schwannomatosis. Am J Hum Genet 80:805–810
214. Hunt SJ, Sharra WG, Abell E (1991) Linear and punctate porokeratosis associated with end-stage liver disease. J Am Acad Dermatol 25:937–939
215. Huson SM, Harper PS, Compston DA (1988) Von Recklinghausen neurofibromatosis: a clinical and population study in south-east Wales. Brain 111:1355–1381
216. Huson SM, Ruggieri M (2006) The neurofibromatoses. In: Harper J, Oranje AP, Prose N (eds) Textbook of pediatric dermatology. Blackwell Publishing, Malden, pp 1467–1490
217. Huter E, Wortham NC, Hartschuh W, Enk A, Jappe U (2008) Single base mutation in the fumarate hydratase gene leading to segmental cutaneous leiomyomatosis. Acta Derm Venereol 88:63–65
218. Hwang LY, Lee JB, Richard G, Uitto JJ, Hsu S (2003) Type 1 segmental manifestation of Hailey-Hailey disease. J Am Acad Dermatol 49:712–714
219. Iacobas I, Burrows PE, Adams DM, Sutton VR, Hollier LH, Chintagumpala MM (2011) Oral rapamycin in the treatment of patients with hamartoma syndromes and *PTEN* mutation. Pediatr Blood Cancer 57:321–323
220. Ireland M, Donnai D, Burn J (1993) Brachmann-de Lange syndrome: delineation of the clinical phenotype. Am J Med Genet 47:959–964
221. Itin PH, Buechner SA (1999) Segmental forms of autosomal dominant skin disorders: the puzzle of mosaicism. Am J Med Genet 85:351–354
222. Izraeli S, Metzker A, Horev G, Karmi D, Merlob P, Farfel Z (1992) Albright hereditary osteodystrophy with hypothyroidism, normocalcemia, and normal Gs protein activity: a family presenting with congenital osteoma cutis. Am J Med Genet 43:764–767
223. Jacobi H, Härtel SL (1996) [Congenital familial plaque-shaped glomus tumors: an unusual variant of multiple regional glomus tumors]. Hautarzt 47:387–390
224. Jansen LH, Driessen FM (1958) Leiomyoma cutis. Br J Dermatol 70:446–451
225. Jassal DS, Qureshi A, Neilan TG, Nomura C, Wittram C, King ME (2006) Pulmonary arteriovenous malformations in hereditary hemorrhagic telangiectasia: an echocardiographic perspective. J Am Soc Echocardiogr 19:229.e225–229.e227
226. Jeong JS, Lee JY, Yoon TY (2011) Unilateral nevoid acanthosis nigricans with a submammary location. Ann Dermatol 23:95–97
227. Jeong KH, Lew BL, Sim WY (2009) Osteoma cutis as the presenting feature of Albright hereditary osteodystrophy associated with pseudopseudohypoparathyroidism. Ann Dermatol 21:154–158
228. Ji Y, Wang X, Lin W, Chen S (2011) Surgical treament of giant plexiform neurofibroma associated with pectus excavatum. J Cardiothorac Surg 6:119

229. Jih MH (2002) Linear unilateral basal-cell nevus syndrome with comedones. Dermatol Online J 8:12
230. John AM, Ruggieri M, Ferner R, Upadhyaya M (2000) A search for evidence of somatic mutations in the NF1 gene. J Med Genet 37:44–49
231. Jokiaho I, Salo A, Niemi KM, Blomstedt GC, Pihkala J (1989) Deletion 3q27→3qter in an infant with mild dysmorphism, parietal meningocele, and neonatal miliaria rubra-like lesions. Hum Genet 83:302–304
232. Jouffrey C, Batard ML, Martin de Lassalle E, Brevière GM, Thomas P, Piette F (2000) Une glomangiomatose segmentaire de type 2. Ann Dermatol Venereol 127:4S85–4S86
233. Jung EG (1988) Segmental neurofibromatosis (NF-5). Neurofibromatosis 1:306–311
234. Jüppner H (2002) The genetic basis of progressive osseous heteroplasia. N Engl J Med 346:128–130
235. Jurenka SB, Van Allen MI (1995) Mixed sclerosing bone dysplasia, small stature, seizure disorder, and mental retardation: a syndrome? Am J Med Genet 57:6–9
236. Kantaputra PN, van den Ouweland A, Sangruchi T, Limwongse C (2012) Severe plexiform facial neurofibromatosis, type 1 with underdeveloped eyes and a novel *NF1* mutation. Am J Med Genet A 158A:1750–1753
237. Kaplan FS, Craver R, MacEwen GD, Gannon FH, Finkel G, Hahn G, Tabas J, Gardner RJ, Zasloff MA (1994) Progressive osseous heteroplasia: a distinct developmental disorder of heterotopic ossification: two new case reports and follow-up of three previously reported cases. J Bone Joint Surg Am 76:425–436
238. Karadaglić DL, Berger S, Janković D, Stefanović Z (1988) Zosteriform porokeratosis of Mibelli. Int J Dermatol 27:589–590
239. Karrer S, Szeimies RM, Hohenleutner U, Heine A, Landthaler M (1995) Unilateral localized basaliomatosis: treatment with topical photodynamic therapy after application of 5-aminolevulinic acid. Dermatology 190:218–222
240. Kern F (1972) Zosteriforme Leiomyomatose. Zeitschr Haut Geschlechtskr 47:536
241. Kikuchi I, Idemori M, Okazaki M (1979) Plaque type syringoma. J Dermatol 6:329–331
242. Kim SE, Heo EP, Yoon TJ, Kim TH (2002) Segmentally distributed neurofibromatosis associated with adenocarcinoma of the colon. J Dermatol 29:350–353
243. Kitoh A, Akiyama H, Yoshida Y, Tanaka T (2003) Multiple piloleiomyomas: do they follow dermatomes or Blaschko lines? J Dermatol 30:851–852
244. Klaassens M, Blom EW, Schrander JJ, Ris-Stalpers C, Nieuwenhuijzen Kruseman AC, van Steensel MA, Schrander-Stumpel CT (2010) Unique skin changes in a case of Albright hereditary osteodystrophy caused by a rare *GNAS1* mutation. Br J Dermatol 162:690–694
245. Klekowski N, Shwayder T (2011) Unilateral acrokeratoelastoidosis–second reported case. Pediatr Dermatol 28:20–22
246. Kluger N, Giraud S, Coupier I, Avril MF, Dereure O, Guillot B, Richard S, Bessis D (2010) Birt-Hogg-Dube syndrome: clinical and genetic studies of 10 French families. Br J Dermatol 162:527–537
247. Kluwe L, Friedrich RE, Mautner VF (1999) Allelic loss of the NF1 gene in NF1-associated plexiform neurofibromas. Cancer Genet Cytogenet 113:65–69
248. Knoth W, Kleinhans D, Boepple D, Lippelt C, Brucker H, Müller D (1974) Systematisierte Glomustumoren. Zeitschr Hautkr 49:189–190
249. Koeppel MC, Sayag J (1990) [Linear mastocytosis of the telangiectasia macularis eruptiva perstans type]. Ann Dermatol Venereol 117:109–111
250. Kohara Y, Takeo T, Oshima Y, Akita Y, Tamada Y, Watanabe D (2011) Linear porokeratosis with nail dystrophy. Eur J Dermatol 21:625–626
251. König A, Happle R (2000) Two cases of type 2 segmental manifestation in a family with cutaneous leiomyomatosis. Eur J Dermatol 10:590–592
252. König A, Hörster S, Vakilzadeh F, Happle R (2000) Type 2 segmental manifestation of Hailey-Hailey disease: poor therapeutic response to dermabrasion is due to severe involvement of adnexal structures. Eur J Dermatol 10:265–268
253. Korf BR (1999) Plexiform neurofibromas. Am J Med Genet 89:31–37
254. Kovalyshyn I, Spencer PS, Busam KJ, Marghoob AA (2009) Multiple tender papules in a 48-year-old man–quiz case. Arch Dermatol 145:829–834
255. Kretzmer E (1911) Ein Beitrag zur Kenntnis der multiplen Dermatomyome (Kutismyome). Arch Dermatol Syph 107:379–385
256. Krieg T, Landthaler M, Ihme A, Wolff HH, Braun-Falco O (1984) [Cutis hyperelastica circumscripta. A case report with a morphologic study]. Hautarzt 35:415–417
257. Krishnaram AS (1991) Unilateral nevoid acanthosis nigricans. Int J Dermatol 30:452–453
258. Kurek KC, Howard E, Tennant LB, Upton J, Alomari AI, Burrows PE, Chalache K, Harris DJ, Trenor CC 3rd, Eng C, Fishman SJ, Mulliken JB, Perez-Atayde AR, Kozakewich HP (2012) PTEN hamartoma of soft tissue: a distinctive lesion in PTEN syndromes. Am J Surg Pathol 36:671–687
259. Kwiatkowski D (2005) TSC1, TSC2, TSC3? Or mosaicism? Eur J Hum Genet 13:695–696
260. Lacour M (2011) Buschke-Ollendorff syndrome, Marfan syndrome, osteogenesis imperfecta, anetodermas and atrophodermas. In: Irvine AD, Hoeger PH, Yan AC (eds) Harper's textbook of pediatric dermatology. Wiley-Blackwell, Oxford. pp 145.1–145.21
261. Lane JE, Allen JH, Lane TN, Lesher JL Jr (2005) Unilateral basal cell carcinomas: an unusual entity

treated with photodynamic therapy. J Cutan Med Surg 9:336–340
262. Launonen V, Vierimaa O, Kiuru M, Isola J, Roth S, Pukkala E, Sistonen P, Herva R, Aaltonen LA (2001) Inherited susceptibility to uterine leiomyomas and renal cell cancer. Proc Natl Acad Sci U S A 98:3387–3392
263. Laymon CW, Peterson WC Jr (1965) Glomangioma (glomus tumor): a clinicopathologic study with special reference to multiple lesions appearing during pregnancy. Arch Dermatol 92:509–514
264. Lázaro C, Ravella A, Gaona A, Volpini V, Estivill X (1994) Neurofibromatosis type 1 due to germ-line mosaicism in a clinically normal father. N Engl J Med 331:1403–1407
265. Lee CH, Hsiao CH, Chiu HC, Tsai TF (2011) Acantholytic dyskeratotic epidermal nevus. Indian J Dermatol Venereol Leprol 77:253
266. Leverkus M, Kluwe L, Röll EM, Becker G, Bröcker EB, Mautner VF, Hamm H (2003) Multiple unilateral schwannomas: segmental neurofibromatosis type 2 or schwannomatosis? Br J Dermatol 148:804–809
267. Levi L, Beneggi M, Crippa D, Sala GP, Calcinati M, Albanese GC (1983) [Multiple eccrine spiradenomas in linear disposition]. G Ital Dermatol Venereol 118:207–210
268. Lim MO, Mukherjee AB, Hansen JW (1981) Dysplastic cutaneous osteomatosis: a unique case of true osteoma. Arch Dermatol 117:797–799
269. Lindhurst MJ, Sapp JC, Teer JK, Johnston JJ, Finn EM, Peters K, Turner J, Cannons JL, Bick D, Blakemore L, Blumhorst C, Brockmann K, Calder P, Cherman N, Deardorff MA, Everman DB, Golas G, Greenstein RM, Kato BM, Keppler-Noreuil KM, Kuznetsov SA, Miyamoto RT, Newman K, Ng D, O'Brien K, Rothenberg S, Schwartzentruber DJ, Singhal V, Tirabosco R, Upton J, Wientroub S, Zackai EH, Hoag K, Whitewood-Neal T, Robey PG, Schwartzberg PL, Darling TN, Tosi LL, Mullikin JC, Biesecker LG (2011) A mosaic activating mutation in AKT1 associated with the Proteus syndrome. N Engl J Med 365:611–619
270. Lindscheid KR, Zabel M (1990) [Zosteriform leiomyomatosis–successful treatment by iontophoresis with tap water]. Z Hautkr 65:923–926
271. Listernick R, Charrow J (2012) The neurofibromatoses. In: Goldsmith LA, Katz SI, Gilchrest BA, Paller AS, Leffell DJ, Wolff K (eds) Fitzpatrick's dermatology in general medicine, vol 2, 8th edn. McGraw-Hill, New York, pp 1680–1690
272. Litani C, Engel G, Piette WW (2009) Tumescent liposuction in the treatment of hemihyperplasia multiple lipomatosis syndrome. Dermatol Surg 35:1147–1151
273. Liu J, Krantz ID (2009) Cornelia de Lange syndrome, cohesin, and beyond. Clin Genet 76:303–314
274. Loffeld A, McLellan NJ, Cole T, Payne SJ, Fricker D, Moss C (2006) Epidermal naevus in Proteus syndrome showing loss of heterozygosity for an inherited *PTEN* mutation. Br J Dermatol 154:1194–1198
275. Löhrer R, Neumann-Acikel A, Eming R, Hartmann K, Rasokat H, Krieg T, Happle R, Eming S (2010) A case of linear porokeratosis superimposed on disseminated superficial actinic porokeratosis. Case Rep Dermatol 2:130–134
276. Longley BJ Jr, Metcalfe DD, Tharp M, Wang X, Tyrrell L, Lu SZ, Heitjan D, Ma Y (1999) Activating and dominant inactivating *c-KIT* catalytic domain mutations in distinct clinical forms of human mastocytosis. Proc Natl Acad Sci U S A 96:1609–1614
277. Lorette G, Fetissoff F, Grangeponte MC, Arbeille B, Despert F, Larrègue M, Laugier J (1984) Érythrodermie ichtyosiforme congénitale bulleuse chez une fille, naevus verruqeux épidermolytique chez son père. Ann Dermatol Venereol 111:858–859
278. Louste LF, Cailliau D (1933) Deuxième cas de dermatomyomes à disposition systématisée. Bull Soc Fr Dermtatol Syphiligr 40:699–702
279. Loveman AB, Gordon AM, Fliegelman MT (1963) Marfan's syndrome: some cutaneous aspects. Arch Dermatol 87:428–435
280. Lucker GP, Steijlen PM (1994) The coexistence of linear and giant porokeratosis associated with Bowen's disease. Dermatology 189:78–80
281. Lun KR, Spelman LJ (2000) Multiple piloleiomyomas. Australas J Dermatol 41:185–186
282. Lustberg H, Gagliardi J, Lawson J (1999) Digital enlargement in tuberous sclerosis. Skeletal Radiol 28:116–118
283. Machino H, Miki Y, Teramoto T, Shiraishi S, Sasaki MS (1984) Cytogenetic studies in a patient with porokeratosis of Mibelli, multiple cancers and a forme fruste of Werner's syndrome. Br J Dermatol 111:579–586
284. Macmillan AL, Roberts SO (1974) Porokeratosis of Mibelli after renal transplantation. Br J Dermatol 90:45–51
285. Maibach HI, Hogan MJ, Tromovitch T (1971) Unilateral "nevoid" basal cell epitheliomas and cataracts. Arch Dermatol 103:564
286. Maini A, Di Carlo A, Donati P, Sacerdoti G, Fazio M (1989) Tumori glomici multipli: osservazione di un caso. Chron Dermatol 20:79–82
287. Mallory SB, Enjolras O, Boon LM, Rogers E, Berk DR, Blei F, Baselga E, Ros AM, Vikkula M (2006) Congenital plaque-type glomuvenous malformations presenting in childhood. Arch Dermatol 142:892–896
288. Marini MA, Saponaro AE, Parra LSR (2004) Siringomas múltiples: forma localizada, variedad clínica "en placa, axilar unilateral". Arch Argent Dermatol 54:7–75
289. Marsh DJ, Kum JB, Lunetta KL, Bennett MJ, Gorlin RJ, Ahmed SF, Bodurtha J, Crowe C, Curtis MA, Dasouki M, Dunn T, Feit H, Geraghty MT, Graham JM Jr, Hodgson SV, Hunter A, Korf BR, Manchester D, Miesfeldt S, Murday VA, Nathanson KL, Parisi M, Pober B, Romano C, Eng C et al (1999) *PTEN* mutation spectrum and genotype-phenotype correlations in Bannayan-Riley-Ruvalcaba syndrome

suggest a single entity with Cowden syndrome. Hum Mol Genet 8:1461–1472
290. Marsh DJ, Trahair TN, Martin JL, Chee WY, Walker J, Kirk EP, Baxter RC, Marshall GM (2008) Rapamycin treatment for a child with germline *PTEN* mutation. Nat Clin Pract Oncol 5:357–361
291. Martinez W, Yebra MT, Arnal F, Casado M, Borbujo J (1992) Multiple linear cylindromas. J Am Acad Dermatol 26:821–824
292. Martínez-García S, Vera-Casaño A, Eloy-García Carrasco C, del Boz-González J, Martínez-Pilar L, Crespo-Erchiga V (2008) Elephantiasis neuromatosa in a patient with neurofibromatosis type 1. J Eur Acad Dermatol Venereol 22:103–105
293. Martinez-Mir A, Glaser B, Chuang GS, Horev L, Waldman A, Engler DE, Gordon D, Spelman LJ, Hatzibougias I, Green J, Christiano AM, Zlotogorski A (2003) Germline fumarate hydratase mutations in families with multiple cutaneous and uterine leiomyomata. J Invest Dermatol 121:741–744
294. Martinez-Mir A, Liu J, Gordon D, Weiner MS, Ahmad W, Fine JD, Ott J, Gilliam TC, Christiano AM (2002) EB simplex superficialis resulting from a mutation in the type VII collagen gene. J Invest Dermatol 118:547–549
295. Martorell-Calatayud A, Blanes M, Requena C, Botella-Estrada R, Guillén-Barona C (2009) [Generalized linear porokeratosis limited to one side of the body]. Actas Dermosifiliogr 100:911–913
296. Mashhood AA, Amjad M (2004) Unilateral tuberous sclerosis complex. J Coll Physicians Surg Pak 14:628–630
297. McCarron KF, Goldblum JR (1998) Plexiform neurofibroma with and without associated malignant peripheral nerve sheath tumor: a clinicopathologic and immunohistochemical analysis of 54 cases. Mod Pathol 11:612–617
298. McGrae JD Jr, Hashimoto K (1996) Unilateral facial angiofibromas–a segmental form of tuberous sclerosis. Br J Dermatol 134:727–730
299. McGrath JA, Bolling MC, Jonkman MF (2010) Lethal acantholytic epidermolysis bullosa. Dermatol Clin 28:131–135
300. McLean WH, Rugg EL, Lunny DP, Morley SM, Lane EB, Swensson O, Dopping-Hepenstal PJ, Griffiths WA, Eady RA, Higgins C et al (1995) Keratin 16 and keratin 17 mutations cause pachyonychia congenita. Nat Genet 9:273–278
301. Menko FH, van Steensel MA, Giraud S, Friis-Hansen L, Richard S, Ungari S, Nordenskjöld M, Hansen TV, Solly J, Maher ER (2009) Birt-Hogg-Dubé syndrome: diagnosis and management. Lancet Oncol 10:1199–1206
302. Menon B, Kumar GS (2010) Giant plexiform neurofibroma presenting with craniovertebral junction instability. J Neurosci Rural Pract 1:129–130
303. Merino Monje E, Guerra Tapia A, Moreno Izquierdo R, Barrio Fuertes A (1978) Tumores glómicos múltiples: forma regional y disseminada en un mismo enfermo. Actas Dermosifiliogr 78:393–398
304. Merks JH, de Vries LS, Zhou XP, Nikkels P, Barth PG, Eng C, Hennekam RC (2003) PTEN hamartoma tumour syndrome: variability of an entity. J Med Genet 40:e111
305. Mibelli V (1893) An uncommon form of keratodermia: "porokeratosis". In: Unna PG, Morris M, Leloir H, Duhring LA (eds) Internationaler Atlas seltener Hautkrankheiten. International atlas of rare skin diseases. Atlas international des maladies rares de la peau. Voss, Hamburg, pp 5–10
306. Miller RM, Sparkes RS (1977) Segmental neurofibromatosis. Arch Dermatol 113:837–838
307. Monrigal E, Gallot D, James I, Hameury F, Vanlieferinghen P, Guibaud L (2009) Venous malformation of the soft tissue associated with blue rubber bleb nevus syndrome: prenatal imaging and impact on postnatal management. Ultrasound Obstet Gynecol 34:730–732
308. Morrison JG, Jones EW, MacDonald DM (1977) Juvenile elastoma and osteopoikilosis (the Buschke–Ollendorff syndrome). Br J Dermatol 97:417–422
309. Morton CA, Shuttleworth D, Douglas WS (1995) Porokeratosis and Crohn's disease. J Am Acad Dermatol 32:894–897
310. Moulin G, Guillaud V, Ferrier E, Marcellin X (1988) [Unilateral basal cell epitheliomatosis]. Ann Dermatol Venereol 115:1188–1190
311. Moulin C, Petiot-Roland A, Chouvet B, Faure M, Claudy A (1996) [Unilateral acro-keratoelastoidosis]. Ann Dermatol Venereol 123:654–656
312. Mounayer C, Wassef M, Enjolras O, Boukobza M, Mulliken JB (2001) Facial "glomangiomas": large facial venous malformations with glomus cells. J Am Acad Dermatol 45:239–245
313. Mukhopadhyay AK (2004) Simultaneous occurrence of disseminated superficial, linear and hypertrophic verrucous forms of porokeratosis in a child. Indian J Dermatol Venereol Leprol 70:364–366
314. Mumm S, Wenkert D, Zhang X, McAlister WH, Mier RJ, Whyte MP (2007) Deactivating germline mutations in *LEMD3* cause osteopoikilosis and Buschke-Ollendorff syndrome, but not sporadic melorheostosis. J Bone Miner Res 22:243–250
315. Munoz C, Bobadilla F, Fuenzalida H, Goldner R, Sina B (2011) Congenital glomangioma of the breast: type 2 segmental manifestation. Int J Dermatol 50:346–349
316. Murata Y, Kumano K, Takai T (2001) Type 2 segmental manifestation of disseminated superficial porokeratosis showing a systematized pattern of involvement and pronounced cancer proneness. Eur J Dermatol 11:191–194
317. Nabai H, Mehregan AH (1979) Porokeratosis of Mibelli: a report of two unusual cases. Dermatologica 159:325–331
318. Nahorski MS, Reiman A, Lim DH, Nookala RK, Seabra L, Lu X, Fenton J, Boora U, Nordenskjöld M, Latif F, Hurst LD, Maher ER (2011) Birt Hogg-Dubé syndrome-associated *FLCN* mutations disrupt protein stability. Hum Mutat 32:921–929

319. Nakagawara G, Asano E, Kimura S, Akimoto R, Miyazaki I (1977) Blue rubber bleb nevus syndrome: report of a case. Dis Colon Rectum 20:421–427
320. Nasir AA, Abdur-Rahman LO, Ibrahim KO, Adegoke MA, Afolabi JK, Adeniran JO (2012) Genitourinary plexiform neurofibroma mimicking sacrococcygeal teratoma. J Surg Tech Case Rep 4:50–52
321. Nath AK, Kumari R, Thappa DM (2009) Eccrine spiradenoma with chondroid syringoma in Blaschkoid distribution. Indian J Dermatol Venereol Leprol 75:600–602
322. Natzel R (1954) Zur Klinik des Leiomyoma cutis. Hautarzt 5:324–325
323. Nazzaro V, Ermacora E, Santucci B, Caputo R (1990) Epidermolytic hyperkeratosis: generalized form in children from parents with systematized linear form. Br J Dermatol 122:417–422
324. Nevin NC, Thomas PS, Davis RI, Cowie GH (1999) Melorheostosis in a family with autosomal dominant osteopoikilosis. Am J Med Genet 82:409–414
325. Niimi Y, Kawana S (2009) Type 2 segmental manifestation of disseminated superficial actinic porokeratosis in a 7-year-old girl. Eur J Dermatol 19:25–28
326. Noda S, Lee JB, Li K, Schwartz L (2012) Unilateral telangiectasia macularis perstans with a clear demarcation in the midline of the trunk. Eur J Dermatol 22:407–408
327. Nomura K, Umeki K, Hatayama I, Kuronuma T (2001) Phenotypic heterogeneity in bullous congenital ichthyosiform erythroderma: possible somatic mosaicism for keratin gene mutation in the mildly affected mother of the proband. Arch Dermatol 137:1192–1195
328. Noto G, Bongiorno MR, Pravatà G, Aricò M (1994) Multiple nevoid spiradenomas. Am J Dermatopathol 16:280–284
329. Ogino A (1976) Linear eccrine poroma. Arch Dermatol 112:841–844
330. Oguzkan S, Cinbis M, Ayter S, Anlar B, Aysun S (2004) Familial segmental neurofibromatosis. J Child Neurol 19:392–394
331. Oh DH, Lane AT, Turk AE, Kohler S (1997) A young boy with a large hemifacial plaque with histopathologic features of trichoepithelioma. J Am Acad Dermatol 37:881–883
332. Oh SW, Lee JS, Kim MY, Kim SC (2007) *COL7A1* mutational analysis in Korean patients with dystrophic epidermolysis bullosa. Br J Dermatol 157:1260–1264
333. Onder M, Adişen E (2009) A new indication of botulinum toxin: leiomyoma-related pain. J Am Acad Dermatol 60:325–328
334. Online Mendelian Inheritance in Man (OMIM). Available from: http//www3.ncbi.nlm.nih.gov/Omim/. Accessed 3 Sept 2013
335. Oranje AP, Vuzevski VD, Kalis TJ, Arts WF, van Joost T, Stolz E (1985) Segmental neurofibromatosis. Br J Dermatol 112:107–112
336. Orloff MS, Eng C (2008) Genetic and phenotypic heterogeneity in the PTEN hamartoma tumour syndrome. Oncogene 27:5387–5397
337. Ortonne JP, Jeune R, Fulton R, Thivolet J (1982) Primary localized gigantism and tuberous sclerosis. Arch Dermatol 118:877–878
338. Ostrowski DM, Gilula LA (1992) Mixed sclerosing bone dystrophy presenting with upper extremity deformities: a case report and review of the literature. J Hand Surg Br 17:108–112
339. Pacifici R, Murphy WA, Teitelbaum SL, Whyte MP (1986) Mixed-sclerosing-bone-dystrophy: 42-year follow-up of a case reported as osteopetrosis. Calcif Tissue Int 38:175–185
340. Palit A, Inamadar AC, Yelikar BR (2004) Unilateral linear annular lesions in a child. Pediatr Dermatol 21:682–683
341. Paller AS, Syder AJ, Chan YM, Yu QC, Hutton E, Tadini G, Fuchs E (1994) Genetic and clinical mosaicism in a type of epidermal nevus. N Engl J Med 331:1408–1415
342. Parreira LM, Sipoli JM, Mercante AM, Orfali RL, Levites J (2009) Case for diagnosis: unilateral multiple piloleiomyoma. An Bras Dermatol 84:197–199
343. Pascual-Castroviejo I, Pascual-Pascual SI, Velázquez-Fragua R, Viaño J, López-Gutiérrez JC (2008) [Segmental neurofibromatosis in children: presentation of 43 patients]. Rev Neurol 47: 399–403
344. Patel N, Stranahan D, Fenske NA (2012) Unilateral telangiectasia macularis eruptiva perstans of the breast. Cutis 90:26–28
345. Patel RC, Zynger DL, Laskin WB (2009) Blue rubber bleb nevus syndrome: novel lymphangiomatosis-like growth pattern within the uterus and immunohistochemical analysis. Hum Pathol 40:413–417
346. Pearson IC, Cliff S (2003) Case 6: plaques extending in a linear pattern from left ankle to hip forming over a 2-year period. Diagnosis: linear porokeratosis with disseminated superficial porokeratosis erupting in pregnancy. Clin Exp Dermatol 28:345–346
347. Peterson WC Jr, Carlson CH, Singer L, Armstrong WD (1963) Analysis of ectopic bone in osteoma cutis. Arch Dermatol 88:540–545
348. Petit A, Lemarchand-Venencie F, Pinquier L, Lebbe C, Bourrat E (2012) [Nevoid acanthosis nigricans or RAVEN (rounded and velvety epidermal nevus): three cases]. Ann Dermatol Venereol 139:183–188
349. Pilarski R, Eng C (2004) Will the real Cowden syndrome please stand up (again)? Expanding mutational and clinical spectra of the PTEN hamartoma tumour syndrome. J Med Genet 41:323–326
350. Piñol Aguadé J, Grimalt F, Mascaró JM (1970) Relaciones entre la poroqueratosis superficial de Respighi y otros cuadros clínicos (hyperkeratosis lenticularis perstans, Flegel; poroqueratosis actínica superficial disseminada, Chernosky; dermatosis psoriasiforme disqueratósica, Hunter y Donald). Medicina Cutánea 4:429–436
351. Pivnick EK, Lobe TE, Fitch SJ, Riccardi VM (1997) Hair whorl as an indicator of a mediastinal plexiform neurofibroma. Pediatr Dermatol 14:196–198
352. Pivnick EK, Riccardi VM (2003) Neurofibromatoses. In: Freedberg IM, Eisen AZ, Wolff K, Austen KF,

Goldsmith LA, Katz SI (eds) Fitzpatrick's dermatology in general medicine, vol 2. McGraw-Hill, New York, pp 1825–1833
353. Poblete-Gutiérrez P, Rübben A, Merk HF, Frank J (2003) Unilateral facial telangiectases suggest type 1 segmental manifestation of Osler-Rendu-Weber syndrome in an 11-year-old boy. Eur J Dermatol 13:537–539
354. Poblete-Gutiérrez P, Wiederholt T, König A, Jugert FK, Marquardt Y, Rübben A, Merk HF, Happle R, Frank J (2004) Allelic loss underlies type 2 segmental Hailey-Hailey disease, providing molecular confirmation of a novel genetic concept. J Clin Invest 114:1467–1474
355. Porneuf M, Ducret JP, Bonvalet D, Perrin P, Duterque M (1993) [Segmental neurilemmomatosis]. Ann Dermatol Venereol 120:824–827
356. Porter AD, Seville RH (1952) Porokeratosis (Mibelli). In: Tenth international congress of dermatology. British Medical Association, London, pp 461–462
357. Price H (1958) Porokeratosis (Mibelli). Arch Dermatol 77:477–478
358. Rahbari H, Fazel Z, Mehregan AH (1995) Destructive facial porokeratosis. J Am Acad Dermatol 33:1049–1050
359. Rallis E, Ragiadakou D (2009) Giant plexiform neurofibroma in a patient with neurofibromatosis type I. Dermatol Online J 15:7
360. Rawlings CE 3rd, Wilkins RH, Cook WA, Burger PC (1987) Segmental neurofibromatosis. Neurosurgery 20:946–949
361. Razack EM, Natarajan M (1977) Ulcerative systematized porokeratosis (Mibelli). Arch Dermatol 113:1583–1584
362. Reddy BS, Thadeus J, Kumar SK, Jaishanker T, Garg BR (1997) Generalized epidermolytic hyperkeratosis in a child born to a parent with systematized epidermolytic linear epidermal nevus. Int J Dermatol 36:198–200
363. Renner R, Sticherling M (2005) [Familial occurrence of a type 2 segmental manifestation of cutaneous leiomyomatosis]. J Dtsch Dermatol Ges 3:695–699
364. Requena L, Farina MC, Robledo M, Sangueza OP, Sanchez E, Villanueva A, Marquina A, Tamarit R (1999) Multiple hereditary infundibulocystic basal cell carcinomas: a genodermatosis different from nevoid basal cell carcinoma syndrome. Arch Dermatol 135:1227–1235
365. Restano L, Cambiaghi S, Brusasco A, Tardini G, Caputo R (1999) A hyperkeratotic linear lesion in a girl with KID syndrome: a further example of early allelic loss? Eur J Dermatol 9:142–143
366. Revencu N, Boon LM, Mulliken JB, Enjolras O, Cordisco MR, Burrows PE, Clapuyt P, Hammer F, Dubois J, Baselga E, Brancati F, Carder R, Quintal JM, Dallapiccola B, Fischer G, Frieden IJ, Garzon M, Harper J, Johnson-Patel J, Labrèze C, Martorell L, Paltiel HJ, Pohl A, Prendiville J, Quere I, Siegel DH, Valente EM, Van Hagen A, Van Hest L, Vaux KK, Vicente A, Weibel L, Chitayat D, Vikkula M (2008) Parkes Weber syndrome, vein of Galen aneurysmal malformation, and other fast-flow vascular anomalies are caused by *RASA1* mutations. Hum Mutat 29:959–965
367. Riccardi VM (1984) Neurofibromatosis heterogeneity. J Am Acad Dermatol 10:518–519
368. Riccardi VM (1992) Neurofibromatosis: phenotype, natural history, and pathogenesis, 2nd edn. Johns Hopkins University Press, Baltimore
369. Rikihisa W, Kiryu H, Furue M (1999) Multiple piloleiomyomas associated with solitary angioleiomyoma. Eur J Dermatol 9:309–310
370. Rodríguez-Martín M, Sánchez González R, Sáez-Rodríguez M, García-Bustinduy M, Martín-Herrera A, Noda-Cabrera A (2009) An unusual case of congenital linear eccrine spiradenoma. Pediatr Dermatol 26:180–183
371. Rongioletti F, Semino MT, Rebora A (1996) Unilateral multiple plaque-like syringomas. Br J Dermatol 135:623–625
372. Rosenfeld SR, Kaplan FS (1995) Progressive osseous heteroplasia in male patients: two new case reports. Clin Orthop Relat Res 317:243–245
373. Roth RR, Martines R, James WD (1987) Segmental neurofibromatosis. Arch Dermatol 123:917–920
374. Rubenstein AE, Bader JL, Aron AA, Wallace S (1983) Familial transmission of segmental neurofibromatosis. Neurology 33:76
375. Ruggieri M (2001) Mosaic (segmental) neurofibromatosis type 1 (NF1) and type 2 (NF2): no longer neurofibromatosis type 5 (NF5). Am J Med Genet 101:178–180
376. Ruggieri M, Huson SM (2001) The clinical and diagnostic implications of mosaicism in the neurofibromatoses. Neurology 56:1433–1443
377. Ruggieri M, Upadhyaya M, Di Rocco C, Gabriele A, Pascual-Castroviejo I (2008) Neurofibromatosis type 1 & related disorders. In: Ruggieri M, Pascual-Castroviejo I, Di Rocco C (eds) Neurocutaneous disorders: phacomatoses and hamartoneoplastic syndromes. Springer, Wien, pp 51–151
378. Ruiz-Villaverde R, Blasco Melguizo J, Menéndez García Estrada AC, Jiménez Cortes MC, Díez García F (2004) [Unilateral type 1 segmental Darier disease]. An Pediatr (Barc) 60:92–94
379. Rushton MA (1956) Some less common bone lesions affecting the jaws. Oral Surg Oral Med Oral Pathol 9:284–304
380. Sahoo B, Handa S, Kumar B (2000) Tuberous sclerosis with macrodactyly. Pediatr Dermatol 17: 463–465
381. Sakuntabhai A, Dhitavat J, Burge S, Hovnanian A (2000) Mosaicism for *ATP2A2* mutations causes segmental Darier's disease. J Invest Dermatol 115:1144–1147
382. Sala GP, Marinaro P, Rossi E, Nicoletti A, Bramati SP, Galbiati G (1990) [Unilateral nevoid syringoma]. Hautarzt 41:272–273
383. Salvitti M, Celestino F, Gerocarni Nappo S, Caione P (2013) Diffuse ganglioneuromatosis and plexiform neurofibroma of the urinary bladder: an uncommon

cause of severe urological disease in an infant. J Pediatr Urol 9:e131–e133
384. Sarma N, Boler AK, Bhattacharya SR (2009) Familial disseminated plaque type porokeratosis with multiple horns and squamous cell carcinoma involving anal skin. Indian J Dermatol Venereol Leprol 75:551
385. Saul RA, Stevenson RG (1984) Segmental neurofibromatosis: a distinct type of neurofibromatosis. Proc Greenwood Genet Cent 3:3–6
386. Savage J (1964) Porokeratosis (Mibelli) and carcinoma. Br J Dermatol 76:489
387. Savage J, Lederer H (1951) Porokeratosis (Mibelli); a case with extensive lesions; results of animal inoculation. Br J Dermatol 63:187–192
388. Schimmel RJ, Pasmans SG, Xu M, Stadhouders-Keet SA, Shore EM, Kaplan FS, Wulffraat NM (2010) *GNAS*-associated disorders of cutaneous ossification: two different clinical presentations. Bone 46:868–872
389. Schirren CG, Wörle B, Kind P, Plewig G (1995) A nevoid plaque with histological changes of trichoepithelioma and cylindroma in Brooke-Spiegler syndrome: an immunohistochemical study with cytokeratins. J Cutan Pathol 22:563–569
390. Schmidt-Holz M (1981) [Leiomyomatosis cutis with paroxysmal pain]. Zeitschr Hautkr 56:1342–1351
391. Scholl OK (1921) Ein halbseitig lokalisierter Fall von Porokeratosis Mibelli. Dermatol Wochenschr 72:1–4
392. Schorr WF, Optiz JM, Reyes CN (1972) The connective tissue nevus-osteopoikilosis syndrome. Arch Dermatol 106:208–214
393. Schulte TL, Liljenqvist U, Görgens H, Hackenberg L, Bullmann V, Tinschert S (2008) Hemihyperplasia-multiple lipomatosis syndrome (HHML): a challenge in spinal care. Acta Orthop Belg 74:714–719
394. Scola N, Skrygan M, Wieland U, Kreuter A, Gambichler T (2012) Altered gene expression in squamous cell carcinoma arising from congenital unilateral linear porokeratosis. Clin Exp Dermatol 37:781–785
395. Segal R (1993) Segmental neurofibromatosis of the sciatic nerve: case report. Neurosurgery 33:948
396. Sertznig P, Wahl R, Mühlhoff C, Megahed M (2009) [Congenital unilateral linear porokeratosis]. Hautarzt 60:949–952
397. Sestini R, Bacci C, Provenzano A, Genuardi M, Papi L (2008) Evidence of a four-hit mechanism involving *SMARCB1* and *NF2* in schwannomatosis-associated schwannomas. Hum Mutat 29:227–231
398. Sethuraman G, Malhotra AK, Khaitan BK, Kumar R, Sharma VK, Kabra M, Singh MK (2006) Osteoma cutis in pseudohypoparathyroidism. Clin Exp Dermatol 31:225–227
399. Shah F, Lester R, Moss C (2001) Buschke-Ollendorff syndrome with associated limb reduction. Br J Dermatol 145:122–129
400. Sharma VK, Khandpur S, Khanna N (2004) An interesting case of unilateral angiofibromas successfully treated with pulsed dye laser. J Eur Acad Dermatol Venereol 18:641–642
401. Sharma S, Sankhyan N, Gulati S, Kumar A, Srinivas M, Shukla B, Mathur SR (2011) Macrodactyly and fibrous hamartoma in a child with tuberous sclerosis complex. J Child Neurol 26:95–98
402. Sharma C, Shekhar S, Sharma M, Rane SU, Aggarwal T (2012) Pre-pubertal isolated plexiform neurofibroma of labium majus without clitoral involvement. Acta Obstet Gynecol Scand 91:1000
403. Sharpe GR, Cox NH (1990) Unilateral naevoid basal-cell-carcinoma syndrome: an individually controlled study of fibroblast sensitivity to radiation. Clin Exp Dermatol 15:352–355
404. Shelley WB, Rawnsley HM, Beerman H (1969) Quadrant distribution of basal cell nevi. Arch Dermatol 100:741–743
405. Shelley WB, Wood MG (1980) A zosteriform network of spiradenomas. J Am Acad Dermatol 2:59–61
406. Shin AY, Garay AA (1997) Unilateral insensate macrodactyly secondary to tuberous sclerosis in a child. Am J Orthop (Belle Mead NJ) 26:30–32
407. Shore EM, Ahn J, Jan de Beur S, Li M, Xu M, Gardner RJ, Zasloff MA, Whyte MP, Levine MA, Kaplan FS (2002) Paternally inherited inactivating mutations of the GNAS1 gene in progressive osseous heteroplasia. N Engl J Med 346:99–106
408. Sidwell RU, Francis N, Grahame R, Pope FM, Bunker CB (2003) Connective tissue naevus (collagenoma) in a patient with benign joint hypermobility syndrome (Ehlers-Danlos syndrome type III). Clin Exp Dermatol 28:323–325
409. Sifaki MK, Krueger-Krasagakis S, Koutsopoulos A, Evangelou GI, Tosca AD (2009) Botulinum toxin type A–treatment of a patient with multiple cutaneous piloleiomyomas. Dermatology 218:44–47
410. Silvestre JF, Bañuls J, Ramón R, Guijarro J, Botella R, Betlloch I (2000) Unilateral multiple facial angiofibromas: a mosaic form of tuberous sclerosis. J Am Acad Dermatol 43:127–129
411. Simpson JR (1974) Proceedings: giant glomangioma. Br J Dermatol 90:229–231
412. Sinha S, Schwartz RA (2007) Juvenile acanthosis nigricans. J Am Acad Dermatol 57:502–508
413. Smith CG, Glaser DA, Leonardi C (1998) Zosteriform multiple leiomyomas. J Am Acad Dermatol 38:272–273
414. Smith JM, Kirk EP, Theodosopoulos G, Marshall GM, Walker J, Rogers M, Field M, Brereton JJ, Marsh DJ (2002) Germline mutation of the tumour suppressor *PTEN* in Proteus syndrome. J Med Genet 39:937–940
415. Smith AD, Waisman M (1960) Connective tissue nevi; familial occurrence and association with osteopoikilosis. Arch Dermatol 81:249–252
416. Soffa DJ, Sire DJ, Dodson JH (1975) Melorheostosis with linear sclerodermatous skin changes. Radiology 114:577–578
417. Sol-Church K, Stabley DL, Demmer LA, Agbulos A, Lin AE, Smoot L, Nicholson L, Gripp KW (2009) Male-to-male transmission of Costello syndrome: G12S *HRAS* germline mutation inherited from a

father with somatic mosaicism. Am J Med Genet A 149A:315–321

418. Sprecher E, Indelman M, Khamaysi Z, Lugassy J, Petronius D, Bergman R (2007) Galli-Galli disease is an acantholytic variant of Dowling-Degos disease. Br J Dermatol 156:572–574
419. Starink TM, Kisch LS, Meijer CJ (1985) Familial multiple trichodiscomas: a clinicopathologic study. Arch Dermatol 121:888–891
420. Starink TM, Woerdeman MJ (1981) Unilateral systematized keratosis follicularis: a variant of Darier's disease or an epidermal naevus (acantholytic dyskeratotic epidermal naevus)? Br J Dermatol 105:207–214
421. Steinmann K, Kluwe L, Friedrich RE, Mautner VF, Cooper DN, Kehrer-Sawatzki H (2009) Mechanisms of loss of heterozygosity in neurofibromatosis type 1-associated plexiform neurofibromas. J Invest Dermatol 129:615–621
422. Stoof TJ, Starink TM (1999) Multipele basaloïde folliculaire hamartomen. Ned Tijdschr Dermatol Venereol 9:131–132
423. Suárez-Amor O, Pereiro-Ferreirós M, Ginarte M, Peteiro C, Toribio J (2007) Coexistence of linear porokeratosis and disseminated superficial actinic porokeratosis: a type 2 segmental manifestation. Acta Derm Venereol 87:363–364
424. Suh DH, Lee HS, Kim SD, Cho KH, Kim KH, Park KC (2000) Coexistence of disseminated superficial porokeratosis in childhood with congenital linear porokeratosis. Pediatr Dermatol 17:466–468
425. Takiguchi PS, Ratz JL (1984) Bilateral dermatomal neurofibromatosis. J Am Acad Dermatol 10:451–453
426. Tan WH, Baris HN, Burrows PE, Robson CD, Alomari AI, Mulliken JB, Fishman SJ, Irons MB (2007) The spectrum of vascular anomalies in patients with *PTEN* mutations: implications for diagnosis and management. J Med Genet 44:594–602
427. Tanaka M, Hara M, Hosokawa M, Miyazawa T, Tagami H (1991) Linear cutaneous neurilemmomas on the forehead. Clin Exp Dermatol 16:247–249
428. Tatnall FM, Leigh IM (1990) Facial leiomyomas. Clin Exp Dermatol 15:296–297
429. Tay YK, Ong BH (1999) Linear warty lesions in a child: linear porokeratosis. Arch Dermatol 135:1544–1545, 1547–1548
430. Tejedor M, Martín-Santiago A, Gómez C, Fiol M, Benítez-Segura I (2010) Congenital plaque-type glomuvenous malformation associated with chylous ascites. Pediatr Dermatol 27:673–675
431. Ter Poorten MC, Barrett K, Cook J (2003) Familial eccrine spiradenoma: a case report and review of the literature. Dermatol Surg 29:411–414
432. Thammaiah S, Manjunath M, Rao K, Uma DH (2011) Intraoral plexiform neurofibroma involving the maxilla – pathognomonic of neurofibromatosis type 1. J Pediatr Neurosci 6:65–68
433. Tijdens EF, Ruiter M (1949) Über "Osteosis cutis". Acta Derm Venereol 29:140–153
434. Tinschert S, Naumann I, Stegmann E, Buske A, Kaufmann D, Thiel G, Jenne DE (2000) Segmental neurofibromatosis is caused by somatic mutation of the neurofibromatosis type 1 (*NF1*) gene. Eur J Hum Genet 8:455–459
435. Titeux M, Mendonça V, Décha A, Moreira E, Magina S, Maia A, Lacaze-Buzy L, Mejia JE, Torrão L, Carvalho F, Eça-Guimarães J, Hovnanian A (2009) Keratitis-ichthyosis-deafness syndrome caused by *GJB2* maternal mosaicism. J Invest Dermatol 129:776–779
436. Tok Celebi J, Chen FF, Zhang H, Ping XL, Tsou HC, Peacocke M (1999) Identification of *PTEN* mutations in five families with Bannayan-Zonana syndrome. Exp Dermatol 8:134–139
437. Tomlinson IP, Alam NA, Rowan AJ, Barclay E, Jaeger EE, Kelsell D, Leigh I, Gorman P, Lamlum H, Rahman S, Roylance RR, Olpin S, Bevan S, Barker K, Hearle N, Houlston RS, Kiuru M, Lehtonen R, Karhu A, Vilkki S, Laiho P, Eklund C, Vierimaa O, Aittomaki K, Hietala M, Sistonen P, Paetau A, Salovaara R, Herva R, Launonen V, Aaltonen LA (2002) Germline mutations in *FH* predispose to dominantly inherited uterine fibroids, skin leiomyomata and papillary renal cell cancer. Nat Genet 30:406–410
438. Torchia D, Palleschi GM, Urso C (2007) Unilateral glomuvenous malformations. J Eur Acad Dermatol Venereol 21:552–553
439. Toro JR, Glenn G, Duray P, Darling T, Weirich G, Zbar B, Linehan M, Turner ML (1999) Birt-Hogg-Dubé syndrome: a novel marker of kidney neoplasia. Arch Dermatol 135:1195–1202
440. Torrelo A, Hadj-Rabia S, Colmenero I, Piston R, Sybert VP, Hilari-Carbonell H, Hernández-Martín A, Ferreres JC, Vañó-Galvan S, Azorín D, de Salamanca JE, Requena L, Bodemer C, Happle R, Garcia-Patos V, Fraitag S (2012) Folliculocystic and collagen hamartoma of tuberous sclerosis complex. J Am Acad Dermatol 66:617–621
441. Torrelo A, Hernández-Martín A, Bueno E, Colmenero I, Rivera I, Requena L, Happle R, González-Sarmiento R (2013) Molecular evidence of type 2 mosaicism in Gorlin syndrome. Br J Dermatol [Epub ahead of print] doi: 10.1111/bjd.12458
442. Traistaru R, Enachescu V, Manuc D, Gruia C, Ghilusi M (2008) Multiple right schwannoma. Rom J Morphol Embryol 49:235–239
443. Trattner A, David M, Hodak E, Ben-David E, Sandbank M (1990) Segmental neurofibromatosis. J Am Acad Dermatol 23:866–869
444. Trattner A, David M, Rothem A, Ben-David E, Sandbank M (1991) Buschke-Ollendorff syndrome of the scalp: histologic and ultrastructural findings. J Am Acad Dermatol 24:822–824
445. Trauner MA, Ruben BS, Lynch PJ (2003) Segmental tuberous sclerosis presenting as unilateral facial angiofibromas. J Am Acad Dermatol 49:S164–S166
446. Tresserra L, Tresserra F, Grases PJ, Badosa J, Tresserra M (1998) Congenital plate-like osteoma cutis of the forehead: an atypical presentation form. J Craniomaxillofac Surg 26:102–106
447. Truffi M (1905) Sur un cas de porokératose systématisée. Ann Dermatol Syphiligr (Paris) 6:521–543

448. Tsoitis G, Kanitakis J, Papadimitriou C, Hatzibougias Y, Asvesti K, Happle R (2001) Cutaneous leiomyomatosis with type 2 segmental involvement. J Dermatol 28:251–255
449. Tsukamoto S, Kido A, Honoki K, Akahane M, Mii Y, Tanaka Y (2010) Type 1 neurofibromatosis with a giant intrathoracic lesion: a case report with 25 years of follow-up. Pathol Res Pract 206:408–410
450. Tsur H, Lipskier E, Fisher BK (1981) Multiple linear spiradenomas. Plast Reconstr Surg 68:100–102
451. Tung HE, Shih SL (2009) Tuberous sclerosis with rare presentation of macrodactyly. Pediatr Radiol 39:878
452. Uhlin SR (1980) Segmental neurofibromatosis. South Med J 73:526–527
453. Urtizberea JA, Testart H, Cartault F, Boccon-Gibod L, Le Merrer M, Kaplan FS (1998) Progressive osseous heteroplasia: report of a family. J Bone Joint Surg Br 80:768–771
454. Vakilzadeh F, Kolde G (1985) Relapsing linear acantholytic dermatosis. Br J Dermatol 112:349–355
455. van Meekeren J (1682) Observationes Medico-Chirurgicae (trans: ex Belgico in latinum translatae ab Abrahamo Blasio MS). Ex Officinâ Henrici & Viduae Theodori Boom, Amstelodami
456. Vandenbroucke I, van Doorn R, Callens T, Cobben JM, Starink TM, Messiaen L (2004) Genetic and clinical mosaicism in a patient with neurofibromatosis type 1. Hum Genet 114:284–290
457. Verbov J (1977) Buschke–Ollendorff syndrome (disseminated dermatofibrosis with osteopoikilosis). Br J Dermatol 96:87–90
458. Verhoef S, Bakker L, Tempelaars AM, Hesseling-Janssen AL, Mazurczak T, Jozwiak S, Fois A, Bartalini G, Zonnenberg BA, van Essen AJ, Lindhout D, Halley DJ, van den Ouweland AM (1999) High rate of mosaicism in tuberous sclerosis complex. Am J Hum Genet 64:1632–1637
459. Verhoef S, Vrtel R, van Essen T, Bakker L, Sikkens E, Halley D, Lindhout D, van den Ouweland A (1995) Somatic mosaicism and clinical variation in tuberous sclerosis complex. Lancet 345:202
460. Verma SB (2009) A rare case of porokeratosis ptychotropica and coexistent linear porokeratosis in a 10-year-old boy. Clin Exp Dermatol 34:e501–e502
461. Vero F (1945) Disseminated congenital osteomas of the skin. J Am Med Ass 129:728
462. Viscasillas XB, Puigdemont GS, Salvador CA (2006) A slowly enlarging, unilateral, erythematous macular lesion. Arch Dermatol 142:641–646
463. Wach F, Hein R, Eckert F, Hohenleutner U, Landthaler M (1993) Multiple Leiomyome der Haut in streifenförmiger Anordnung: Therapie mit dem CO_2-Laser. Zeitschr Haut Geschlechtskr 68:168–172
464. Wada T, Shirakata Y, Takahashi H, Murakami S, Iizuka H, Suzuki H, Hashimoto K (2003) A Japanese case of segmental Darier's disease caused by mosaicism for the *ATP2A2* mutation. Br J Dermatol 149:185–188
465. Waite KA, Eng C (2002) Protean *PTEN*: form and function. Am J Hum Genet 70:829–844
466. Walker GF (1964) Mixed sclerosing bone dystrophies. Two case reports. J Bone Joint Surg Br 46:546–552
467. Ward S, Sugo E, Verge CF, Wargon O (2011) Three cases of osteoma cutis occurring in infancy: a brief overview of osteoma cutis and its association with pseudo-pseudohypoparathyroidism. Australas J Dermatol 52:127–131
468. Watabe H, Kashima M, Baba T, Mizoguchi M (2000) A case of unilateral dermatomal cavernous haemangiomatosis. Br J Dermatol 143:888–891
469. Webb DW, Clarke A, Fryer A, Osborne JP (1996) The cutaneous features of tuberous sclerosis: a population study. Br J Dermatol 135:1–5
470. Wedgwood D, Skene-Smith HS (1973) Tuberose sclerosis with unilateral involvement of the maxilla and mandible: a case report. Br J Oral Surg 11:126–130
471. Weidman FD, Anderson NP, Ayres S (1933) Juvenile elastoma. Arch Dermatol Syphilol 28:182–189
472. Weintraub R, Pinkus H (1977) Multiple fibrofolliculomas (Birt-Hogg-Dubé) associated with a large connective tissue nevus. J Cutan Pathol 4:289–299
473. Weleber RG, Zonana J (1983) Iris hamartomas (Lisch nodules) in a case of segmental neurofibromatosis. Am J Ophthalmol 96:740–743
474. White JM, Calonje E (2008) Painful red spots on the cheek of a young man–quiz case. Nevoidal pilar leiomyoma. Arch Dermatol 144:1217–1222
475. White JM, Short K, Salisbury JR, Fuller LC (2006) A novel case of linear syringomatous hamartoma. Clin Exp Dermatol 31:222–224
476. Whyte MP, Murphy WA, Fallon MD, Hahn TJ (1981) Mixed-sclerosing-bone-dystrophy: report of a case and review of the literature. Skeletal Radiol 6:95–102
477. Wilkin JK (1980) Unilateral dermatomal cavernous hemangiomatosis. Dermatologica 161:347–354
478. Wilms NA, Douglass MC (1981) An unusual case of preponderantly right-sided syringomas. Arch Dermatol 117:308
479. Wright S, Ryan J (1990) Multiple familial eccrine spiradenoma with cylindroma. Acta Derm Venereol 70:79–82
480. Wu M, Wang Y, Zhang D, Jia G, Bu W, Fang F, Zhao L (2011) A case of giant primary osteoma cutis successfully treated with tissue expansion and surgical excision. Indian J Dermatol Venereol Leprol 77:79–81
481. Yadegari M, Whyte MP, Mumm S, Phelps RG, Shanske A, Totty WG, Cohen SR (2010) Buschke-Ollendorff syndrome: absence of *LEMD3* mutation in an affected family. Arch Dermatol 146:63–68
482. Yang CC, Happle R, Chao SC, Yu-Yun Lee J, Chen W (2008) Giant café-au-lait macule in neurofibromatosis 1: a type 2 segmental manifestation of neurofibromatosis 1? J Am Acad Dermatol 58:493–497

483. Yang KH, Rhim H, Cho OK, Ko BH, Kim Y, Lee HW, Hong EK (2002) Segmental colonic involvement of plexiform neurofibroma in neurofibromatosis type 1. J Comput Assist Tomogr 26:129–131
484. Yeh GL, Mathur S, Wivel A, Li M, Gannon FH, Ulied A, Audi L, Olmstead EA, Kaplan FS, Shore EM (2000) *GNAS1* mutation and *Cbfa1* misexpression in a child with severe congenital platelike osteoma cutis. J Bone Miner Res 15:2063–2073
485. Yen A, Raimer SS (1996) Multiple painful blue nodules. Multiple glomus tumors (glomangiomas). Arch Dermatol 132:704–705, 707–708
486. Yoon TY, Lee HT, Chang SH (1999) Giant congenital multiple patch-like glomus tumors. J Am Acad Dermatol 40:826–828
487. Yoshida A, Takahashi K, Maeda F, Akasaka T (2010) Multiple vascular eccrine spiradenomas: a case report and published work review of multiple eccrine spiradenomas. J Dermatol 37: 990–994
488. Youn M, Hann SK, Moon TK, Lee MG (1998) Acantholytic dyskeratotic epidermal nevus induced by ultraviolet B radiation. J Am Acad Dermatol 39:301–304
489. Yung CW, Soltani K, Bernstein JE, Lorincz AL (1981) Unilateral linear nevoidal syringoma. J Am Acad Dermatol 4:412–416
490. Zankl A, Rampa A, Schinzel A (2003) Brachmann-de Lange syndrome (BDLS) with asymmetry and skin pigmentary anomalies: a result of mosaicism for a putative *BDLS* gene mutation? Am J Med Genet A 118A:358–361
491. Zeller J, Friedmann D, Clerici T, Revuz J (1995) The significance of a single periungual fibroma: report of seven cases. Arch Dermatol 131: 1465–1466
492. Zhou X, Hampel H, Thiele H, Gorlin RJ, Hennekam RC, Parisi M, Winter RM, Eng C (2001) Association of germline mutation in the PTEN tumour suppressor gene and Proteus and Proteus-like syndromes. Lancet 358:210–211
493. Zhou XP, Marsh DJ, Hampel H, Mulliken JB, Gimm O, Eng C (2000) Germline and germline mosaic *PTEN* mutations associated with a Proteus-like syndrome of hemihypertrophy, lower limb asymmetry, arteriovenous malformations and lipomatosis. Hum Mol Genet 9:765–768
494. Ziprkowski L, Krakowski A (1961) Contribution to eccrine spiradenoma. Arch Dermatol 84:792–797

Revertant Mosaicism

11

Contents

In both autosomal dominant and recessive skin disorders, a thorough clinical examination may reveal some patches of healthy skin reflecting revertant mutations. Moreover, in some Mendelian disorders such as ichthyosis in confetti (ichthyosis variegata) or Kindler syndrome, multiple events of mitotic recombination or slipped mispairing may give rise to numerous patches of healthy skin [5, 9, 11]. In these traits, manifold reversion mechanisms appear to be rather the rule than an exceptional event.

11.1 Revertant Mosaicism in Autosomal Dominant Skin Disorders

Patches of clinically healthy skin with molecular evidence of revertant mosaicism were documented in epidermolysis bullosa of the Dowling-Meara type [16] and in ichthyosis variegata (ichthyosis with confetti) (Fig. 11.1) [2–5]. Moreover, clinical evidence for revertant mosaicism without molecular examination was described in epidermolytic ichthyosis of Brocq [6] and Darier disease [7, 15] (see Sects. 3.1.1.6, 8.1.3, and 8.1.4).

11.2 Revertant Mosaicism in Autosomal Recessive Skin Disorders

Patches of skin that did not blister have been noted in non-Herlitz junctional epidermolysis bullosa (Fig. 11.2) [13, 14], recessive dystrophic

R. Happle, *Mosaicism in Human Skin*,
DOI 10.1007/978-3-642-38765-4_11, © Springer-Verlag Berlin Heidelberg 2014

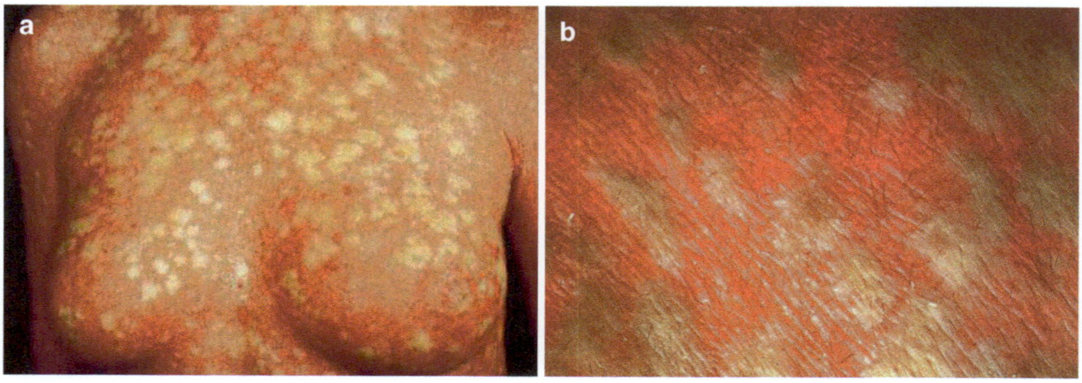

Fig. 11.1 (**a**) Ichthyosis variegata (ichthyosis with confetti) showing thousands of islands of revertant mosaicism [3] (Reprinted with permission from the British Association of Dermatologists); (**b**) Close-up showing hypertrichosis of the areas affected by the forward mutation [2] (Reprinted with permission from Karger Publishers, Basel, Switzerland)

epidermolysis bullosa [1, 14], and Kindler syndrome [9, 11] (Fig. 11.3) (see Sect. 3.1.1.6). In all of these cases, molecular analysis revealed diverse events of revertant mosaicism.

In autosomal recessive forms of epidermolysis bullosa, even two or three different spontaneous back-mutations were found in the same individual [12]. Such cases of "natural gene therapy" may serve as a model to develop new strategies of treating severe hereditary skin disorders [8, 10]. Perhaps, a safe gene therapeutic approach will be developed by expanding a normally functioning cell clone derived from the islands of healthy skin that appear to be noted rather frequently in such diseases when sought for with a prepared mind.

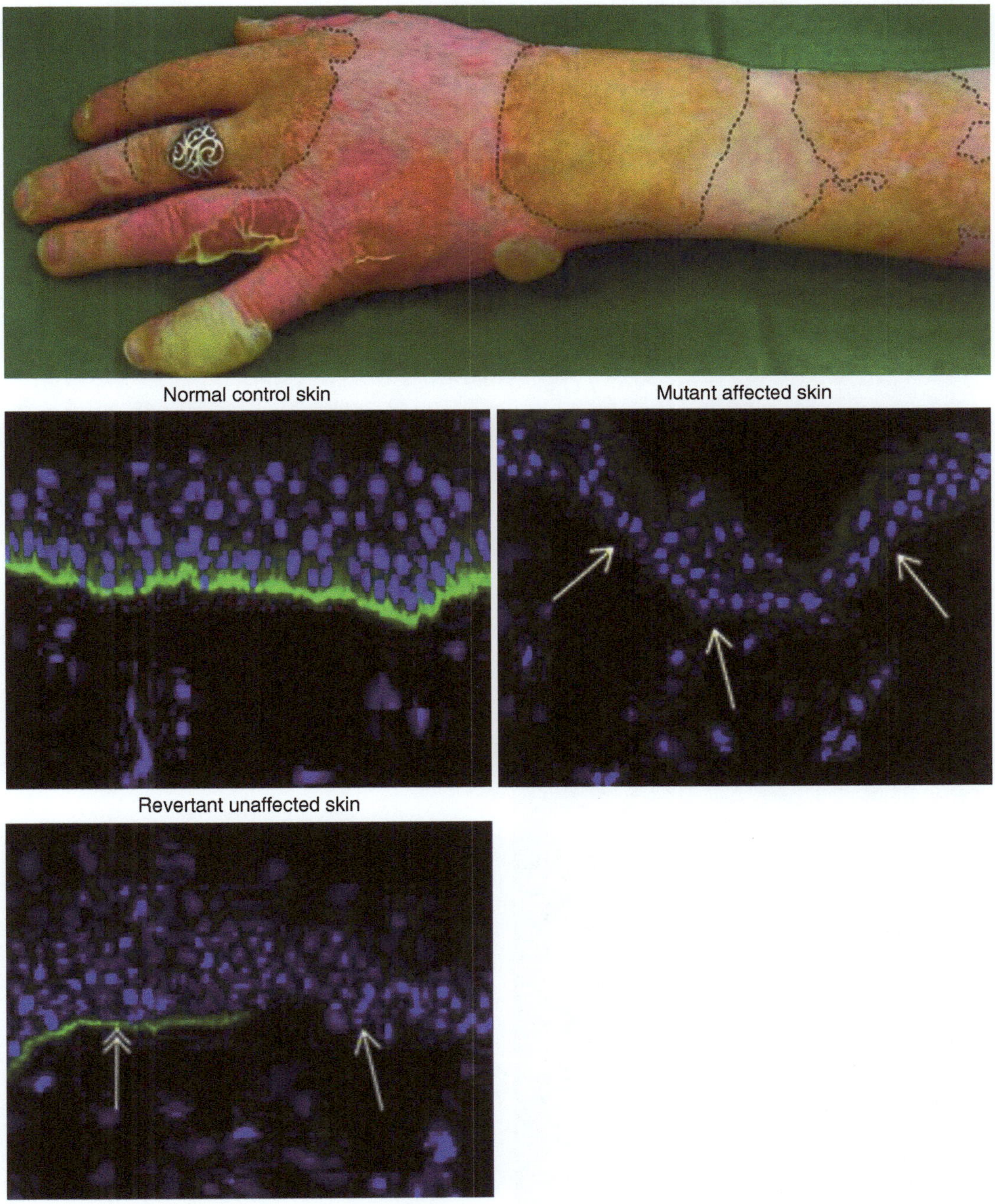

Fig. 11.2 Revertant mosaicism in a case of recessive junctional epidermolysis bullosa. Biopsies showed absence of collagen 17 staining (*green*) in the mutant affected skin, whereas patchy re-expression of the protein was noted in the revertant unaffected areas [14]. Note that the patient prefers to wear her ring on a finger with revertant unaffected skin (Reprinted with permission from Discovery Medicine, Publisher of Johns Hopkins CME Series)

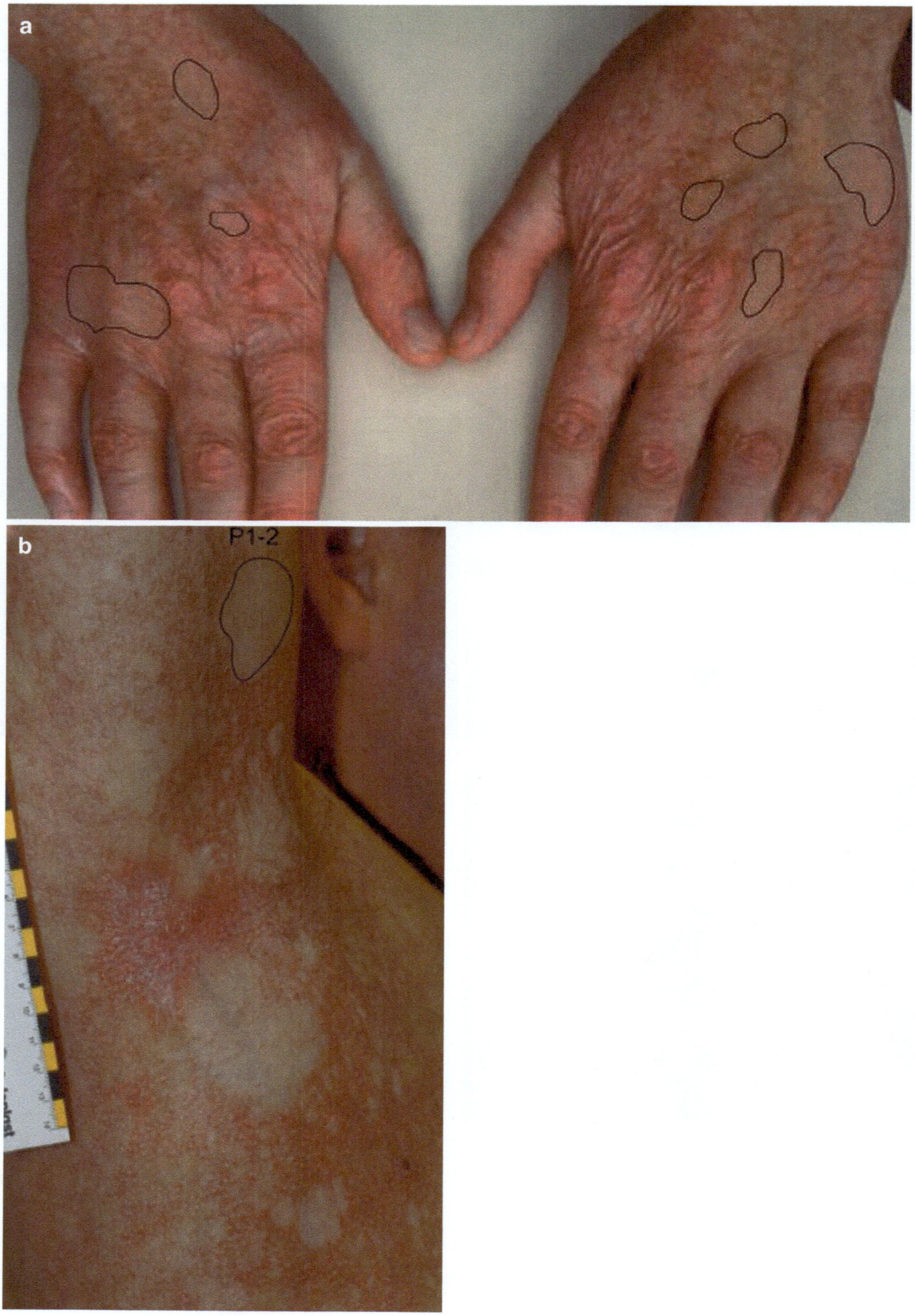

Fig. 11.3 Islands of revertant mosaicism (**a**) on the hands and (**b**) around the axillary region of a woman with Kindler syndrome. These patches of normal skin were caused by either slipped mispairing or mitotic recombination [9] (Reprinted with permission from the American Society for Clinical Investigation)

References

1. Almaani N, Nagy N, Liu L, Dopping-Hepenstal PJ, Lai-Cheong JE, Clements SE, Techanukul T, Tanaka A, Mellerio JE, McGrath JA (2010) Revertant mosaicism in recessive dystrophic epidermolysis bullosa. J Invest Dermatol 130:1937–1940
2. Brusasco A, Tadini G, Cambiaghi S, Ermacora E, Grimalt R, Caputo R (1994) A case of congenital reticular ichthyosiform erythroderma – ichthyosis 'en confettis'. Dermatology 188:40–45
3. Brusasco A, Cambiaghi S, Tadini G, Berti E, Caputo R (1998) Unusual hyperpigmentation developing in congenital reticular ichthyosiform erythroderma (ichthyosis variegata). Br J Dermatol 139: 893–896
4. Burger B, Spoerri I, Schubert M, Has C, Itin PH (2012) Description of the natural course and clinical manifestations of ichthyosis with confetti caused by a novel *KRT10* mutation. Br J Dermatol 166:434–439
5. Choate KA, Lu Y, Zhou J, Choi M, Elias PM, Farhi A, Nelson-Williams C, Crumrine D, Williams ML, Nopper AJ, Bree A, Milstone LM, Lifton RP (2010) Mitotic recombination in patients with ichthyosis causes reversion of dominant mutations in KRT10. Science 330:94–97
6. Happle R, König A (1999) Dominant traits may give rise to paired patches of either excessive or absent involvement. Am J Med Genet 84:176–177
7. Itin PH, Happle R (2002) Darier disease with paired segmental manifestation of either excessive or absent involvement: a further step in the concept of twin spotting. Dermatology 205:344–347
8. Jonkman MF, Pasmooij AM (2009) Revertant mosaicism – patchwork in the skin. N Engl J Med 360:1680–1682
9. Kiritsi D, He Y, Pasmooij AM, Onder M, Happle R, Jonkman MF, Bruckner-Tuderman L, Has C (2012) Revertant mosaicism in a human skin fragility disorder results from slipped mispairing and mitotic recombination. J Clin Invest 122:1742–1746
10. Lai-Cheong JE, McGrath JA, Uitto J (2011) Revertant mosaicism in skin: natural gene therapy. Trends Mol Med 17:140–148
11. Lai-Cheong JE, Moss C, Parsons M, Almaani N, McGrath JA (2012) Revertant mosaicism in Kindler syndrome. J Invest Dermatol 132:730–732
12. Pasmooij AM, Pas HH, Deviaene FC, Nijenhuis M, Jonkman MF (2005) Multiple correcting *COL17A1* mutations in patients with revertant mosaicism of epidermolysis bullosa. Am J Hum Genet 77:727–740
13. Pasmooij AM, Pas HH, Bolling MC, Jonkman MF (2007) Revertant mosaicism in junctional epidermolysis bullosa due to multiple correcting second-site mutations in *LAMB3*. J Clin Invest 117:1240–1248
14. Pasmooij AM, Jonkman MF, Uitto J (2012) Revertant mosaicism in heritable skin diseases: mechanisms of natural gene therapy. Discov Med 14:167–179
15. Rodríguez-Pazos L, Gomez-Bernal S, Loureiro M, Toribio J (2011) Type 2 segmental Darier disease with twin spot phenomenon. J Eur Acad Dermatol Venereol 25:496–497
16. Smith FJ, Morley SM, McLean WH (2004) Novel mechanism of revertant mosaicism in Dowling-Meara epidermolysis bullosa simplex. J Invest Dermatol 122:73–77

Nevoid Skin Disorders 12

Contents

The group of nevoid conditions includes all skin disorders that are reminiscent of a nevus, but do not fulfill the criteria of a true nevus [83] (see Chap. 7). On the other hand, this group comprises some X-linked "true nevi" such as the cutaneous lesions of focal dermal hypoplasia that, by convention, are categorized as "nevoid."

12.1 Cutaneous Lesions Reflecting Functional X-Chromosome Mosaicism

The skin lesions reflecting lyonization in women with X-linked traits are taken, by convention, as nevoid conditions although they fulfill, in principle, the criteria of true nevi [83]. There is,

R. Happle, *Mosaicism in Human Skin*,
DOI 10.1007/978-3-642-38765-4_12, © Springer-Verlag Berlin Heidelberg 2014

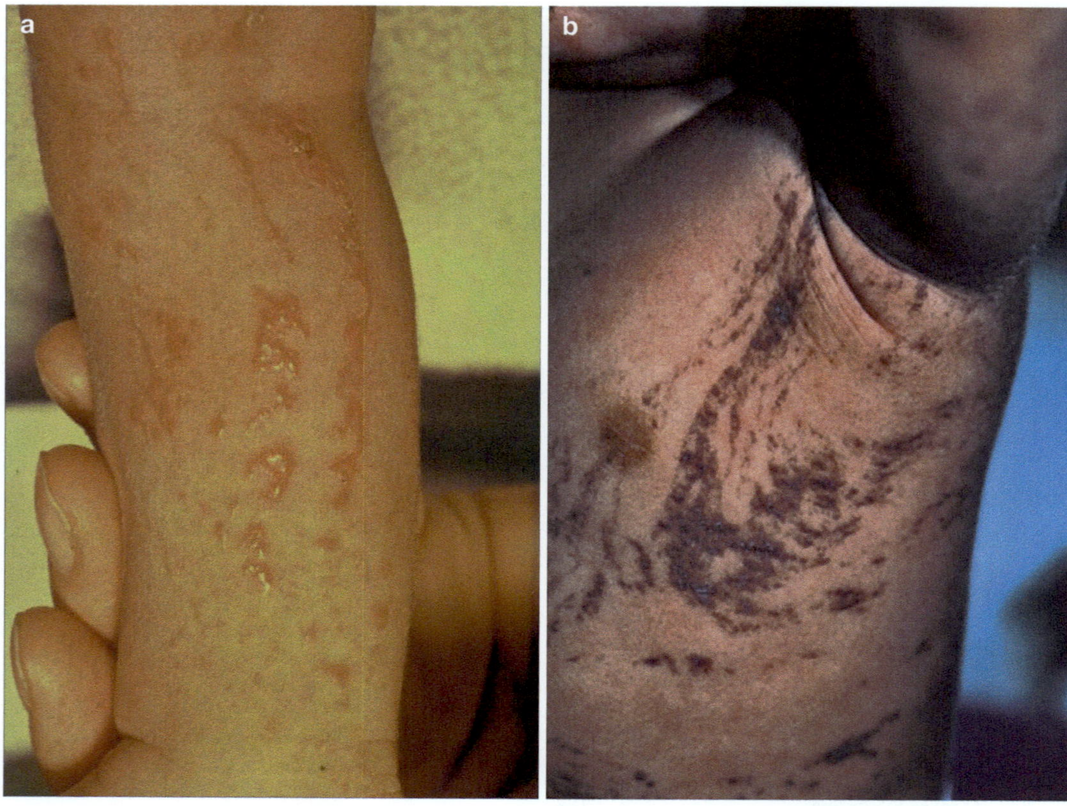

Fig. 12.1 Incontinentia pigmenti. (**a**) First stage of in the form of linear blistering; (**b**) third stage characterized by linear hyperpigmentation [19] (Reprinted with permission from John Wiley & Sons, USA)

however, one exception from this rule. The CHILD nevus reflects functional X-chromosome mosaicism but is categorized as a true nevus (see Sect. 7.3.1.9 and Table 7.1).

12.1.1 Incontinentia Pigmenti

This X-linked multisystem birth defect is caused by *NEMO* mutations [227]. The disorder is characterized by linear skin lesions that develop in four stages. During the first months of life, an episode of inflammation and blistering of the skin occurs, being arranged along Blaschko's lines (Fig. 12.1). Later, these lesions are replaced by hyperkeratotic verrucous areas, whereas in older girls a linear or patchy hyperpigmentation is noted (Figs. 3.14 and 12.1b). In adult women, this hyperpigmentation often fades away and gives rise to hypopigmented linear lesions that are slightly atrophic and hairless. This fourth stage tends to be most conspicuous on the calves.

The early inflammatory stage apparently reflects an attempt of the functionally normal cells to eliminate the functionally aberrant clone. This elimination process, however, is not always complete because exceptional eruptions reiterating a stage 1 may occur in older children [6, 50, 197] or even in adult women [14].

Hypomorphic NEMO alleles may cause a quite different phenotype in the form of ectodermal dysplasia of Zonana (see Chap. 4).

12.1.2 Focal Dermal Hypoplasia

Focal dermal hypoplasia (Goltz syndrome) is an X-linked dominant, male-lethal disorder caused by *PORCN* mutations [77]. The syndrome includes multiple congenital defects of the mesoderm and the ectoderm, reflecting lyonization. A cutaneous hallmark are linear lesions of thinned or absent dermis (Fig. 12.2), giving rise to herniation of the underlying fatty tissue [82]. Longitudinal

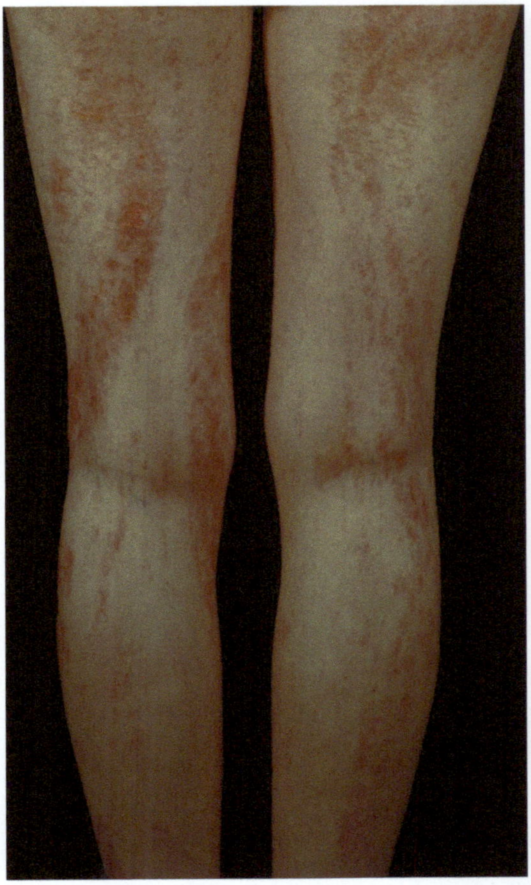

Fig. 12.2 Focal dermal hypoplasia showing a systematized linear arrangement (Courtesy of Dr. Regina Föster-Holst, Kiel, Germany)

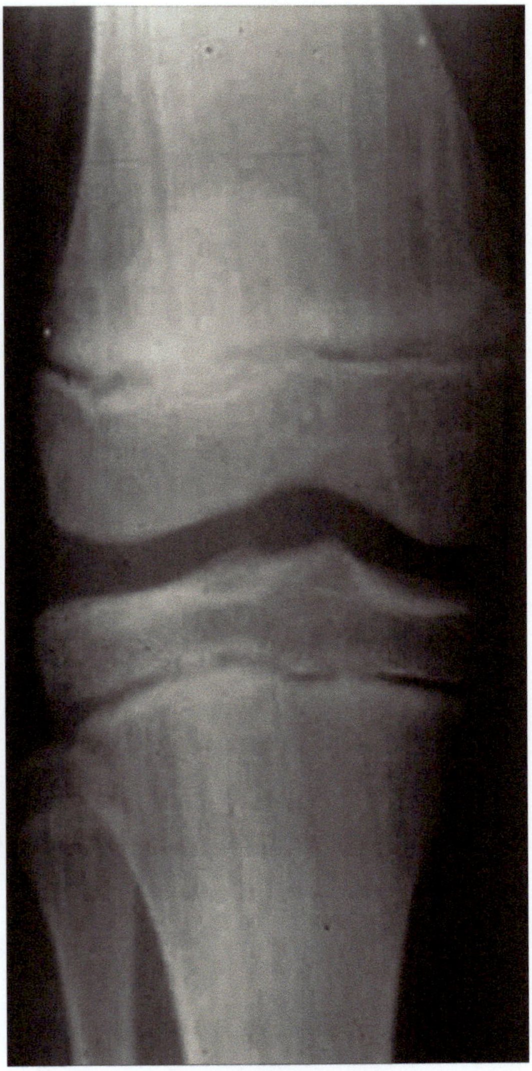

Fig. 12.3 Longitudinal striation of long bones reflecting functional X-chromosome mosaicism in a patient with focal dermal hypoplasia

striation of the long bones as noted on x-rays is likewise very characteristic (Fig. 12.3) [97].

12.1.3 Conradi-Hünermann-Happle Syndrome

This X-linked dominant, male-lethal phenotype is caused by *EBP* mutations [102]. The syndrome is also called X-linked dominant chondrodysplasia punctata and includes asymmetric bone defects, sectorial cataracts, and skin lesions following Blaschko's lines. Epiphyseal stippling is found on x-rays taken shortly after birth. During the first weeks of life, the skin shows linear areas of inflammation and hyperkeratosis (Fig. 12.4). This stage is followed by linear areas of atrophy and hypopigmentation with rather mild scaling. Lesions of follicular atrophoderma are predominantly found on the scalp, the wrists, and the dorsal and lateral aspects of the hands (Fig. 12.5).

A girl with an *EBP* mutation and classical features of the syndrome had a mother who was phenotypically healthy with the exception of a very small patch of erythematous and slightly hyperkeratotic skin present on one of her legs. The mutation was absent in the mother's blood lymphocytes but present in DNA obtained from her tiny skin lesion, providing evidence that she carried an *EBP* mutation in the form of postzygotic mosaicism [165].

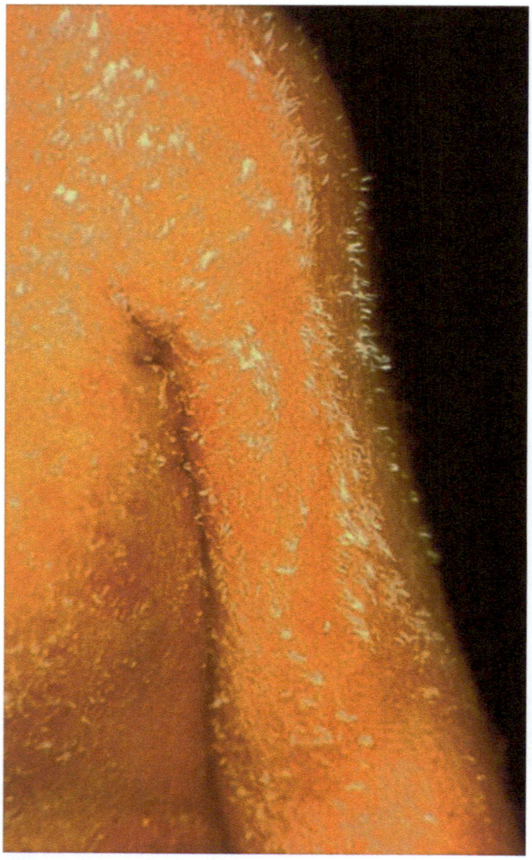

Fig. 12.4 Characteristic feather-like hyperkeratoses showing a linear arrangement in a newborn girl with Conradi-Hünermann-Happle syndrome (Courtesy of Dr. Elzo Folkers, Zaandam, The Netherlands)

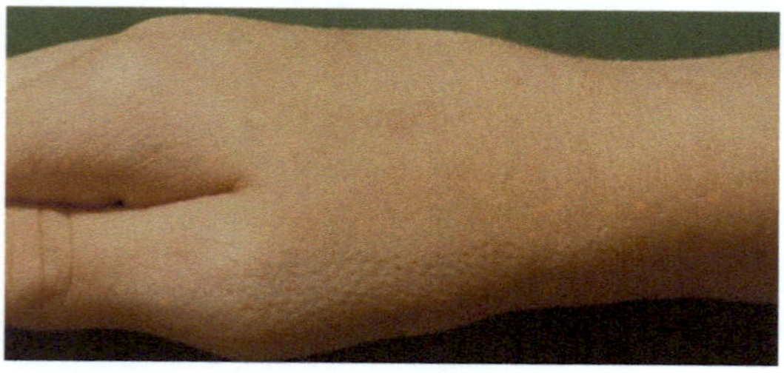

Fig. 12.5 Linear arrangement of follicular atrophoderma in an adolescent girl with Conradi-Hünermann-Happle syndrome

12.1.4 MIDAS Syndrome

MIDAS is an acronym for *mi*crophthalmia, *d*ermal *a*plasia, and *s*clerocornea [36, 92]. The disorder is also called MLS (microphthalmia with linear skin defects) syndrome [255]. It is caused by male-lethal mutations in one of the X-linked genes, holocytochrome c-type synthetase [198] or *COX7B* [113]. Oncocytic cardiomyopathy may cause early death [21]. For unknown reasons, the cutaneous lesions tend to involve almost exclusively the face and the neck (Fig. 12.6) [179, 263].

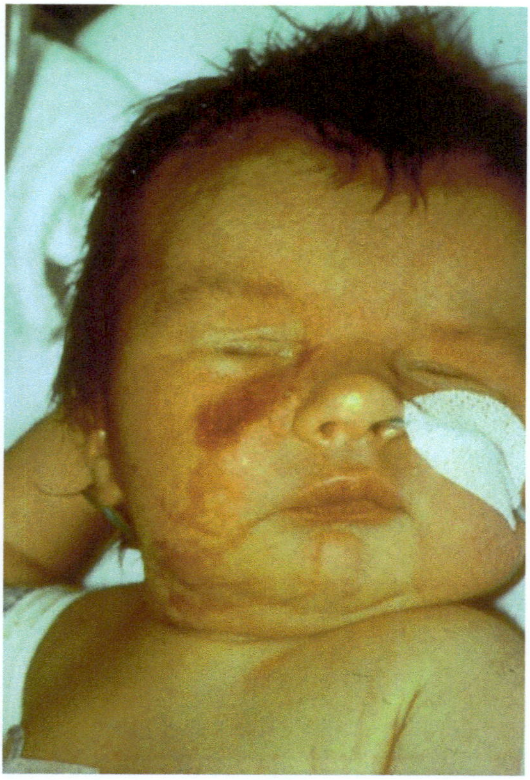

Fig. 12.6 MIDAS syndrome in a newborn girl affected with bilateral microphthalmia, blepharophimosis, saddle nose, and unilateral linear areas of dermal aplasia [92] (Reprinted with permission from John Wiley & Sons, USA)

12.1.5 Oral-Facial-Digital Syndrome Type 1

This X-linked, male-lethal trait is caused by *CXORF5* mutations. It includes gingival frenula and multilobulated tongue (Fig. 12.7), median cleft of upper lip, asymmetric cleft palate, malformed digits, and mild mental deficiency. Multiple milia involve the face and the

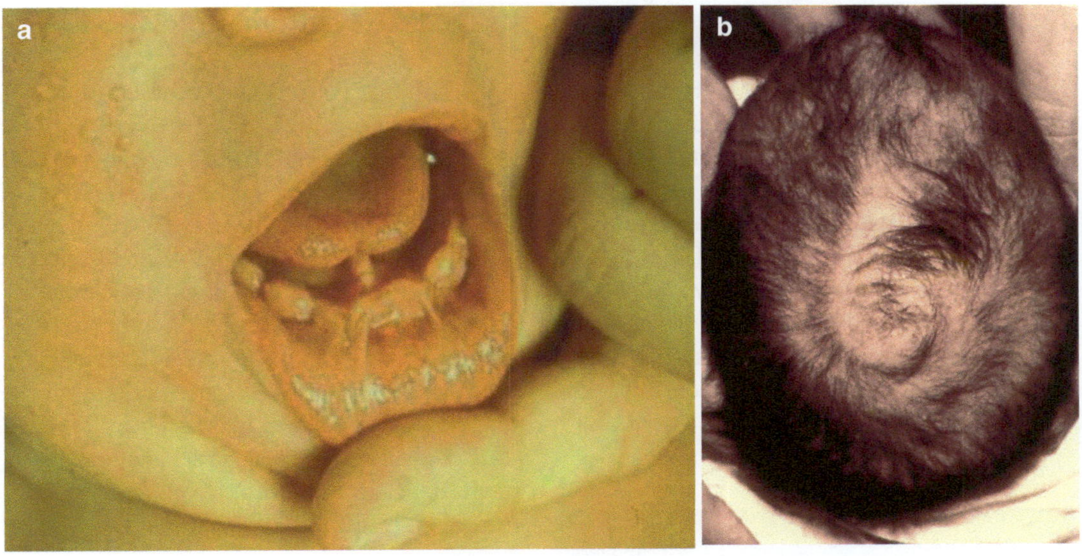

Fig. 12.7 Oral-facial-digital syndrome type 1. (**a**) Note gingival frenula and lobulated tongue; (**b**) spiral pattern of hairlessness reflecting lyonization [95]

ears but disappear after infancy. On the scalp a spiral pattern of congenital alopecia, reflecting lyonization, is noted (Fig. 12.7) [25, 95].

12.1.6 Christ-Siemens-Touraine Syndrome

With few exceptions, the full-blown phenotype of hypohidrosis is observed in male patients only, whereas heterozygous female individuals tend to be mildly affected. Sweat testing of carrier females reveals hypohidrotic areas arranged along Blaschko's lines (Fig. 12.8) [15, 94].

12.1.7 X-Linked Dyskeratosis Congenita

Male patients with X-linked dyskeratosis congenita show a diffusely arranged reticular hyperpigmentation and atrophy of skin, dystrophic nails, mucosal leukoplakia, progressive failure of bone marrow, and increased proneness to develop malignancies. Heterozygous women are, as a rule, only mildly affected, and their skin lesions follow Blaschko's lines (Fig. 12.9) [54].

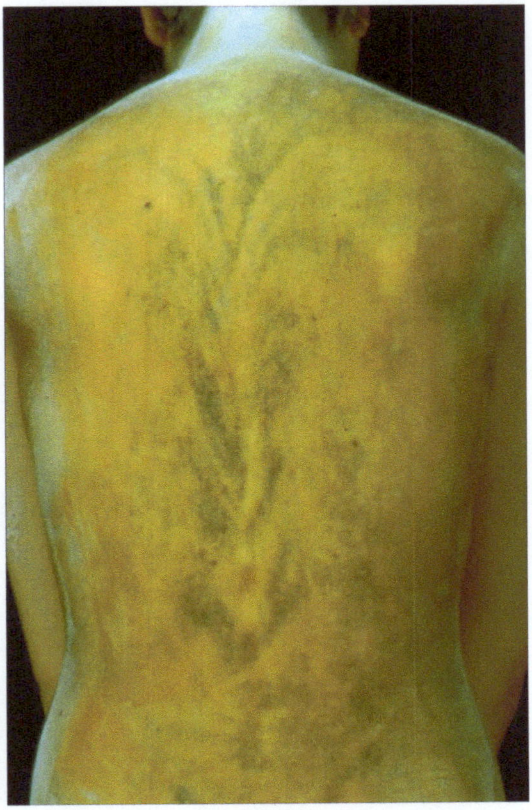

Fig. 12.8 Minor's sweat testing in a female carrier of Christ-Siemens-Touraine syndrome [94] (Reprinted with permission from John Wiley & Sons, USA)

12.1.8 Menkes Syndrome

This neurodegenerative disorder is caused by mutations in the gene encoding Cu(2+)-transporting ATPase, alpha polypeptide (*ATP7A*). Major features include twisting and kinking of scalp hair, mental deficiency, seizures, bone defects, and hypothermia. Affected boys usually die during early infancy. Heterozygous women may show a linear or otherwise segmental pattern of cutaneous hypopigmentation (Fig. 12.10) [147, 253] and a mixture of anomalous and normal scalp hair [40, 49].

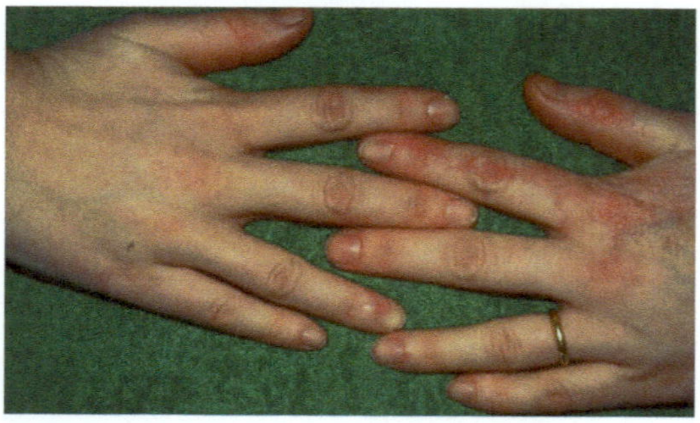

Fig. 12.9 Linear involvement of left hand in a female carrier of dyskeratosis congenita [54] (Courtesy of Dr. Hugo Degreef, Leuven, Belgium)

12.1.9 IFAP Syndrome

IFAP (ichthyosis follicularis, atrichia, photophobia) syndrome is caused by mutations in *MBTPS2*, an intramembrane zinc metalloprotease controlling cholesterol homeostasis and stress response of the endoplasmatic reticulum [178]. Boys are diffusely and severely affected (Fig. 12.11a), whereas female carriers show a linear pattern of follicular ichthyosis, atrophoderma, asymmetric hairlessness (Fig. 12.11b), and diminished sweating [56, 137]. Hyperkeratotic streaks involving the feet have likewise been reported [240]. Unilateral acne lesions are another mosaic manifestation as noted in a 14-year-old heterozygous girl (Fig. 12.11c).

12.1.10 Aicardi Syndrome

This X-linked dominant, male-lethal disorder is characterized by chorioretinal lacunae, agenesis of corpus callosum, and infantile spasms [209]. Associated features include microphthalmia, coloboma, hypoplasia of the optic nerve, and various cerebral defects [231]. A 3-month-old blind girl with Aicardi syndrome had a "trichogephyrosis" in the form of hypertrichosis bridging

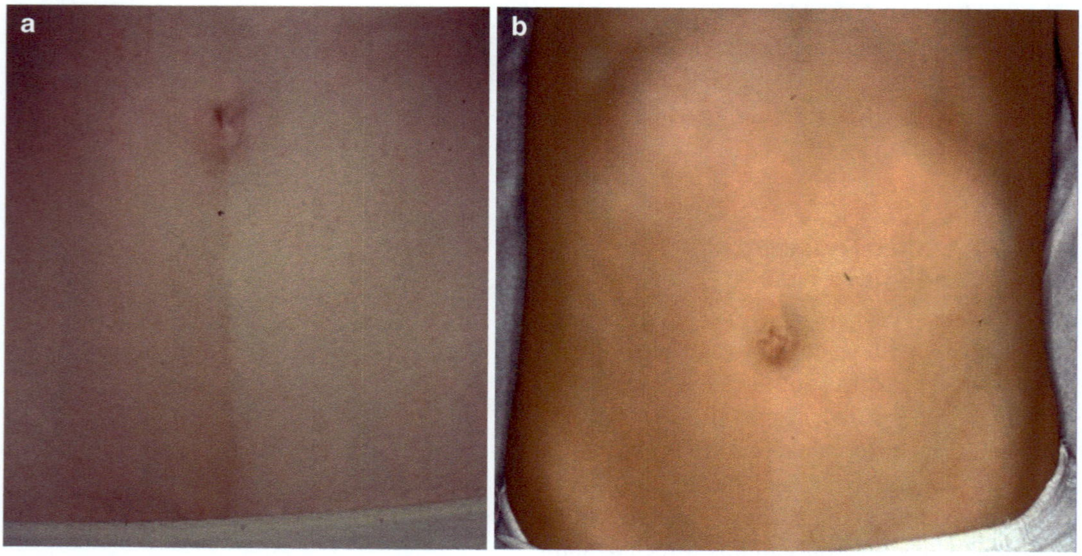

Fig. 12.10 (**a**, **b**) Unilateral arrangement of abdominal hypopigmentation in two sisters of a boy with Menkes syndrome [147] (Courtesy of Dr. Gérard Lorette, Tours, France)

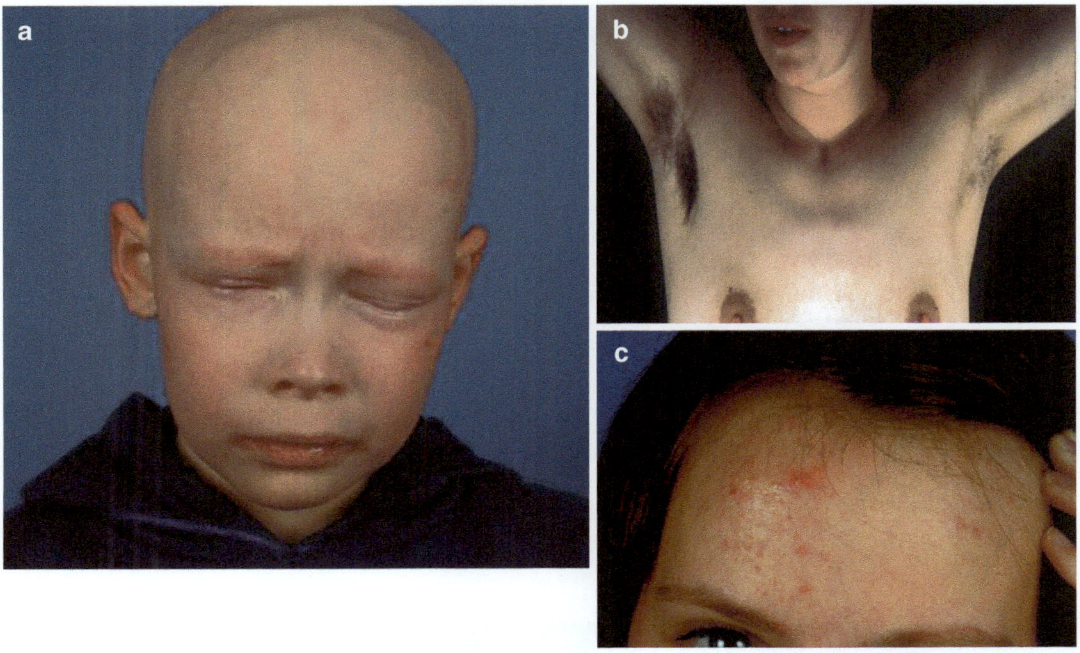

Fig. 12.11 IFAP syndrome. (**a**) A boy with complete baldness and pronounced photophobia [137]; (**b**) unilateral axillary hypotrichosis in his heterozygous mother; (**c**) asymmetrical acne lesions in his heterozygous sister

the gap between the right eyebrow and scalp hair (Fig. 12.12) [96]. Additional clinical reports are needed to determine whether the unilateral involvement reflects lyonization.

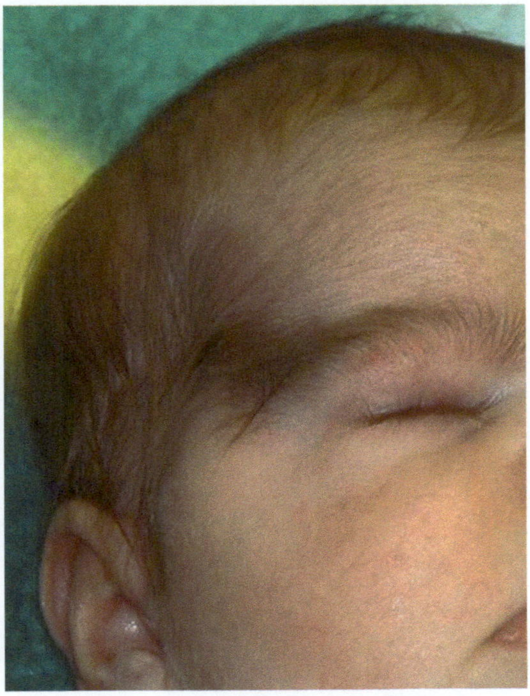

Fig. 12.12 Unilateral nevoid hypertrichosis in the form of "trichogephyrosis" in an infant girl with Aicardi syndrome [96] (Reprinted with permission from John Libbey Eurotext, Montrouge, France)

12.1.11 Reticulate Pigmentary Disorder of Partington

The full-blown syndrome occurs in male individuals who show diffuse reticulate dyspigmentation (Fig. 12.13a), hypohidrosis, coarse hair, failure to thrive, recurrent pneumonia, colitis, and corneal dyskeratosis with photophobia. Female carriers have linear lesions of dyspigmentation (Fig. 12.13b) [9, 131, 190, 191]. The underlying gene has been assigned to Xp22–p21 but is so far unknown [117].

12.1.12 X-linked Albinism-Deafness Syndrome

In 1962, an unusual X-linked form of albinism with deafness was described in two reports on

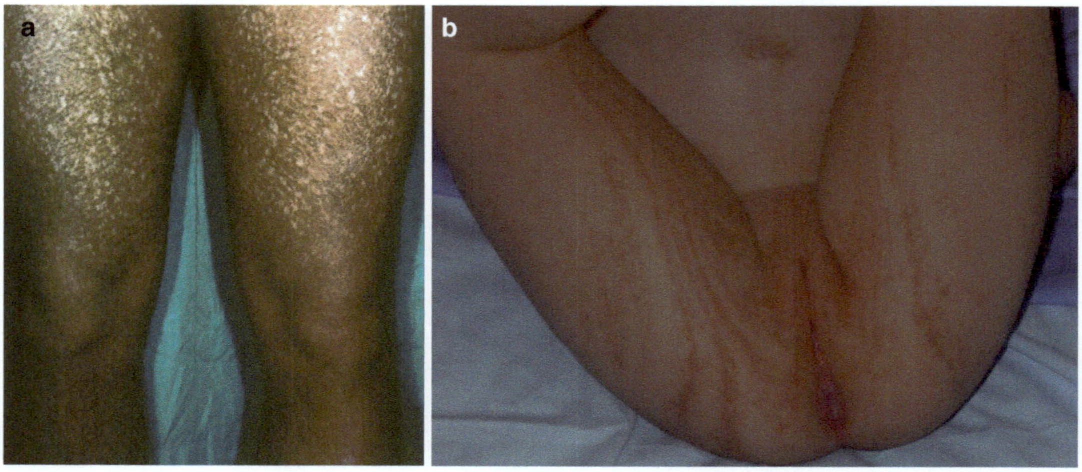

Fig. 12.13 Reticulate pigmentary disorder of Partington. (**a**) Diffuse involvement in a man; (**b**) linear lesions in his daughter (Courtesy of Dr. Eulalia Baselga, Barcelona, Spain)

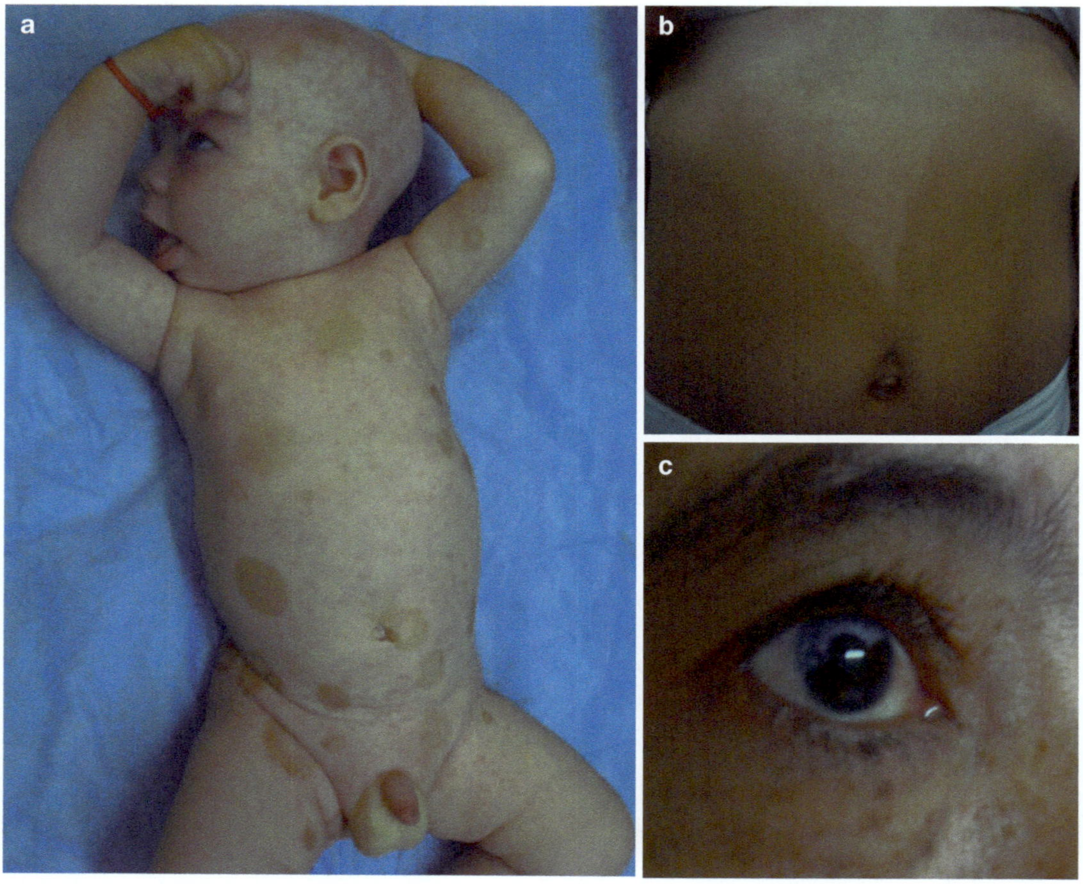

Fig. 12.14 X-linked albinism-deafness syndrome. (**a**) Diffuse involvement in a boy; (**b**) segmental hypopigmentation in his mother who suffers from hearing loss; (**c**) sectorial depigmentation of the iris in the boy's grandmother who has no hearing loss (Courtesy of Dr. Silvina Sartori, Santa Fe, Argentina)

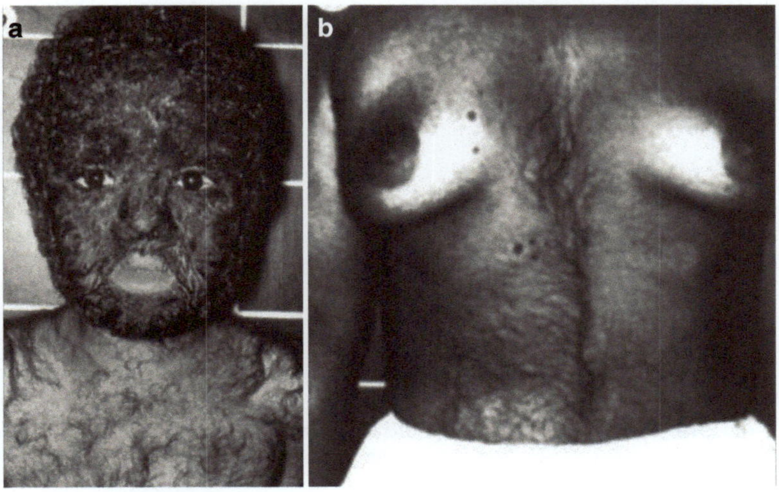

Fig. 12.15 X-linked dominant hypertrichosis. (**a**) Diffuse involvement in a boy; (**b**) segmental pattern of lyonization in a woman [63] (Reprinted with permission from Nature Publishing Group)

the same family [152, 261]. Initially this was taken as an X-linked recessive trait, but female members of the family were later found to be likewise affected [66]. The trait has been assigned to Xq26.3–q27.1 [223]. Male patients are deaf and blue eyed and show peculiar patches of hyperpigmentation being diffusely distributed within the albinotic skin (Fig. 12.14a), whereas affected women may have difficulties of hearing or may show segmental depigmentated areas involving the skin (Fig. 12.14b) or the irides (Fig. 12.14c). Remarkably, the cutaneous pattern of lyonization does not follow Blaschko's lines.

12.1.13 X-linked Dominant Hypertrichosis

Affected males have a severe and diffusely distributed hypertrichosis, whereas heterozygous females show asymmetrical patches of increased hairiness arranged in a checkerboard pattern, reflecting lyonization (Fig. 12.15) [63, 150]. In a large Chinese family, Zhu et al. [260] confirmed that heterozygous women tend to be more mildly affected. The authors assumed that a palindrome-mediated insertion near *SOX3* may be involved in the etiology. A syndromic form of this X-linked hair disorder has likewise been documented [233].

12.2 Congenital Autosomal Disorders Representing Non-Nevi

The following skin disorders are either autosomal sporadic tumor syndromes or other non-nevi that have so far been mistaken as nevi either because they look like nevi or simply because someone has erroneously categorized them as nevi.

12.2.1 Benign Skin Tumors Reflecting Lethal Autosomal Mutations Surviving by Mosaicism

The following sporadic phenotypes cannot be categorized among the group of nevi because they are characterized by the presence benign neoplasias.

12.2.1.1 Syringocystadenoma Papilliferum

Multiple tumors of this type show rather frequently an arrangement along Blaschko's lines (Fig. 12.16) [139, 154, 174, 187, 194, 256, 258]. All cases so far reported were sporadic.

12.2.1.2 Maffucci Syndrome

In this syndrome multiple enchondromas, most frequently affecting the fingers and toes, are found to be associated with multiple hemangiomas that

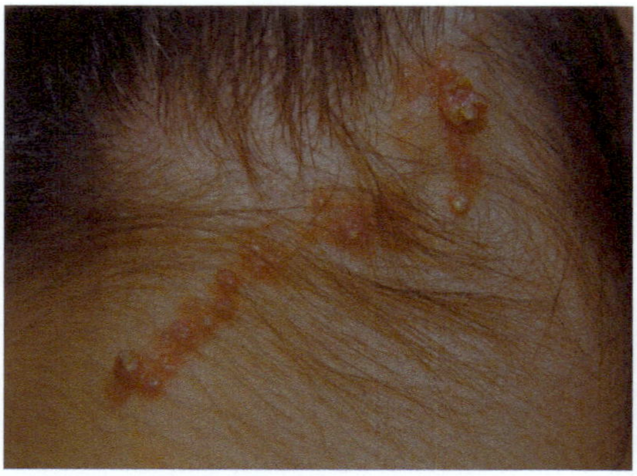

Fig. 12.16 Syringocystadenoma papilliferum showing a linear arrangement [154] (Reprinted with permission from Creative Commons)

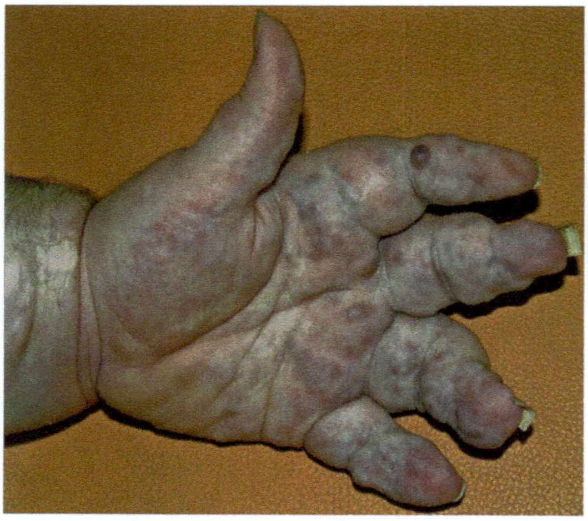

Fig. 12.17 Maffucci syndrome [205] (Reprinted with permission from the American Society of Clinical Oncology)

predominantly involve the skin (Fig. 12.17) [120, 126] but may also be found in the oral cavity, esophagus, or intestinal tract [222]. The syndrome always occurs sporadically and the lesions show an asymmetrical arrangement. Therefore, the phenotype was proposed to be caused by a lethal mutation surviving by mosaicism [84]. This etiological concept was corroborated in recent molecular studies. Pansuriya et al. [188] found postzygotic heterozygous mutations in *IDH1* or *IDH2* within the lesional tissue of patients with Maffucci syndrome and a related disorder, Ollier disease.

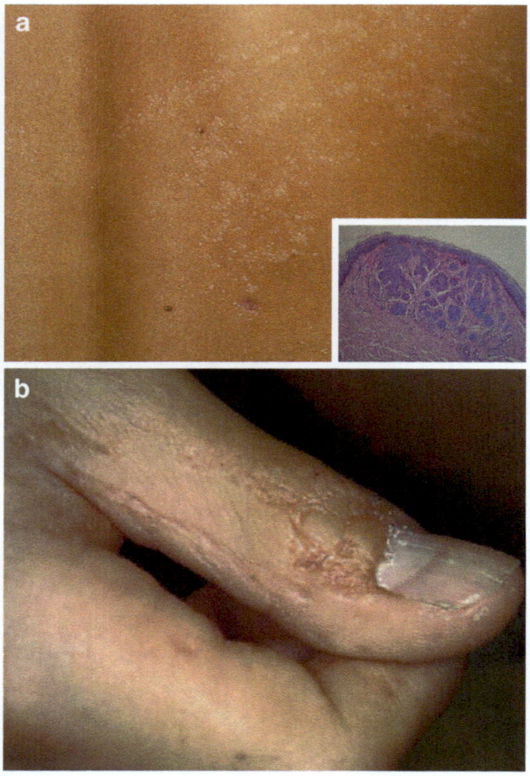

Fig. 12.18 Happle-Tinschert syndrome. (**a**) Linear arrangement of multiple follicular basaloid hamartomas in a 7-year-old boy; (**b**) ipsilateral linear atrophoderma involving the thumb [116] (Reprinted with permission from S. Karger AG, Basel, Switzerland)

12.2.1.3 Happle-Tinschert Syndrome

This birth defect includes multiple, segmentally arranged basaloid follicular hamartomas (Fig. 12.18a), linear atrophoderma (Fig. 12.18b), and osseous, dental, ocular, or cerebral anomalies [22, 99, 116]. A case described by Bedi [17] as "nevus unius lateris" is another typical example of this disorder.

It is unlikely that the phenotype represents a type 2 segmental manifestation of hereditary multiple basaloid follicular hamartoma because so far this autosomal dominant trait has neither been observed in patients with Happle-Tinschert syndrome nor in their relatives [99]. Admittedly, however, it is so far difficult to determine whether cases of multiple congenital, segmentally arranged and tightly packed basaloid

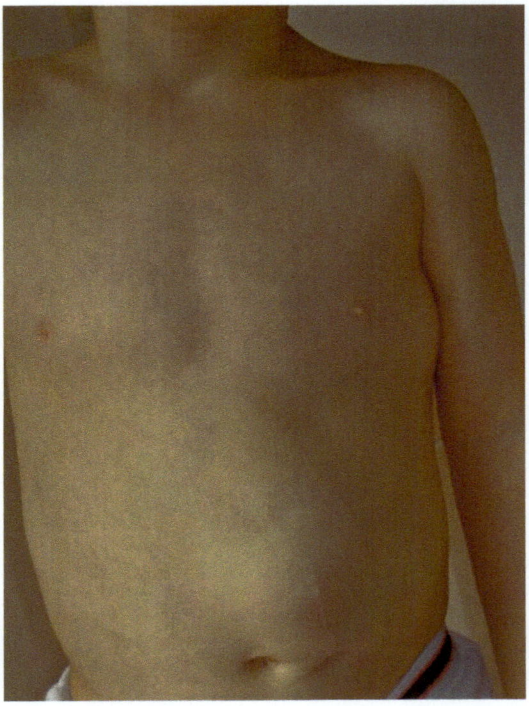

Fig. 12.19 Hemihyperplasia-multiple lipomata syndrome in a 3-year-old boy. Note strictly unilateral arrangement of the lipomas [30] (Reprinted with permission from Elsevier Limited, Oxford, UK)

follicular hamartomas [8, 100, 110] may represent a minimal form of this syndrome or a superimposed manifestation of autosomal dominant multiple basaloid follicular hamartoma (see Sect. 10.1.4).

12.2.2 Hemihyperplasia-Multiple Lipomata Syndrome: A Nevoid Disorder of Unknown Origin

This phenotype was delineated in 1998 by Biesecker et al. [20]. It is characterized by a rather stable overgrowth of one half of the body and ipsilateral development of multiple lipomas (Fig. 12.19) [30, 215]. The term "multiple lipomatosis" [20] is redundant because the word lipomatosis implies multiplicity of the tumors. The syndrome has always occurred sporadically. Its molecular basis is so far unknown. Possible mechanisms are discussed in OMIM under the entry 235000 ("hemihyperplasia, isolated") [182].

Practical aspect: Liposuction appears to be a suitable therapeutic option to alleviate the burden of large lipomas [144].

12.2.3 Other Autosomal Non-Nevi

Some other skin lesions are called nevi without being nevi. The following disorders refer to such historical misnomers.

12.2.3.1 Salmon Patch ("Unna's Nevus,""Median Nevus Flammeus,""Nevus Simplex")

Congenital telangiectatic macules involving the glabella ("angels kiss") or the occipital or nuchal region ("stork bite") (Fig. 12.20) or the sacral area have so far been categorized as median vascular nevi [122, 193]. These lesions do not fulfill, however, the criteria of a nevus because they do not reflect mosaicism. Frontal salmon patches are noted in 30–40 % of newborns, whereas the prevalence of the nuchal erythema is even higher [250]. If we would include such median vascular lesions into the group of nevi, this term would lose any specific meaning.

12.2.3.2 White Sponge Hyperplasia of the Mucosa ("White Sponge Nevus")

This autosomal dominant trait is caused by mutations of keratin 4 or 13 and has so far been called "white sponge nevus" (Fig. 12.21) [146]. This name, however, is an absurdity. With the same right one could categorize palmoplantar keratoderma as a "hyperkeratotic nevus." The symmetric oral leukoplakia does not look like a nevus, nor does it reflect mosaicism.

12.2.3.3 "Basal Cell Nevus"

This term should no longer be used since it is both incorrect and ambiguous [99]. It is incorrect because it refers to non-melanocytic neoplasias and thus not to nevi. It is ambiguous because it

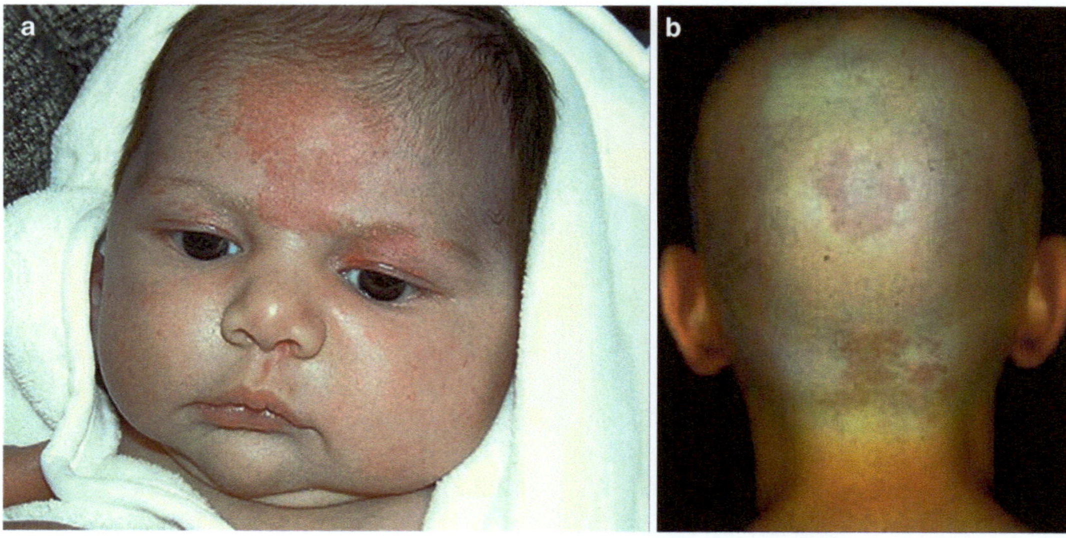

Fig. 12.20 Salmon patch, a frequently occurring nonnevus. (**a**) Rather extensive glabellar involvement [122]; (**b**) occipital involvement in a adult patient with alopecia totalis (**a**: Reprinted with permission from Elsevier Limited, Oxford, UK)

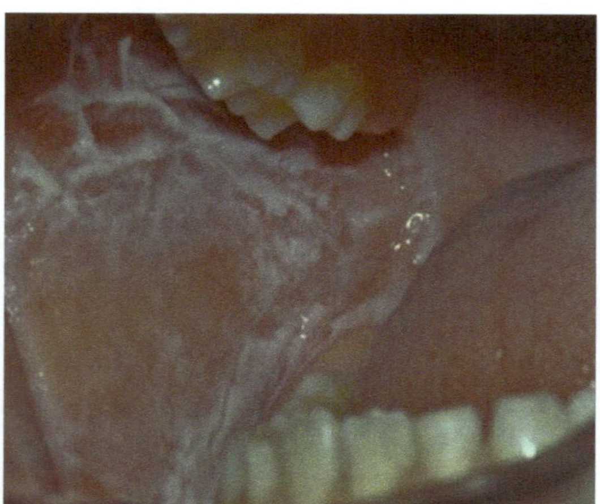

Fig. 12.21 White sponge hyperplasia of the mucosa [146] (Reprinted with permission from John Wiley & Sons, USA)

has been used in the past to describe Gorlin syndrome [32] and Happle-Tinschert syndrome [7, 116] and mosaic manifestations of either Gorlin syndrome [221] or nonsyndromic basal cell carcinomas [31] or nonsyndromic basaloid follicular hamartomas [44].

12.2.3.4 "Blue Rubber Bleb Nevus"

Blue rubber bleb "nevi" are not nevi but represent true angiomas. Hence, the so-called blue rubber bleb nevus syndrome is in fact a blue rubber bleb angiomatosis (see Sect. 10.1.11).

12.3 Nevoid Arrangement of Acquired Skin Disorders

This group includes some disorders that tend to occur exclusively in a segmental arrangement, and mosaic manifestations of many common skin disorders with a polygenic background.

12.3.1 Lichen Striatus

This transient disorder is preponderantly noted in children and consists of one or more linear areas showing inflammatory papular lesions (Fig. 12.22). The distribution follows Blaschko's lines. Histopathologically, a lichenoid infiltrate with lymphocytic exocytosis and spongiosis is noted [229]. Spontaneous resolution tends to occur after some weeks or months. The cause of the disorder is so far unknown. Some cases of lichen striatus being associated with atopic dermatitis [55, 192, 243] may in fact represent examples of superimposed linear atopic dermatitis [245].

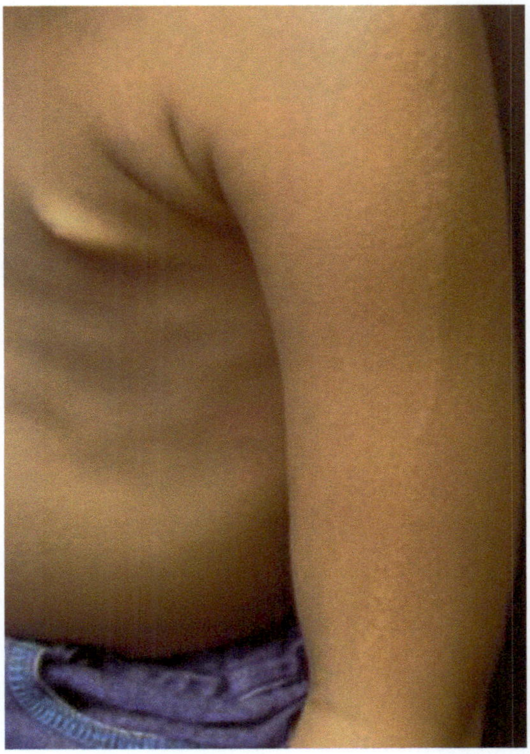

Fig. 12.22 Lichen striatus in a 3-year-old girl [200] (Reprinted with permission from John Wiley & Sons, USA)

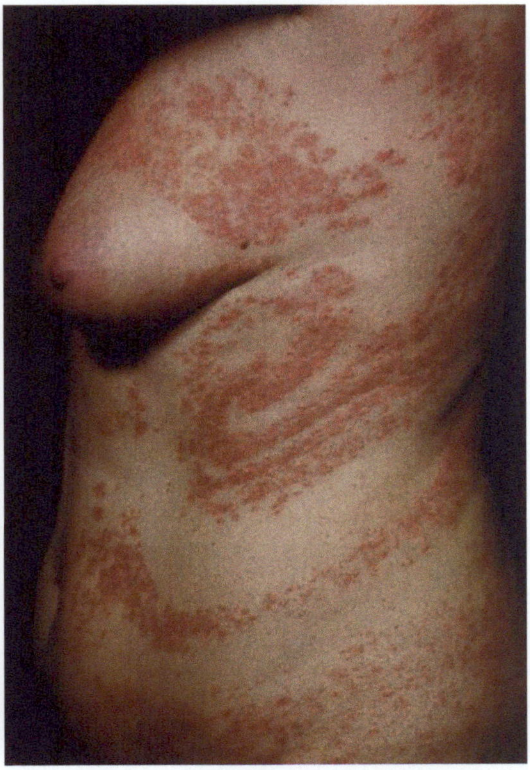

Fig. 12.23 Unilateral systematized "Blaschko dermatitis" in a 44-year-old woman [158] (Reprinted with permission from Elsevier Limited, Oxford, UK)

12.3.2 "Blaschkitis": No Entity, but Either a Variant of Lichen Striatus or an Umbrella Term Including the Linear Manifestation of Various Acquired Inflammatory Skin Disorders

This term was proposed by Grosshans and Marot [76] for a transient linear skin disorder that resembled lichen striatus but showed a more pronounced inflammation and was preponderantly noted in adults (Fig. 12.23). A clear-cut separation from lichen striatus, however, appears to be rather difficult [161, 235], and "pediatric blaschkitis" was also reported [53, 130]. Many authors believe that blaschkitis is merely a variant of lichen striatus [79, 108, 163, 169, 201, 241]. In 2006, Grosshans himself agreed that "blaschkitis probably belongs to the spectrum of lichen striatus" [236].

In a broad sense, the term "blaschkitis" includes the isolated or superimposed linear manifestation of many acquired inflammatory skin disorders such as atopic dermatitis, lichen planus, various types of lupus erythematosus, dermatomyositis, or drug eruptions (see Sect. 12.3.7.1).

12.3.3 Lichen Aureus

Lichen aureus is usually arranged in a segmental pattern (Fig. 12.24) [142, 162] and can, therefore, be categorized as a mosaic disorder. Synonyms include unilateral linear capillaritis [149], unilateral linear capillaropathy [259], and linear progressive pigmentary purpura [106]. It is a controversial issue whether there is a relationship between this disorder and the nonsegmental lesions of Schamberg disease. As a rule of thumb, a nonsegmental involvement that usually occurs in adults tends to be called "Schamberg disease," whereas the name "lichen aureus" is preferred for the segmental lesions that are preponderantly noted in children [119]. Remarkably, however, cases of superimposed segmental involvement have so far not been reported.

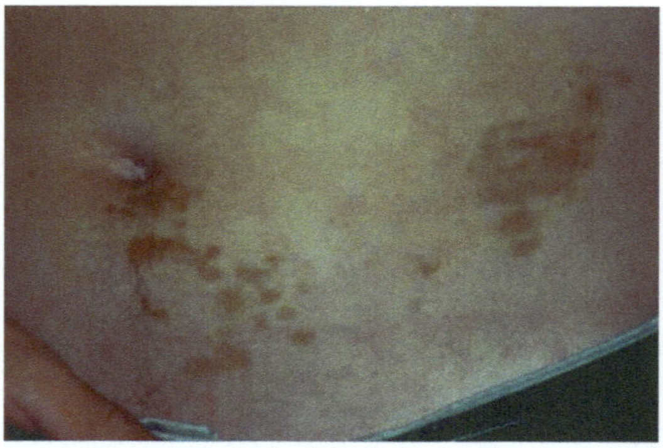

Fig. 12.24 Lichen aureus showing a characteristic segmental distribution [162] (Reprinted with permission from the Australasian College of Dermatologists © 2010 the Australasian College of Dermatologists)

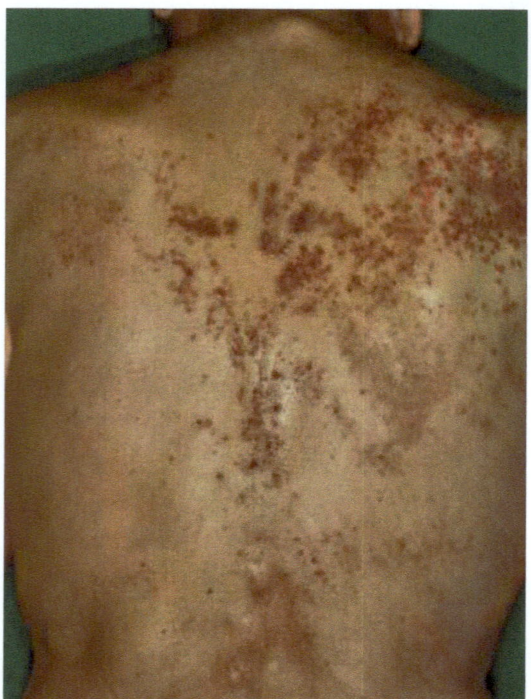

Fig. 12.25 Grover disease arranged along Blaschko's lines [12] (Reprinted with permission from S. Karger AG, Basel, Switzerland)

12.3.4 Linear Grover Disease

Grover disease is an acquired disorder characterized by multiple disseminated papular lesions resembling those of Darier disease both clinically and histopathologically [189]. Contrasting with Darier disease, however, the disorder is rather pruritic

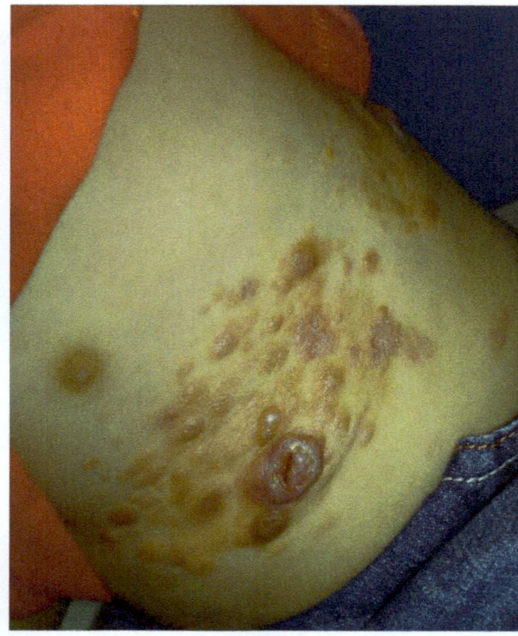

Fig. 12.26 Segmentally arranged juvenile xanthogranuloma [176] (Reprinted with permission from John Wiley & Sons, USA)

[12]. It tends to develop later in life and usually disappears spontaneously after some months. A systematized linear arrangement of the disorder (Fig. 12.25) was reported by Asahina et al. [12].

12.3.5 Linear Juvenile Xanthogranuloma

The small yellowish plaques of juvenile xanthogranuloma are a rather common disorder of infancy. On rare occasions, a pronounced linear arrangement may occur [68, 133]. This segmental involvement may include large nodules that even may show ulceration (Fig 12.26) [176]. Such cases may herald the presence of one or more genes predisposing to juvenile xanthogranuloma.

12.3.6 Linear Atrophoderma of Moulin

This acquired disorder consists of atrophic and hyperpigmented bands arranged along Blaschko's lines (Fig. 12.27) [16, 166]. As a rule, no initial inflammatory stage is reported, and sclerotic skin

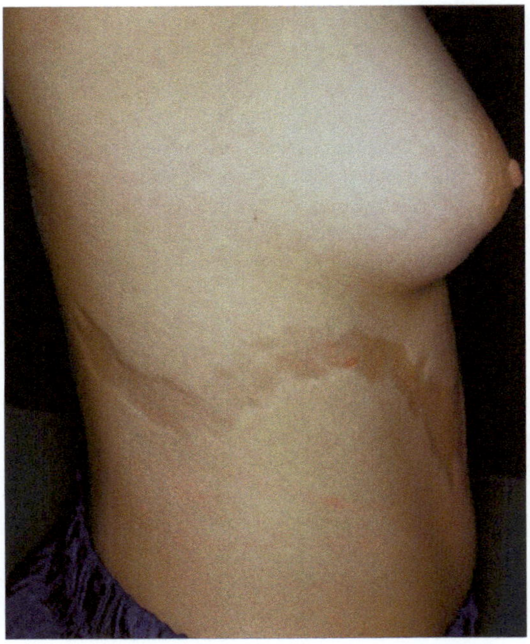

Fig. 12.27 Linear atrophoderma of Moulin [208] (Reprinted with permission from John Libbey Eurotext, Montrouge, France)

changes are absent [208]. The disorder may reflect an early postzygotic mutation giving rise to mosaicism [48].

12.3.7 Superimposed Segmental Manifestation of Common Polygenic Skin Disorders

Numerous acquired skin disorders such as psoriasis or atopic dermatitis have a polygenic background. Such diseases sometimes show a linear or otherwise segmental arrangement that may either occur as an isolated manifestation or be superimposed on the ordinary nonsegmental phenotype [86, 88] (see Sect. 3.1.3). The following paragraphs deal with both types of mosaicism but shall focus on the superimposed segmental manifestation of polygenic skin disorders.

12.3.7.1 Acquired Inflammatory Disorders

In common inflammatory skin disorders, many examples of mosaic involvement have been reported.

Psoriasis Vulgaris

Linear psoriasis may occur as an isolated disorder [33, 38, 226]. The similarities between, and the distinguishing criteria of, linear psoriasis and ILVEN have fervently been debated [3, 73, 109, 202]. In my view, the two disorders can usually be distinguished by application of clinical and histochemical criteria [85].

Historical cases of superimposed linear psoriasis [90] have been reviewed elsewhere [86, 88]. More recently published reports (Fig. 12.28) [11, 41, 134, 217, 219] corroborate the rule that the superimposed linear lesions tend to appear much earlier than the nonsegmental psoriatic plaques (Table 12.1). A novel phenomenon is the unmasking of linear involvement by effective treatment

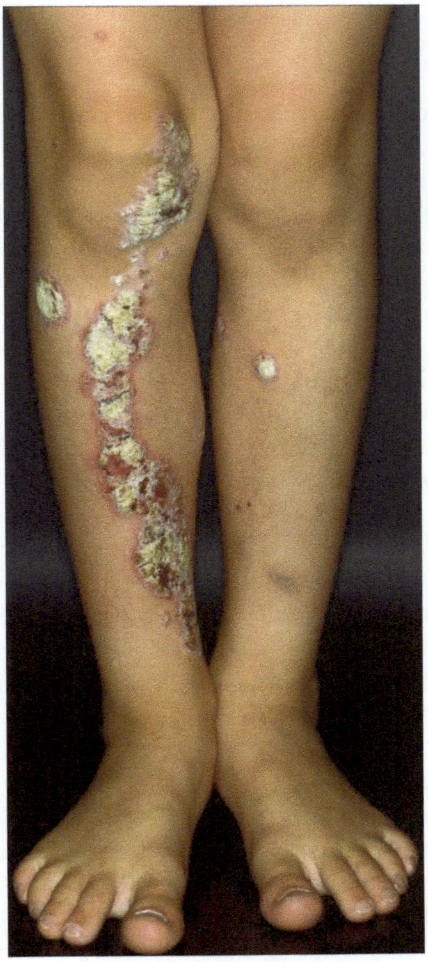

Fig. 12.28 Superimposed linear psoriasis in a 7-year-old boy. Note additional nonsegmental lesions on both legs [167]

Table 12.1 Recently published reports on superimposed linear psoriasis

Authors	Age	Age at onset of segmental lesions	Age at onset of nonsegmental lesions	Remarks
Seitz et al. [217]	19	Birth	Adolescence	One grandfather had psoriasis vulgaris Linear lesions were resistant to topical treatment with steroids and dithranol
Sfia et al. [219]	29	?	12 years	Linear lesions were unmasked by treatment with infliximab
Arnold et al. [11]	50	?	Adolescence	Linear lesions were unmasked by treatment with adalimumab
Kira and Katayama [134]	18	3 years	17 years	Linear lesion showed epidermal expression of involucrin, thus excluding "ILVEN" It was resistant to topical treatment with calcipotriol and clobetasol
Colombo et al. [41]	67	65	Prior to the onset of linear lesions (!)	Linear lesions were less responsive to treatment with etanercept
Özdemir et al. [184]	28	3	Later in life	Linear lesions described as "ILVEN" were unresponsive to treatment with etanercept
Özdemir et al. [183]	42	7	Later in life	Linear lesions described as "ILVEN" were less responsive to treatment with adalimumab

of nonlinear psoriasis with biologicals [11, 41, 183, 219].

Linear psoriasis is not arranged in a "zosteriform" pattern. It should be distinguished from true zosteriform psoriasis occurring as a Köbner phenomenon in a dermatome that had previously been involved by herpes zoster.

Psoriasis Pustulosa

There are four reports on linear psoriasis pustulosa being superimposed on the ordinary nonsegmental disorder (Fig. 12.29) [124, 151, 185, 211]. In three of the patients, the pustular psoriasis was preceded by plaques of psoriasis vulgaris. In one patient the linear lesions were unmasked by oral etretinate treatment [124]. Such cases indicate that generalized pustular psoriasis is not always inherited as a Mendelian trait caused by a homozygous *IL36RN* mutation [153].

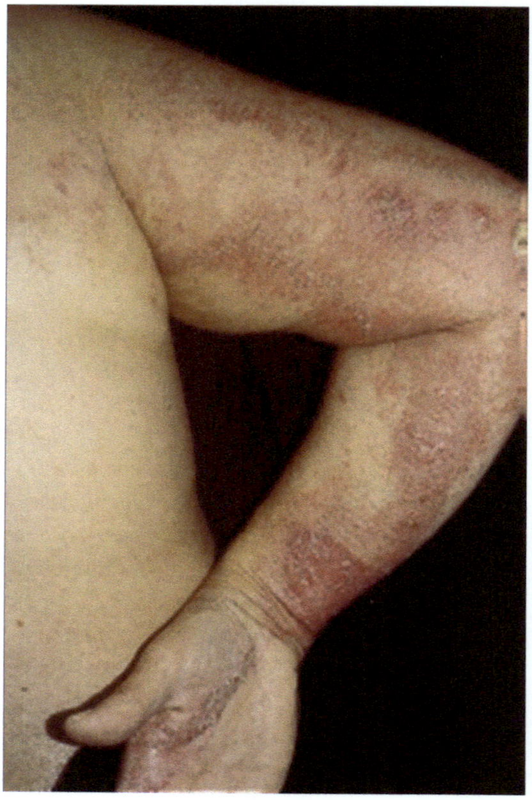

Fig. 12.29 Linear pattern of pustular psoriasis in a 57-year-old woman who had previously developed nonsegmental lesions of both psoriasis vulgaris and pustular psoriasis [185] (Reprinted with permission from Elsevier Limited, Oxford, UK)

Atopic Dermatitis

Some convincing examples of linear atopic dermatitis superimposed on nonlinear eczematous lesions have been documented (Fig. 12.30) [27, 86, 107, 234, 245]. These cases indicate that keratinocytic epitopes play a primary role in the pathogenesis of atopic dermatitis. In another report, an 18-year-old woman had itchy papules

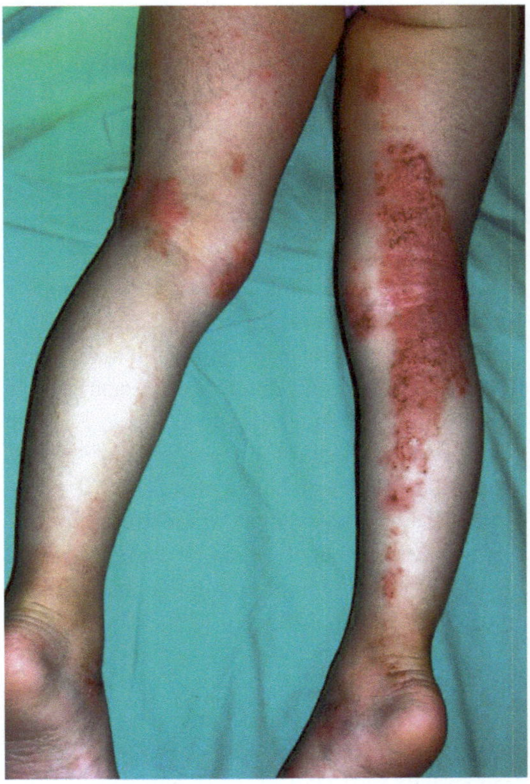

Fig. 12.30 Superimposed linear atopic dermatitis. Note nonlinear involvement of the left leg (Courtesy of Dr. Antonio Torrelo, Madrid, Spain)

in a linear distribution down the inner aspect of her right leg [248]. Nonsegmental skin lesions were absent, but she had atopic features in the form of hay fever: positive prick tests to house dust mite, feathers, and grass; and raised IgE. Removal of split skin in an itchy area did not yield lasting relief, whereas excision of full thickness skin with subsequent grafting of the defect by split skin from an unaffected donor area resulted in permanent relief.

Moreover, some cases of "lichen striatus" associated with atopic dermatitis [55, 192, 229, 243, 245] may in fact be categorized as examples of superimposed linear atopic dermatitis.

Chronic Prurigo

Kawachi et al. [128] described a 65-year-old man with a 20-year history of chronic prurigo and a 3-year history of a pronounced linear pruritic lesion involving his left leg (Fig. 12.31). It is not clear whether the authors wanted to separate the disorder from atopic dermatitis, but by all means this case represents an impressive example of superimposed segmental manifestation of a chronic pruriginous disorder.

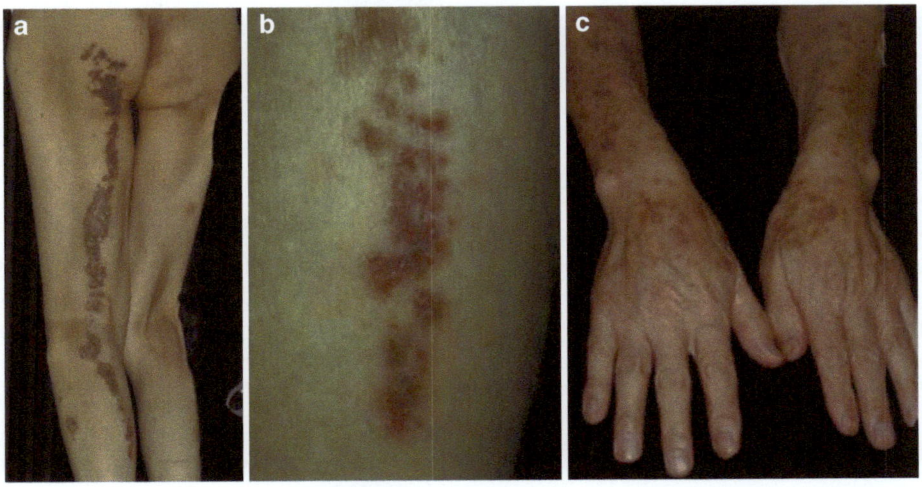

Fig. 12.31 Superimposed linear pruriginous dermatitis in a 65-year-old man. (**a**, **b**) Segmental involvement present for 3 years; (**c**) nonsegmental prurigo present for 20 years [128] (Reprinted with permission from John Libbey Eurotext, Montrouge, France)

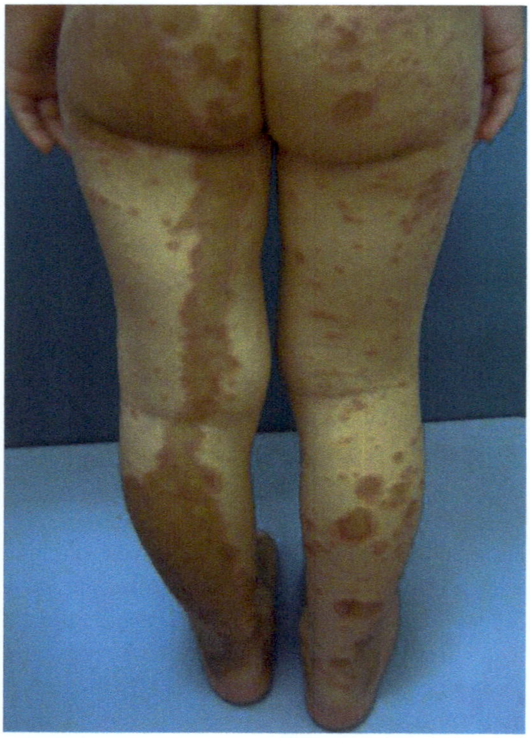

Fig. 12.32 Superimposed linear lichen planus in a 4-year-old girl [181] (Reprinted with permission from John Wiley & Sons, USA)

Table 12.2 Cases of superimposed linear lichen planus preceding the appearance of nonlinear lesions

Authors	Age at onset of linear lesions	Onset of nonlinear lesions
Davis [52]	42 years	45 years
Irgang [115]	46 years	4 months later
Nagy et al. [172]	68 years	Some weeks later
Fink-Puches et al. [64]	42 years	"Later"
Gunning and Turiansky [78]	54 years	4 months later[a]
Kanwar and De [125] Case 1	Childhood	"Subsequently"
Kanwar and De [125] Case 2	Childhood	"Subsequently"
Hamade et al. [80]	5 years	Some weeks or months later
Onder et al. [181]	4 years	"During the following 3 months"

[a]Patient had developed a rash of nonsegmental lichen planus 5 years earlier

Lichen Planus

Isolated forms of linear involvement are often described [28, 132, 145, 214, 218, 230, 249]. By contrast, cases of superimposed linear lichen planus (Fig. 12.32) are more rarely published [86]. Sometimes their arrangement is erroneously described as being "zosteriform" [195]. From the reports documenting the sequence of events, it is obvious that the linear lesions tend to appear earlier than the nonlinear ones (Table 12.2).

In a report on 100 children with lichen planus, Kanwar and De [125] diagnosed a linear type in 12 patients. Subsequently two of these children developed, in addition, nonsegmental lesions. From other studies on lichen planus in childhood, one has the impression that the authors still follow the simple scheme of "classic" versus "linear" involvement, without accepting the possibility of "mixed" cases [81, 148, 173, 177, 220].

Lichen Planopilaris

An isolated mosaic manifestation is rather frequently reported. Cases of linear facial lesions predominate [70, 71, 138, 257], but other regions of the body may likewise be affected [13].

A superimposed linear involvement was described in a 29-year-old man who had itchy papules arranged in a linear distribution on the right side of his scalp, neck, and trunk (Fig. 12.33a–c) [35]. On his abdomen multiple, disseminated, nonsegmental lesions of lichen planopilaris giving rise to hairlessness were noted (Fig. 12.33d).

Lichen Nitidus

A few cases of isolated linear lichen nitidus have been documented [196, 199]. In another report, a 6-year-old boy had pronounced lesions of linear lichen nitidus on his left hand, giving rise to disturbed nail growth [23]. Moreover, the patient developed bilateral disseminated lesions of lichen nitidus (Fig. 12.34). Hence, this appears to be a classical example of superimposed linear lichen nitidus [87].

Acne Vulgaris

The following case reports can today be categorized as examples of superimposed segmental

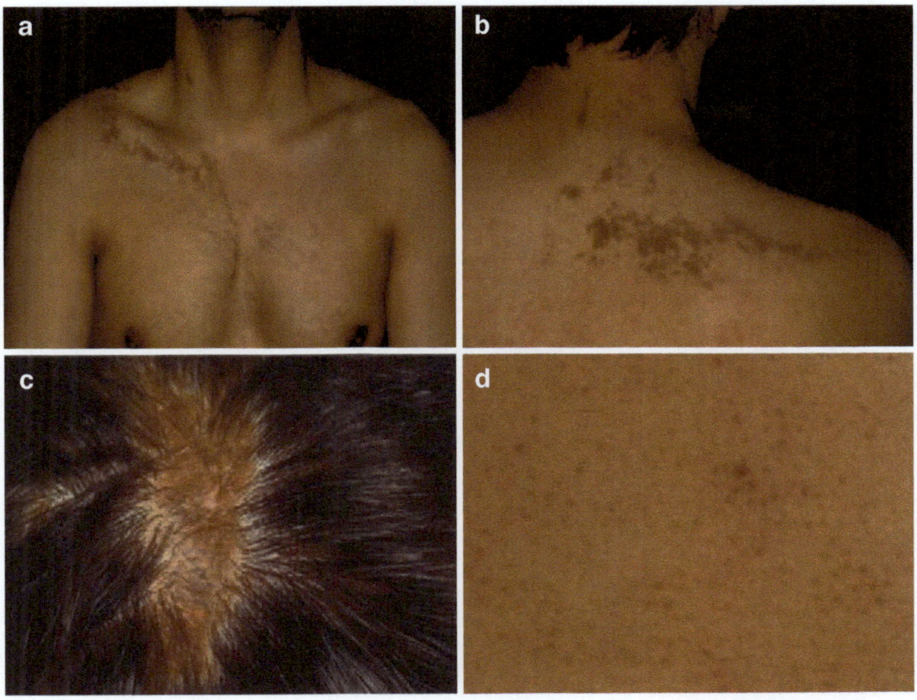

Fig. 12.33 Superimposed linear lichen planopilaris. (**a**, **b**) Linear lesions involving the trunk; (**c**) linear lesion on the scalp; (**d**) disseminated lesions on the abdomen [35] (Reprinted with permission from John Libbey Eurotext, Montrouge, France)

manifestation of acne vulgaris. Hughes and Cunliffe [111, 112] described a 17-year-old boy who had, since 2 years, bilateral acne lesions involving his face, back, and chest. In addition, a patchy area of severe acne involved the left side of his chest (Fig. 12.35). This "acne nevus" was said to have been present since birth. Similarly, González-Hermosa et al. [72] described a 26-year-old man with a 9-year history of recalcitrant acne lesions involving the right half of the back, whereas only three individual lesions were found on the left side of his body. In a 26-year-old woman with "grade 2 acne on the left side and grade 4 on the right side," Cooper et al. [42] found that the sebum excretion rate was higher on the right-hand side, and that conversion of dehydroepiandrosterone to androstenedione and 5α-steroids was greater in the biopsy from the right side.

An interesting counterpart of superimposed segmental acne vulgaris was described in the form of an "acne-free nevus" [46]. A 17-year-old boy had mild facial acne and severe papulopustular acne on his back and chest. On the back, four well-demarcated patches being completely free of acne were noted. In these acne-free areas, the sebaceous glands were smaller and produced less sebum when compared to acne-bearing skin. Moreover, the percentage of squalene was lower in the acne-free patches.

Discoid Lupus Erythematosus

An isolated linear involvement is rather frequently reported in children [57, 75, 129, 203], adolescents [47], and adults [2, 4, 29, 67, 123, 216, 242]. The lesions are predominantly noted on the face. Sometimes antinuclear antibodies are found in the blood [1, 75, 212], which would raise the question of a "superimposed" manifestation. A convincing case of superimposed linear manifestation of the disorder

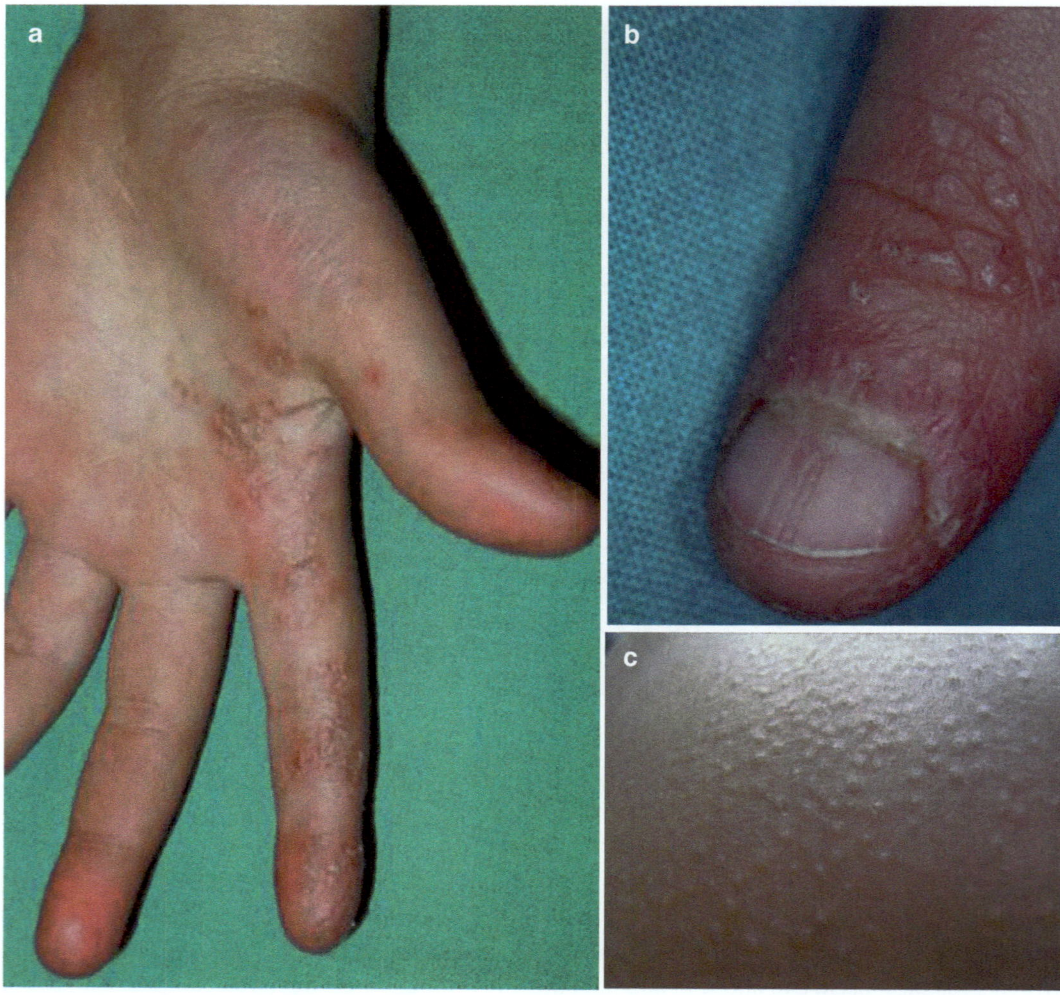

Fig. 12.34 Superimposed linear lichen nitidus. (**a**) Pronounced linear involvement of the left hand; (**b**) disturbed growth of left thumb nail; (**c**) disseminated papules on the abdomen [23] (Reprinted with permission from *European Journal of Pediatric Dermatology*, Bari, Italy)

(Fig. 12.36) was reported by Verma and Wollina [252].

Lupus Erythematosus Profundus

Linear forms of lupus panniculitis have sometimes been reported [141, 155, 232, 238]. The scalp is rather often involved (Fig. 12.37) [37, 135, 171, 204, 247]. Antinuclear antibodies have rarely been found [141, 171, 247]. Two of the patients later developed classical features of systemic lupus erythematosus [62, 155]. Such cases can be taken as examples superimposed segmental manifestation of SLE.

Subacute Cutaneous Lupus Erythematosus

In a 42-year-old woman, typical polycyclic lesions of subacute cutaneous lupus erythematosus arranged in a strictly unilateral, systematized, linear pattern were noted (Fig. 12.38) [206]. Oral treatment with corticosteroids and antimalarial drugs resulted in remission.

Practical Aspects

The linear forms of cutaneous lupus erythematosus tend to respond rather quickly to established therapeutic approaches such as topical steroids or

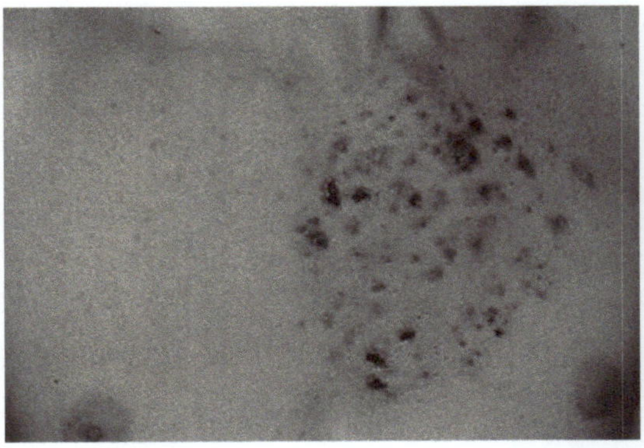

Fig. 12.35 Superimposed segmental acne in a 17-year-old boy who developed nonsegmental lesions of acne vulgaris at the age of 15 years, whereas the "acne nevus" was present since birth [111, 112] (Reprinted with permission from John Wiley & Sons, USA)

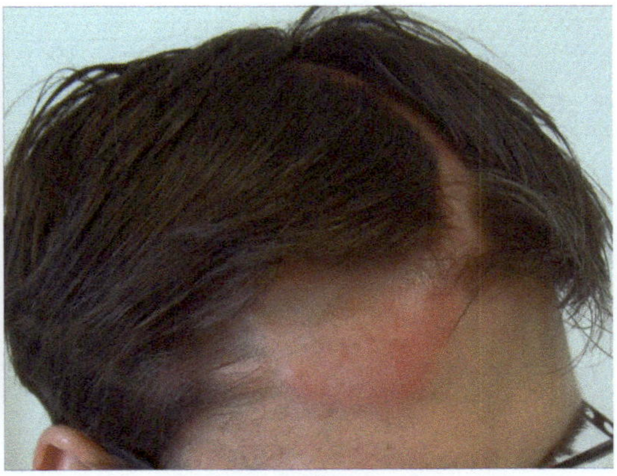

Fig. 12.37 Linear lupus erythematosus profundus in 32-year-old man who did not show any nonsegmental lesions [135]

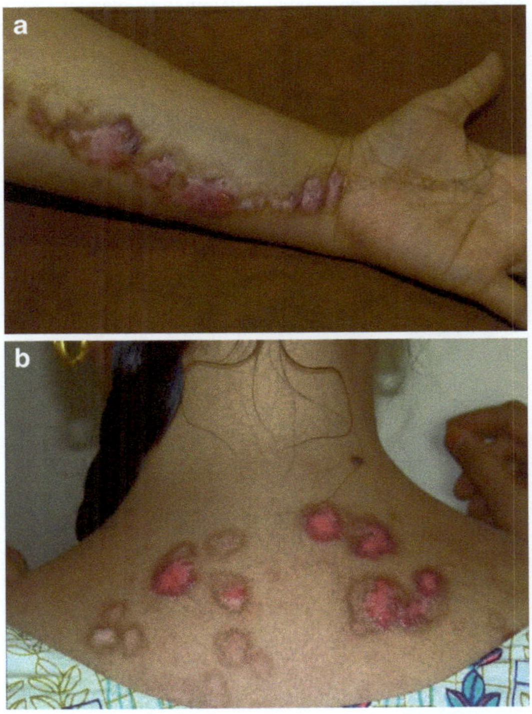

Fig. 12.36 Superimposed linear manifestation of discoid lupus erythematosus. (**a**) Linear involvement of the left arm; (**b**) nonlinear lesions on the back [252] (Courtesy of Dr. Shyam B. Verma, Vadodara, India)

tacrolimus or systemic administration of steroids [206, 239], chloroquine [37, 141, 206], or dapsone [51].

Systemic Lupus Erythematosus

A 9-year-old girl with typical clinical and laboratory findings of systemic lupus erythematosus (SLE) developed papules and bullae arranged in a linear configuration on her right forearm and hand [207]. A severe general exacerbation gave rise to acute deterioration of these linear skin lesions as well as to nonlinear blistering of the lips and the oropharynx. Another patient had linear lupus erythematosus profundus of the face and scalp [155]. She later developed clinical symptoms and laboratory changes of systemic lupus erythematosus, which is why this case can likewise be categorized as an example of superimposed segmental manifestation of SLE.

Dermatomyositis

An 8-year-old boy had an atrophic and indurated streak involving his left leg from the buttock to the heel [24]. The linear lesions were partly ulcerated and intermittently showed extrusion of calcified material. Soon after the onset of the linear disease, the patient developed typical nonsegmental signs and symptoms of juvenile dermatomyositis including malar rash, bilateral Gottron papules, arthralgia, proximal weakness, and general malaise. Hence, the linear disorder was interpreted as a superimposed segmental manifestation of dermatomyositis (Fig. 12.39). Similar cases were

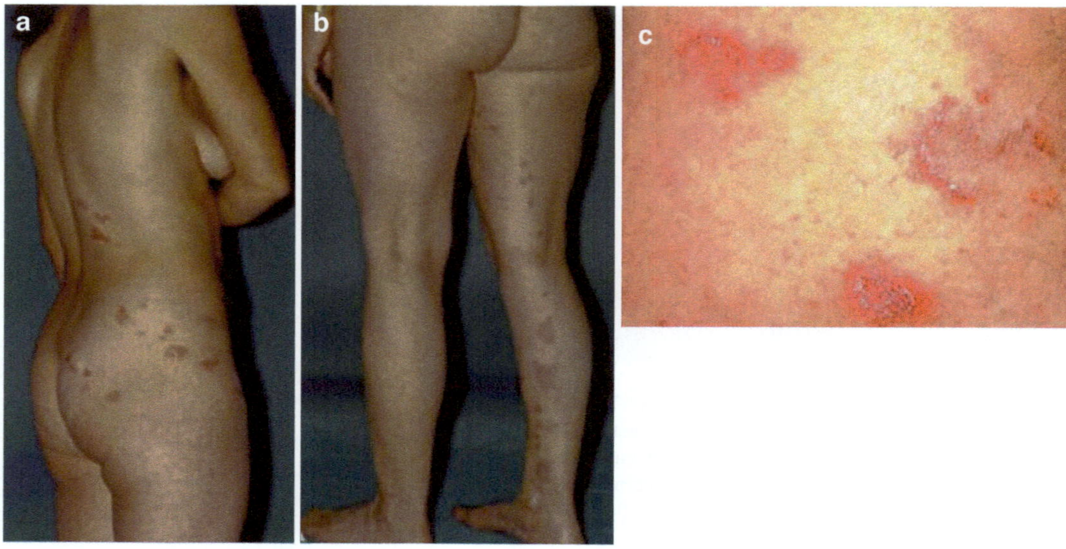

Fig. 12.38 Linear subacute lupus erythematosus in a 42-year-old woman. (**a**) Unilateral lesions on the trunk and buttock; (**b**) involvement of the ipsilateral leg; (**c**) close-up view of polycyclic inflammatory plaques [206] (Reprinted with permission from John Libbey Eurotext, Montrouge, France)

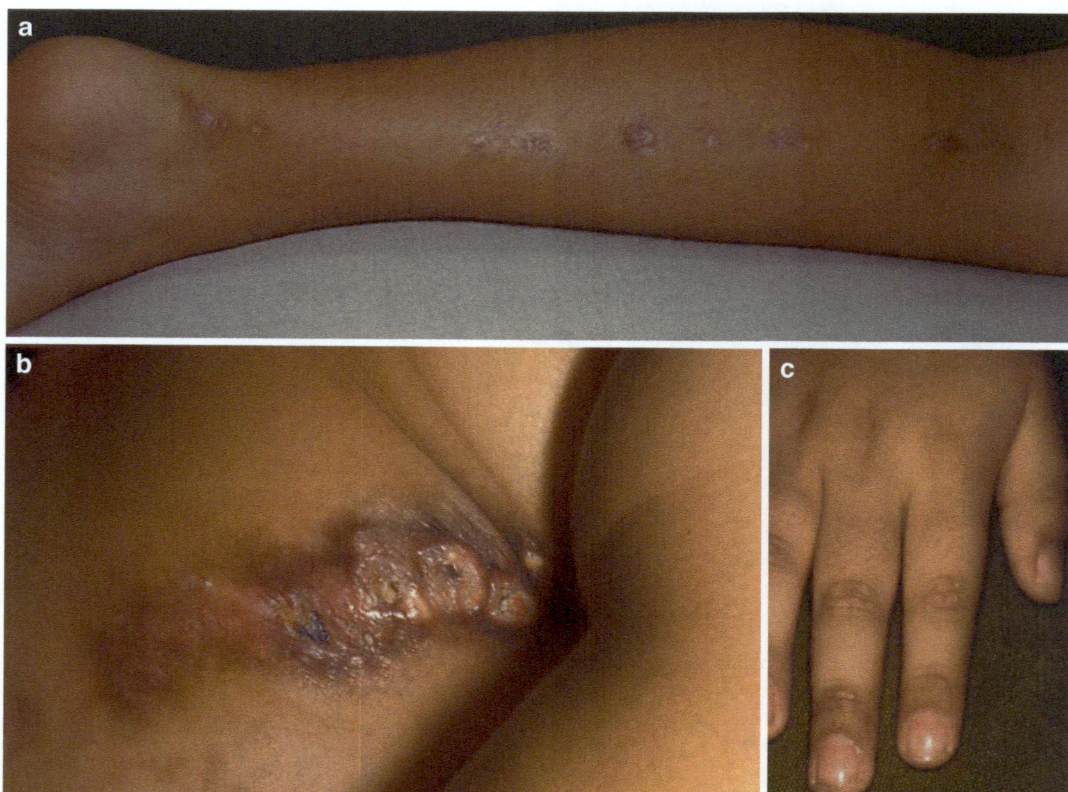

Fig. 12.39 Superimposed linear manifestation of dermatomyositis. (**a**) Linear arrangement of lesions showing extrusion of calcified deposits; (**b**) close-up of proximal part of linear involvement; (**c**) subsequent nonsegmental involvement in the form of bilateral Gottron papules [24] (Reprinted with permission from S. Karger AG, Basel, Switzerland)

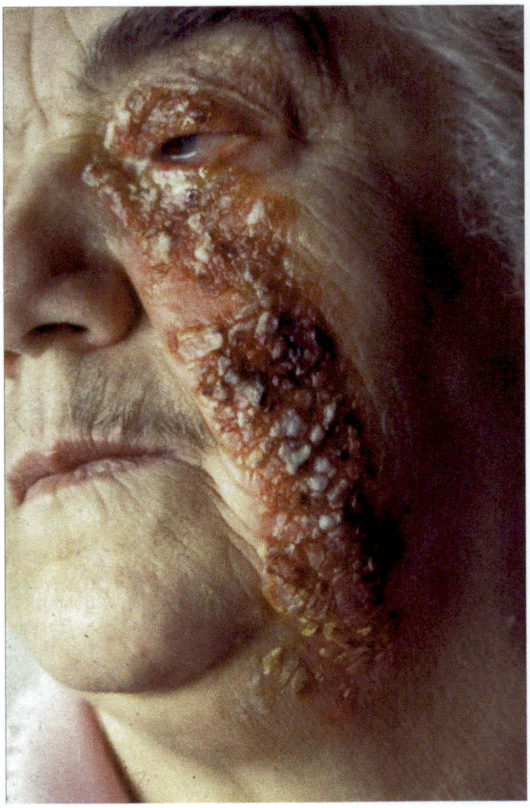

Fig. 12.40 Superimposed linear pemphigus vulgaris in a 64-year-old woman [45] (Reprinted with permission from S. Karger AG, Basel, Switzerland. Color photograph kindly provided by Dr. Franco Crovato, Genoa, Italy)

documented by Lipsker and Lenormand [143] and by Torchia [244] who revisited a case of “blaschkitis” as noted in a child with dermatomyositis [237].

A large segmental calcinosis of the muscular loges of the right shoulder and upper arm was described in a 37-year-old man who had, in addition, classical nonsegmental features of dermatomyositis [65]. The calcified masses constituted a major handicap and had to be surgically removed. Most likely they represented a superimposed segmental manifestation [89].

Pemphigus Vulgaris

A Blaschko-linear arrangement of facial pemphigus vegetans (Fig. 12.40) was documented in an elderly woman who had, in addition, bilateral nonlinear lesions of pemphigus vulgaris [45]. The nonsegmental lesions responded well to oral steroid treatment, whereas the linear pemphigus vegetans remained unchanged. This case can be taken as an example of superimposed linear pemphigus vulgaris [86].

Hasson et al. [103] described a 64-year-old woman with pronounced, unilateral, linear lesions of pemphigus vulgaris being superimposed on disseminated, less-severe lesions. The authors argued that the band-like distribution might be due to a mastectomy scar. On the other hand, they stated that the linear arrangement “can be closely correlated with Blaschko’s lines… However, as the scar grossly follows this pattern, it is difficult to decide which factor might be implicated.”

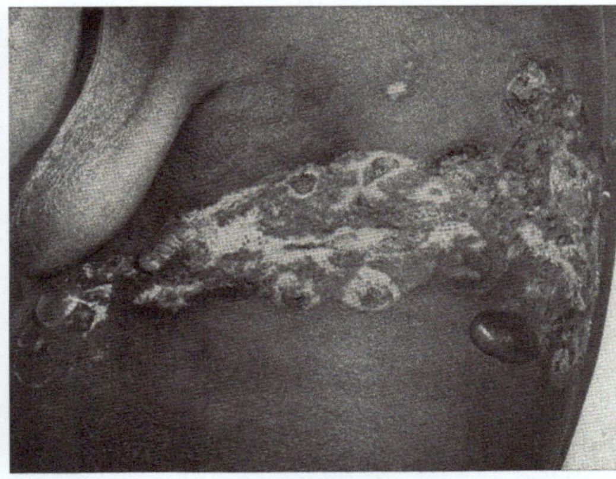

Fig. 12.41 Superimposed segmental manifestation of bullous pemphigoid in a 64-year-old woman [114] (Reprinted with permission from John Wiley & Sons, USA)

Bullous Pemphigoid

A 64-year-old woman had scattered cutaneous and oral lesions of bullous pemphigoid involving both sides of her body [114]. At an early stage of the disease she developed, on the left side of her chest, an eruption that was “massive and zosteriform in type.” The author considered this segmental eruption (Fig. 12.41) to be unique and suggested the possibility of a “posterior root ganglion involvement.” The clinical and histopathological features, however, do not appear to be compatible with a concomitant zoster infection. A diagnosis of superimposed segmental manifestation of bullous pemphigoid is more likely.

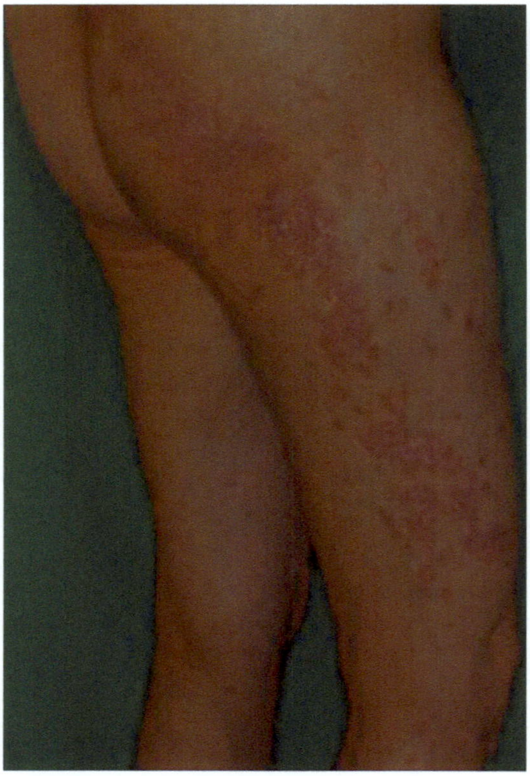

Fig. 12.42 Superimposed linear graft-versus-host disease in a 50-year-old woman [180] (Reprinted with permission from John Libbey Eurotext, Montrouge, France)

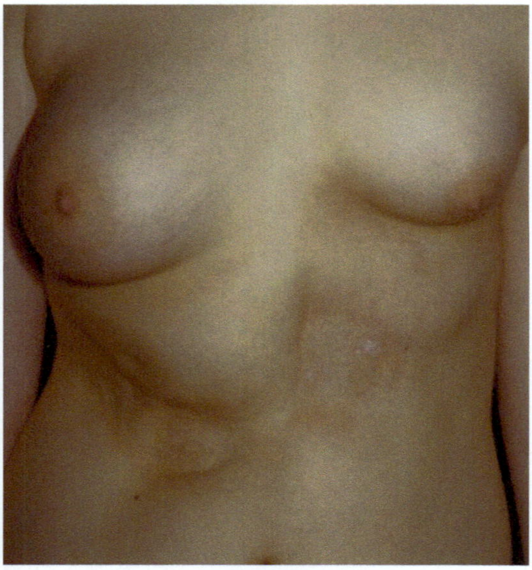

Fig. 12.43 Systematized linear morphea with unilateral atrophy of breast (Courtesy of Dr. Antonio Torrelo, Madrid, Spain)

Graft-Versus-Host Disease

A case of "lichen striatus" occurring after allogeneic stem cell transplantation [170] may best be explained as an example of isolated linear graft-versus-host disease.

A superimposed linear graft-versus-host reaction being followed by disseminated nonsegmental lesions (Fig. 12.42) has so far been described in seven recipients of allogeneic bone-marrow transplantation [39, 86, 180]. In such cases, the linear eruption was neither "dermatomal" nor "zosteriform" [39] but followed Blaschko's lines. By way of exception, however, a true dermatomal arrangement reflecting a Köbner phenomenon caused by herpes zoster infection may also occur [213].

Morphea

Morphea may sometimes be arranged in a segmental distribution suggesting mosaicism (Fig. 12.43). Several authors assumed that such lesions follow Blaschko's lines [254], but convincing evidence was only presented, so far, in one case [104] that was reminiscent of linear atrophoderma of Moulin [48]. In other reports [121, 127, 210, 225, 228] the neutral term "segmental" is more appropriate [26], and the statement that "morphea follows Blaschko's lines" [254] seems unjustified. Remarkably, several authors mentioned "mixed" cases of linear scleroderma associated with nonsegmental morphea [105, 262]. According to Zulian et al. [262], in most patients (64 %) the linear lesions appeared before, or at the same time as, the plaque lesions. Possibly, such cases represent an additional example of superimposed segmental manifestation of a polygenic skin disorder.

Remarkably, Alain Taïeb from Bordeaux made a diagnosis of eosinophilic fasciitis (Shulman syndrome [10]) being confirmed histopathologically in a man who had, since the age of 38 years, a linear scleroderma-like induration of his right pectoralis region and arm (Fig. 12.44a). Three years later, the patient developed bilaterally disseminated lesions of morphea (Fig. 12.44b). Dr. Taïeb inferred that the patient's linear disorder could be taken as a superimposed mosaic

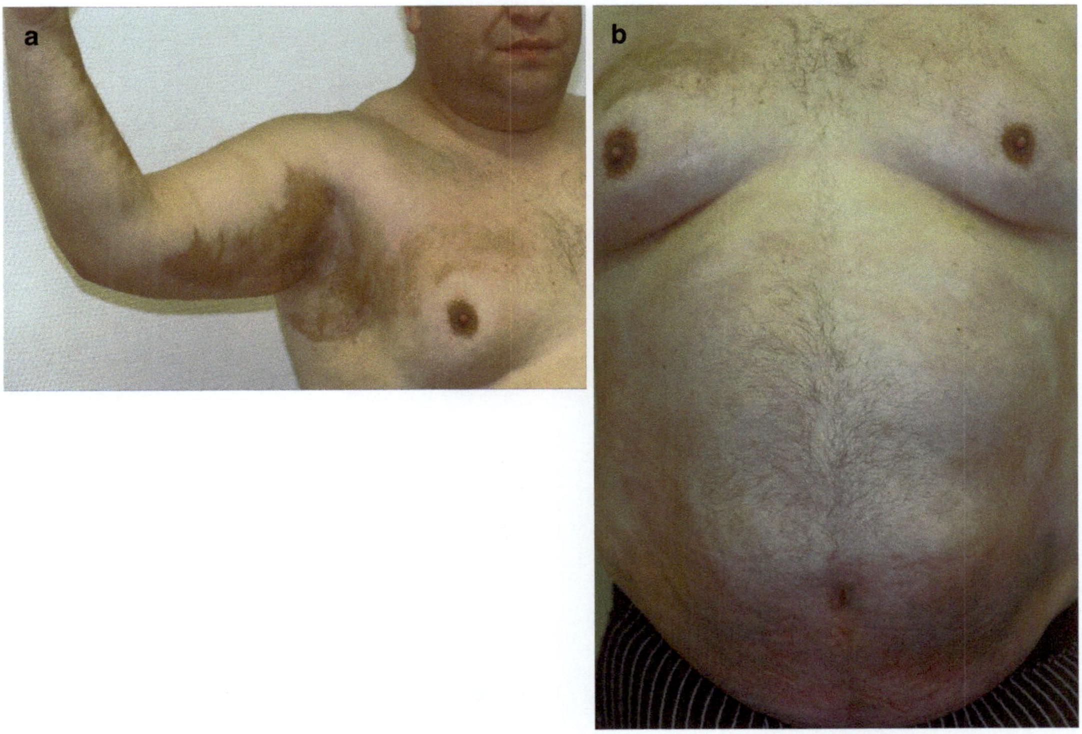

Fig. 12.44 Coexistence of Shulman syndrome and morphea in a 42-year-old man. (**a**) Unilateral linear lesions of eosinophilic fasciitis were first noted at the age of 38 years; (**b**) bilateral patches of morphea appeared 3 years later (Courtesy of Dr. Alain Taïeb, Bordeaux, France)

manifestation of disseminated morphea. A close relationship between eosinophilic fasciitis and morphea had previously been assumed by other authors [159].

Granuloma Annulare

Two cases of isolated linear granuloma annulare have been documented [101, 157].

Two other reports described superimposed segmental granuloma annulare [164, 246]. In one of these patients [164], the skin lesions were arranged in a systematized pattern (Fig. 12.45a and b) and preceded the appearance of disseminated nonsegmental granuloma annulare (Fig. 12.45c).

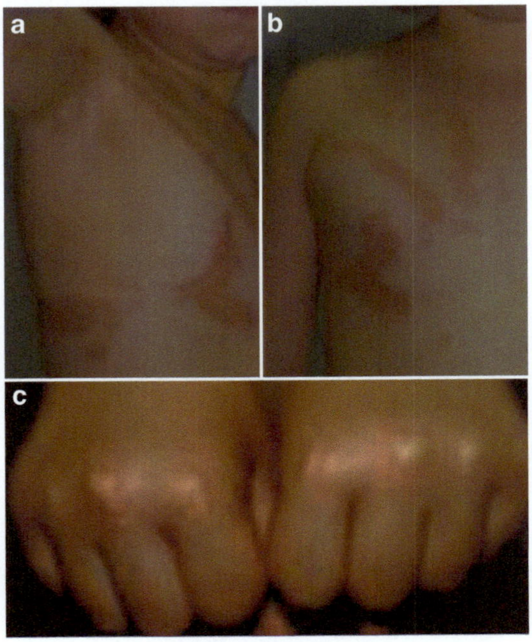

Fig. 12.45 Superimposed linear granuloma annulare in a 7-year-old girl. (**a, b**) Pronounced linear lesions present since infancy; (**c**) nonsegmental lesions that appeared later during childhood [164] (Reprinted with permission from John Wiley & Sons, USA)

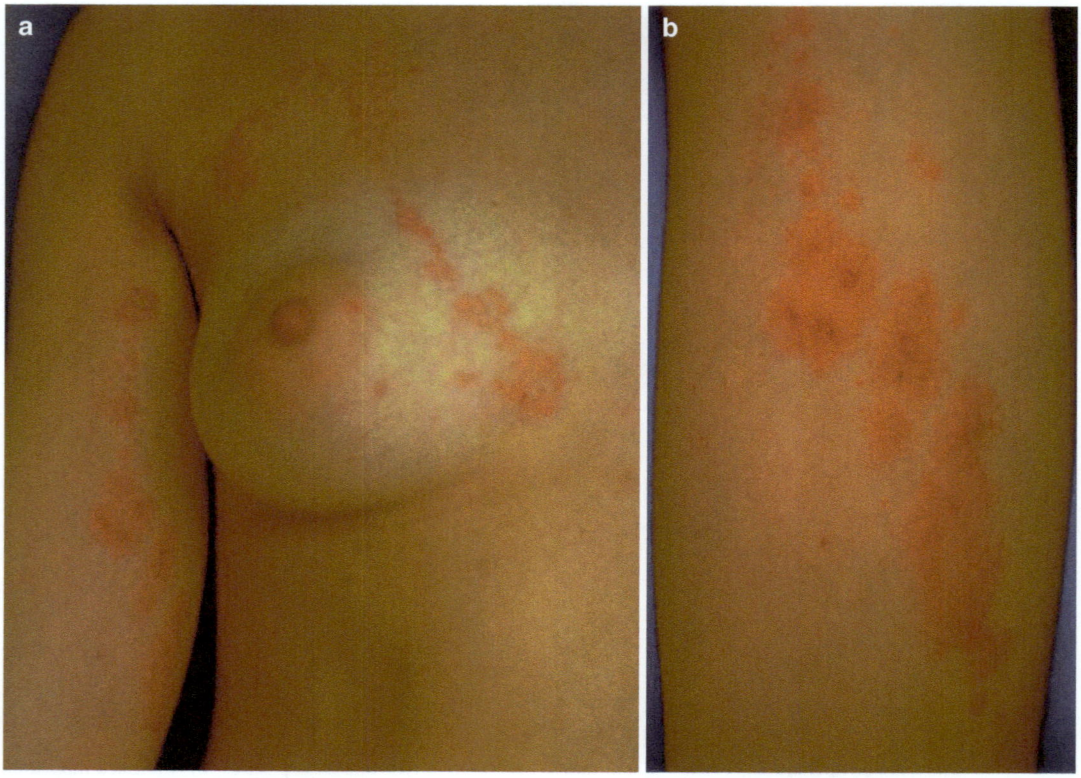

Fig. 12.46 Isolated linear erythema multiforme in a 20-year-old woman [160]. (**a**) General view; (**b**) close-up showing bullous target lesions on the right arm (Reprinted with permission from John Wiley & Sons)

Erythema Multiforme

Isolated linear erythema multiforme was described in a patient with herpes labialis (Fig. 12.46) [160]. A case of superimposed linear erythema multiforme was documented in a review on "acute linear dermatoses" [18].

Common Drug Eruption

Two cases of isolated linear lichenoid drug eruption induced by the antihypertensive valsartan were documented (Fig. 12.47) [69].

A 19-year-old woman developed linear inflammatory skin lesions on the right clavicular region and on her right arm after intake of ibuprofen [5]. A few days later a bilateral diffuse rash appeared. This case was interpreted as an example of superimposed linear drug eruption [93]. Two cases of "blaschkitis" induced by metronidazole [34] or lenalidomide [74] can likewise be reclassified as examples of superimposed linear drug eruption since the photographic documentation (Fig. 12.48) clearly shows additional nonsegmental lesions that were not mentioned in the these reports.

Fixed Drug Eruption

An isolated segmental manifestation was described in a 25-year-old woman who had repeated attacks of fixed drug eruption involving her right arm in a linear pattern (Fig. 12.49) [186]. Oral challenge with a low dose of cotrimoxazole gave rise to a severe reactivation of the linear eruption.

A convincing example of superimposed linear fixed drug eruption has likewise been documented [224]. The lesions involved the right leg in a Blaschko-linear arrangement. Moreover, the patient had a few nonsegmental lesions on his other leg, the hands, the upper lip, and the scrotum.

Superimposed Lateralized Exanthem of Childhood

"Asymmetric periflexural exanthem of childhood" or "unilateral laterothoracic exanthem" is

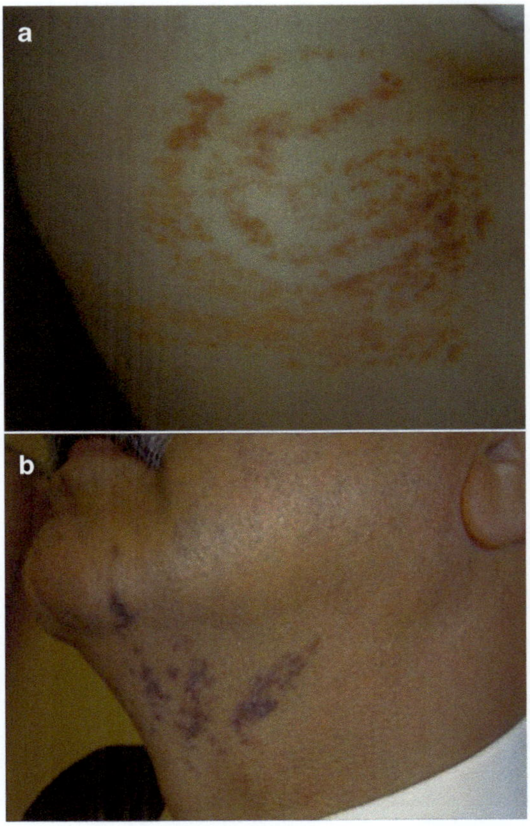

Fig. 12.47 Isolated form of linear lichenoid drug eruption induced by valsartan in (**a**) a 46-year-old woman; (**b**) a 57-year-old man [69] (Reprinted with permission from John Wiley & Sons, USA)

most likely caused by viral infections [43]. The disease begins with a unilateral involvement, but subsequently some contralateral lesions tend to develop in most cases [156]. This phenomenon (Fig. 12.50) can perhaps be explained by the concept of superimposed lateralized exanthem of childhood [98].

12.3.7.2 Vitiligo

As a general rule, segmental vitiligo does not follow Blaschko's lines, nor is it arranged in a dermatomal pattern. Isolated forms occur rather frequently [61, 136]. On the other hand, numerous cases of superimposed segmental vitiligo (Fig. 12.51) have likewise been documented, mostly under the term "mixed vitiligo" [58, 60, 61, 168, 251]. In such cases, the superimposed segmental component usually precedes the appearance of nonsegmental lesions by many months and sometimes even by more than 2 years [60, 175]. If a patient with segmental vitiligo has also halo nevi or leukotrichia, his risk to develop other nonsegmental lesions of vitiligo is particularly high [59].

Remarkably, superimposed segmental vitiligo has been reported to be more prone to local recurrences after suction blister grafting when compared to the isolated involvement [60, 140].

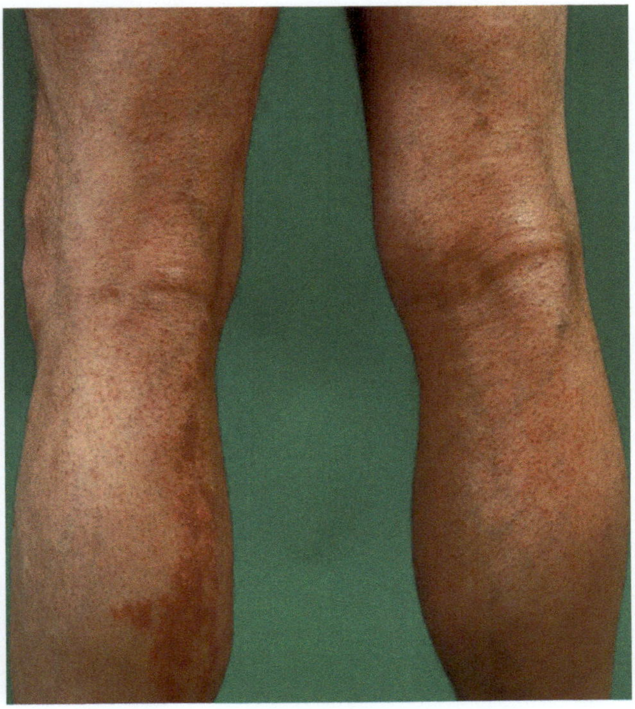

Fig. 12.48 Superimposed linear manifestation of a papular drug eruption induced by lenalidomide [74]

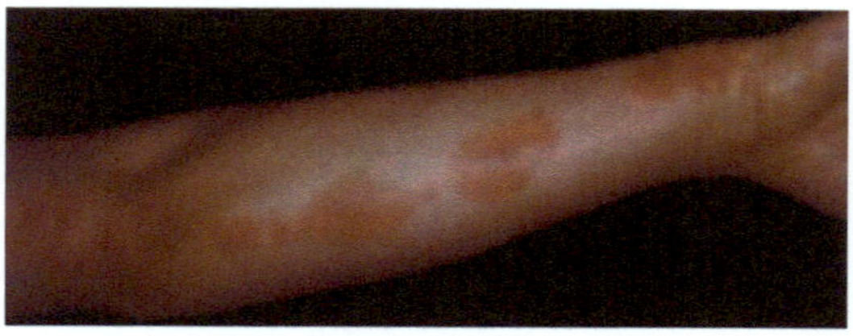

Fig. 12.49 Linear fixed drug eruption induced by cotrimoxazole in a 25-year-old woman [186] (Reprinted with permission from John Wiley & Sons, USA)

Because vitiligo is a polygenic disorder, the term "type 2 mosaicism" [60] is not appropriate [91].

12.3.7.3 Cherry Angiomas

Multiple cherry angiomas ("senile angiomas") usually develop during the second half of life. They are so common that a polygenic background is very likely. A 62-year-old woman had multiple cherry angiomas that had appeared eruptively at

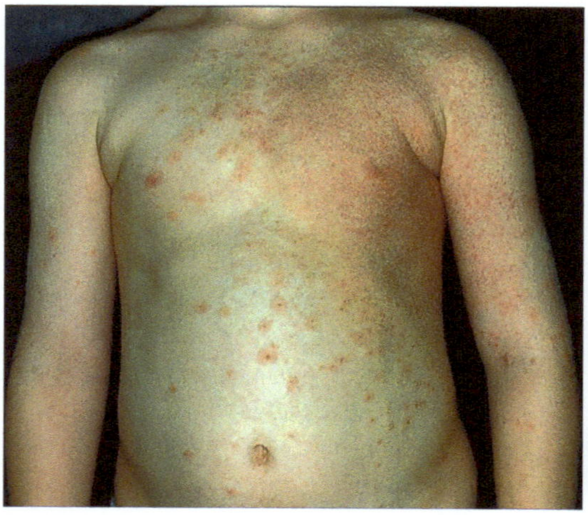

Fig. 12.50 Superimposed lateralized exanthem of childhood in a 6-year-old boy. Note presence of rather mild contralateral lesions (Courtesy of Dr. Thomas Ruzicka, Munich, Germany)

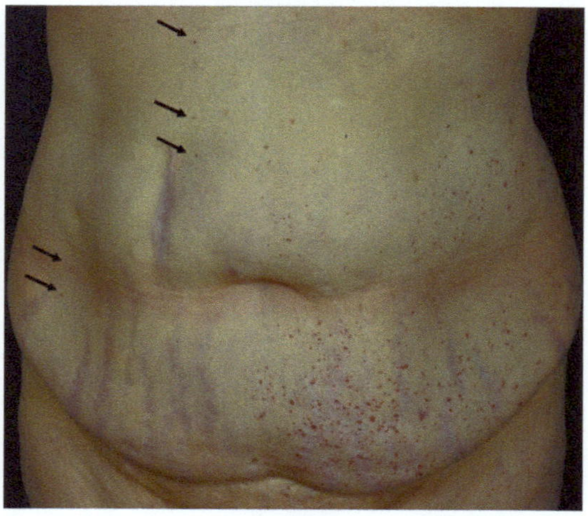

Fig. 12.52 Superimposed segmental manifestation of multiple cherry angiomas. Note paucity of disseminated lesions (*arrows*) [118] (Courtesy of Dr. WenChieh Chen, Munich, Germany)

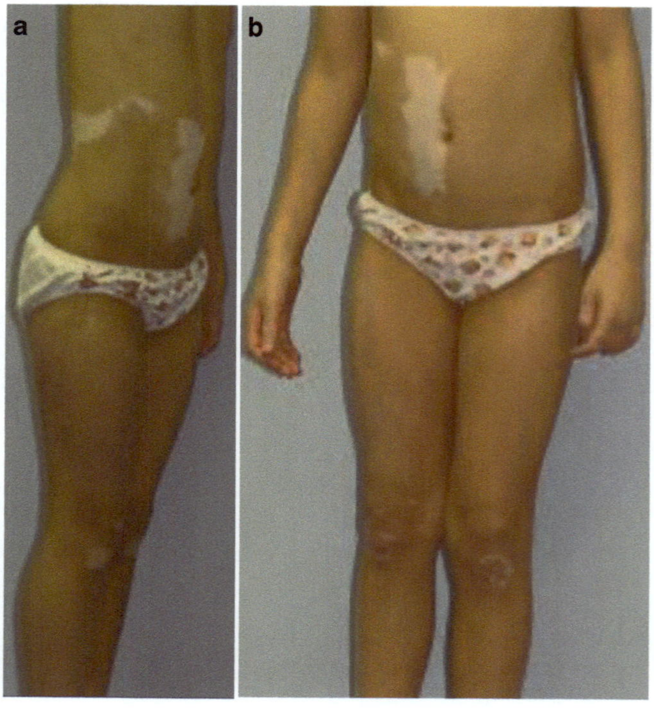

Fig. 12.51 Superimposed segmental vitiligo in a 6-year-old boy. (**a**) Segmental lesions developed at 4 years of age, whereas bilateral nonsegmental vitiligo appeared at 5 years of age; (**b**) nonsegmental vitiligo responded to narrow band UVB therapy, whereas the segmental lesions remained unchanged [60] (Reprinted with permission from Elsevier Limited, Oxford, UK)

the age of 30 years in a systematized segmental arrangement involving the left side of her trunk and the right thigh [118]. Additional nonsegmental lesions developed later, after menopause (Fig. 12.52). The authors interpreted this case as a further example of superimposed segmental manifestation of a polygenic trait.

References

1. Abe M, Ishikawa O, Miyachi Y (1998) Linear cutaneous lupus erythematosus following the lines of Blaschko. Br J Dermatol 139:307–310
2. Abe M, Ohnishi K, Ishikawa O (2000) Guess what? Linear cutaneous lupus erythematous (LCLE): relationship with Blaschko's lines. Eur J Dermatol 10:229–231
3. Akelma AZ, Cizmeci MN, Kanburoglu MK, Mete E (2013) A diagnostic dilemma: inflammatory linear verrucous epidermal nevus versus linear psoriasis. J Pediatr 162:879
4. Alcantára-González J, Fernandez-Guarino M, Carrillo-Gijon R, Jáen-Olasolo P (2011) Linear cutaneous lupus erythematosus. Indian J Dermatol Venereol Leprol 77:717–719
5. Alfonso R, Belinchon I (2001) Linear drug eruption. Eur J Dermatol 11:122–123
6. Alikhan A, Lee AD, Swing D, Carroll C, Yosipovitch G (2010) Vaccination as a probable cause of incontinentia pigmenti reactivation. Pediatr Dermatol 27:62–64
7. Aloi FG, Tomasini CF, Isaia G, Grazia Bernengo M (1989) Unilateral linear basal cell nevus associated with diffuse osteoma cutis, unilateral anodontia, and abnormal bone mineralization. J Am Acad Dermatol 20:973–978
8. Anderson TE, Best PV (1962) Linear basal-cell naevus. Br J Dermatol 74:20–23
9. Anderson RC, Zinn AR, Kim J, Carder KR (2005) X-linked reticulate pigmentary disorder with systemic manifestations: report of a third family and literature review. Pediatr Dermatol 22:122–126
10. Antic M, Lautenschlager S, Itin PH (2006) Eosinophilic fasciitis 30 years after - what do we really know? Report of 11 patients and review of the literature. Dermatology 213:93–101
11. Arnold AW, Happle R, Itin PH (2010) Superimposed linear psoriasis unmasked by therapy with adalimumab. Eur J Dermatol 20:573–574
12. Asahina A, Ishiko A, Saito I, Hasegawa K, Sawamura D, Nakano H (2012) Grover's disease following multiple bilateral Blaschko lines: a rare clinical presentation with genetic and electron microscopic analyses. Dermatology 225:183–187
13. Baker K, Pehr K (2006) Linear lichen planopilaris of the trunk: first report of a case. J Cutan Med Surg 10:136–138
14. Barnes CM (1978) Incontinentia pigmenti: report of a case with persistent activity into adult life. Cutis 22:621–624
15. Bartstra HL, Hulsmans RF, Steijlen PM, Ruige M, de Die-Smulders CE, Cassiman JJ (1994) Mosaic expression of hypohidrotic ectodermal dysplasia in an isolated affected female child. Arch Dermatol 130:1421–1424
16. Baumann L, Happle R, Plewig G, Schirren CG (1994) Atrophodermia linearis Moulin. A new disease picture, following the Blaschko lines. Hautarzt 45:231–236
17. Bedi TR (1979) Nevus unius lateralis [sic] exhibiting intraepidermal basaloid cell formation simulating superficial basal cell epithelioma. Dermatologica 159:340–345
18. Beek CH, Jantzen J (1955) On acute linear dermatoses. Acta Derm Venereol 35:318–326
19. Berlin AL, Paller AS, Chan LS (2002) Incontinentia pigmenti: a review and update on the molecular basis of pathophysiology. J Am Acad Dermatol 47: 169–187
20. Biesecker LG, Peters KF, Darling TN, Choyke P, Hill S, Schimke N, Cunningham M, Meltzer P, Cohen MM Jr (1998) Clinical differentiation between Proteus syndrome and hemihyperplasia: description of a distinct form of hemihyperplasia. Am J Med Genet 79:311–318
21. Bird LM, Krous HF, Eichenfield LF, Swalwell CI, Jones MC (1994) Female infant with oncocytic cardiomyopathy and microphthalmia with linear skin defects (MLS): a clue to the pathogenesis of oncocytic cardiomyopathy? Am J Med Genet 53:141–148
22. Boccaletti V, Accorsi P, Pinelli L, Ungari M, Giordano L, Neri I, De Panfilis G (2011) Congenital systematized basaloid follicular hamartoma with microphthalmia and hemimegalencephaly. Pediatr Dermatol 28:555–560
23. Boccaletti V, Salsi B, Ricci R, De Panfilis G (2007) Palmar lichen nitidus following Blaschko's lines with nail involvement in a child. J Pediatr Dermatol 17:145–148
24. Boente MC, Nadra G, Asial R, Happle R (2009) Pronounced linear calcinosis in a boy with mild dermatomyositis: a further possible example of superimposed segmental manifestation of a polygenic disorder. Dermatology 219:155–157
25. Boente MC, Primc N, Veliche H, Rosales S, Carrero-Valenzuela R, Saleme C, Asial R (1999) A mosaic pattern of alopecia in the oral-facial-digital syndrome type I (Papillon-Léage and Psaume syndrome). Pediatr Dermatol 16:367–370
26. Bolognia JL, Orlow SJ, Glick SA (1994) Lines of Blaschko. J Am Acad Dermatol 31:157–190, quiz 190–152
27. Bonifazi E (2012) Segmental atopic dermatitis. Eur J Pediatr Dermatol 22:161
28. Bordel MT (2006) Unilateral Blaschkoid linear lichen planus. Actas Dermosifiliogr 97:225–226
29. Bouzit N, Grézard P, Wolf F, Balme B, Perrot H (1999) Linear cutaneous lupus erythematosus in an adult. Dermatology 199:60–62

30. Boybeyi O, Alanay Y, Kayikçioğlu A, Karnak I (2010) Hemihyperplasia-multiple lipomatosis syndrome: an underdiagnosed entity in children with asymmetric overgrowth. J Pediatr Surg 45:E19–E23
31. Brailey LL, Davis T, Kolker SE, Murry TC, Thomas D, Bale AE, Ruhoy SM (2007) Congenital linear unilateral basal cell nevus: a case report with patched gene molecular studies. J Cutan Pathol 34:65–70
32. Bree AF, Shah MR (2011) Consensus statement from the first international colloquium on basal cell nevus syndrome (BCNS). Am J Med Genet A 155A: 2091–2097
33. Brinca A, Santiago F, Serra D, Andrade P, Vieira R, Figueiredo A (2011) Linear psoriasis - a case report. Case Rep Dermatol 3:8–12
34. Brinkmeier T, Herbst RA, Schaller J, Kuegler K, Pirker C, Beiteke U, Grosshans E, Frosch PJ (2004) Drug-induced blaschkitis. Acta Derm Venereol 84:314–315
35. Bussmann C, Happle R, Baar W, Bieber T, Haidl G, Novak N (2010) Superimposed linear lichen planopilaris: another polygenic disorder exemplifying a new genetic concept. Eur J Dermatol 20:269–270
36. Carman KB, Yakut A, Sabuncu I, Yarar C (2009) MIDAS (microphthalmia, dermal aplasia, sclerocornea) syndrome with central nervous system abnormalities. Clin Dysmorphol 18:234–235
37. Chen YA, Hsu CK, Lee JY, Yang CC (2012) Linear lupus panniculitis of the scalp presenting as alopecia along Blaschko's lines: a distinct variant of lupus panniculitis in East Asians? J Dermatol 39:385–388
38. Chien P Jr, Rosenman K, Cheung W, Wang N, Sanchez M (2009) Linear psoriasis. Dermatol Online J 15:4
39. Cohen PR, Hymes SR (1994) Linear and dermatomal cutaneous graft-versus-host disease. South Med J 87:758–761
40. Collie WR, Moore CM, Goka TJ, Howell RR (1978) Pili torti as marker for carriers of Menkes disease. Lancet 1:607–608
41. Colombo L, Marconi M, Mapelli ET, Crespi E, Gualandri L, Menni S, Cerri A (2011) Superimposed linear psoriasis: low effectiveness of biologic therapy. G Ital Dermatol Venereol 146:311–313
42. Cooper MF, Hay JB, McGibbon D, Shuster S (1976) Androgen metabolism and sebaceous activity in clonal acne. J Invest Dermatol 66:261
43. Coustou D, Léauté-Labrèze C, Bioulac-Sage P, Labbé L, Taïeb A (1999) Asymmetric periflexural exanthem of childhood: a clinical, pathologic, and epidemiologic prospective study. Arch Dermatol 135:799–803
44. Crivellato E, Trevisan G, Grandi G, Lepore L (1983) Bilateral follicular basal cell nevus with comedo-like lesions. Acta Derm Venereol 63:77–79
45. Crovato F, Desirello G, Nazzari G, De Marchi R (1989) Linear pemphigus vulgaris after X-ray irradiation. Dermatologica 179:135–136
46. Cunliffe WJ, Ead RD, Perera WH, Hodgins MB (1977) An acne-free naevus. Br J Dermatol 96:287–290
47. Daldon PE, Lage R (2011) Linear chronic discoid lupus erythematosus following the lines of Blaschko. An Bras Dermatol 86:553–556
48. Danarti R, Bittar M, Happle R, König A (2003) Linear atrophoderma of Moulin: postulation of mosaicism for a predisposing gene. J Am Acad Dermatol 49:492–498
49. Danks DM, Campbell PE, Stevens BJ, Mayne V, Cartwright E (1972) Menkes's kinky hair syndrome: an inherited defect in copper absorption with widespread effects. Pediatrics 50:188–201
50. Darne S, Carmichael AJ (2007) Isolated recurrence of vesicobullous incontinentia pigmenti in a schoolgirl. Br J Dermatol 156:600–602
51. Davies MG, Newman P (2001) Linear cutaneous lupus erythematosus in association with ipsilateral submandibular myoepithelial sialadenitis. Clin Exp Dermatol 26:56–58
52. Davis MI (1938) Zosteriform lichen planus. Arch Dermatol 38:615–618
53. Denk K, Flux K (2011) Blaschkitis in children - a new entity? J Dtsch Dermatol Ges 9:48–49
54. Devriendt K, Matthijs G, Legius E, Schollen E, Blockmans D, van Geet C, Degreef H, Cassiman JJ, Fryns JP (1997) Skewed X-chromosome inactivation in female carriers of dyskeratosis congenita. Am J Hum Genet 60:581–587
55. Di Lernia V, Ricci G, Bonci A, Patrizi A (1991) Lichen striatus and atopy. Int J Dermatol 30:453–454
56. Ding YG, Wang JY, Qiao JJ, Mao XH, Cai SQ (2010) A novel mutation in *MBTPS2* causes ichthyosis follicularis, alopecia and photophobia (IFAP) syndrome in a Chinese family. Br J Dermatol 163:886–889
57. Engelman DE, Kotz EA 3rd, Maize JC Sr (2007) Linear cutaneous lupus erythematosus in the lines of Blaschko. Pediatr Dermatol 24:125–129
58. Ezzedine K, Diallo A, Léauté-Labrèze C, Mossalayi D, Gauthier Y, Bouchtnei S, Cario-Andre M, Sénéschal J, Boralévi F, Jouary T, Taïeb A (2011) Multivariate analysis of factors associated with early-onset segmental and nonsegmental vitiligo: a prospective observational study of 213 patients. Br J Dermatol 165:44–49
59. Ezzedine K, Diallo A, Léauté-Labrèze C, Séneschal J, Prey S, Ballanger F, Alghamdi K, Cario-André M, Jouary T, Gauthier Y, Taieb A (2012) Halo naevi and leukotrichia are strong predictors of the passage to mixed vitiligo in a subgroup of segmental vitiligo. Br J Dermatol 166:539–544
60. Ezzedine K, Gauthier Y, Léauté-Labrèze C, Marquez S, Bouchtnei S, Jouary T, Taïeb A (2011) Segmental vitiligo associated with generalized vitiligo (mixed vitiligo): a retrospective case series of 19 patients. J Am Acad Dermatol 65:965–971
61. Ezzedine K, Lim H, Suzuki T, Katayama I, Hamzavi I, Lan C, Goh B, Anbar T, de Castro CS, Lee A, Parsad D, van Geel N, Le Poole I, Oiso N, Benzekri L, Spritz R, Gauthier Y, Hann S, Picardo M, Taieb A (2012) Revised classification/nomenclature of vitiligo and related issues: The Vitiligo Global Issues

Consensus Conference. Pigment Cell Melanoma Res 25:E1–E13
62. Fernandes S, Santos S, Freitas I, Salgado M, Afonso A, Cardoso J (2012) Linear lupus erythematosus profundus as an initial manifestation of systemic lupus erythematosus in a child. Pediatr Dermatol (Epub ahead of print) doi: 10.1111/pde.12014
63. Figuera LE, Pandolfo M, Dunne PW, Cantú JM, Patel PI (1995) Mapping of the congenital generalized hypertrichosis locus to chromosome Xq24-q27.1. Nat Genet 10:202–207
64. Fink-Puches R, Hofmann-Wellenhof R, Smolle J (1996) Zosteriform lichen planus. Dermatology 192:375–377
65. Fleury J, Guillet G, Anyfantakis V, Huguier V (2010) Unusual calcinosis of muscular loges in a 37 year old patient with a history of juvenile dermatomyositis. Eur J Dermatol 20:647–648
66. Fried K, Feinmesser M, Tsitsianov J (1969) Hearing impairment in female carriers of the sex-linked syndrome of deafness with albinism. J Med Genet 6:132–134
67. Gaitanis G, Nomikos K, Chaniotakis I, Stergiopoulou C, Zioga A, Bassukas I (2009) Linear cutaneous lupus erythematosus: a subset of childhood cutaneous lupus erythematosus. Lupus 18:759–761
68. Garay M, Moreno S, Apreea G, Pizzi-Parra N (2004) Linear juvenile xanthogranuloma. Pediatr Dermatol 21:513–515
69. Gencoglan G, Ceylan C, Kazandi AC (2009) Linear lichenoid drug eruption induced by valsartan. Clin Exp Dermatol 34:e334–e335
70. Gerritsen MJ, de Jong EM, van de Kerkhof PC (1998) Linear lichen planopilaris of the face. J Am Acad Dermatol 38:633–635
71. Giménez-García R, Lázaro-Cantalejo TE, Sánchez-Ramón S, Velasco Fernandez C (2005) Linear lichen planopilaris of the face. J Eur Acad Dermatol Venereol 19:770–772
72. González-Hermosa MR, Escario E, De las Heras C, Casado M (1990) Acne naevus. Clin Exp Dermatol 15:154–155
73. Goujon C, Pierini AM, Thivolet J (1981) Does linear psoriasis exist? Ann Dermatol Venereol 108:643–650
74. Grape J, Frosch P (2011) Papular drug eruption along the lines of Blaschko caused by lenalidomide. Hautarzt 62:618–620
75. Green JJ, Baker DJ (1999) Linear childhood discoid lupus erythematosus following the lines of Blaschko: a case report with review of the linear manifestations of lupus erythematosus. Pediatr Dermatol 16: 128–133
76. Grosshans E, Marot L (1990) Blaschkitis in adults. Ann Dermatol Venereol 117:9–15
77. Grzeschik KH, Bornholdt D, Oeffner F, König A, Boente CM, Enders H, Fritz B, Hertl M, Grasshoff U, Höfling K, Oji V, Paradisi M, Schuchardt C, Szalai Z, Tadini G, Traupe H, Happle R (2007) Deficiency of *PORCN*, a regulator of Wnt signaling, is associated with focal dermal hypoplasia. Nat Genet 39:833–835
78. Gunning ST, Turiansky GW (2003) Successive linear, generalized, and oral lichen planus in a patient with chronic hepatitis C infection. J Am Acad Dermatol 49:1190–1191
79. Hafner C, Landthaler M, Vogt T (2006) Lichen striatus (blaschkitis) following varicella infection. J Eur Acad Dermatol Venereol 20:1345–1347
80. Hamade Y, Arita K, Toyonaga E, Inokuma D, Hamasaka K, Shimizu H (2012) Lichen planus in childhood showing various cutaneous features. Acta Derm Venereol 92:386–387
81. Handa S, Sahoo B (2002) Childhood lichen planus: a study of 87 cases. Int J Dermatol 41:423–427
82. Happle R (1985) Lyonization and the lines of Blaschko. Hum Genet 70:200–206
83. Happle R (1995) What is a nevus? A proposed definition of a common medical term. Dermatology 191:1–5
84. Happle R (2002) Dohi Memorial Lecture. New aspects of cutaneous mosaicism. J Dermatol 29:681–692
85. Happle R (2006) Linear psoriasis and ILVEN: is lumping or splitting appropriate? Dermatology 212:101–102
86. Happle R (2007) Superimposed segmental manifestation of polygenic skin disorders. J Am Acad Dermatol 57:690–699
87. Happle R (2007) Superimposed linear lichen nitidus: a recent report revisited. Eur J Pediatr Dermatol 17:239
88. Happle R (2009) Superimposed segmental manifestation of both rare and common cutaneous disorders: a new paradigm. Actas Dermosifiliogr 100:77–85
89. Happle R (2010) Superimposed segmental dermatomyositis: an emerging new paradigm. Eur J Dermatol 20:551
90. Happle R (2011) Superimposed linear psoriasis: a historical case revisited. J Dtsch Dermatol Ges 9:1027–1028, discussion 1029
91. Happle R (2012) Superimposed segmental vitiligo should not be categorized as "type 2 mosaicism". J Am Acad Dermatol 66:852
92. Happle R, Daniëls O, Koopman RJ (1993) MIDAS syndrome (microphthalmia, dermal aplasia, and sclerocornea): an X-linked phenotype distinct from Goltz syndrome. Am J Med Genet 47:710–713
93. Happle R, Effendy I (2001) Coexisting linear and disseminated drug eruption: a clinical clue to the understanding of the genetic basis of drug eruptions. Eur J Dermatol 11:89
94. Happle R, Frosch PJ (1985) Manifestation of the lines of Blaschko in women heterozygous for X-linked hypohidrotic ectodermal dysplasia. Clin Genet 27:468–471
95. Happle R, Fuhrmann-Rieger A, Fuhrmann W (1984) How are the Blaschko lines arranged on the scalp? Hautarzt 35:366–369
96. Happle R, Kroll P (2013) Nevoid hypertrichosis of the face in a 3-month-old girl with Aicardi syndrome. Eur J Dermatol 23 (Epub ahead of print) doi: 101684/ejd.2013.2082

97. Happle R, Lenz W (1977) Striation of bones in focal dermal hypoplasia: manifestation of functional mosaicism? Br J Dermatol 96:133–135
98. Happle R, Nuñez-Giralda AI (2012) Superimposed lateralized exanthema of childhood: a proposed explanation for an enigmatic disorder. Eur J Dermatol 22:542–544
99. Happle R, Tinschert S (2008) Segmentally arranged basaloid follicular hamartomas with osseous, dental and cerebral anomalies: a distinct syndrome. Acta Derm Venereol 88:382–387
100. Harman M, Inaloz HS, Akdeniz S, Inaloz SS, Aslan A (1999) Congenital non-familial unilateral basaloid follicular hamartoma. J Eur Acad Dermatol Venereol 13:210–213
101. Harpster EF, Mauro T, Barr RJ (1989) Linear granuloma annulare. J Am Acad Dermatol 21:1138–1141
102. Has C, Bruckner-Tuderman L, Muller D, Floeth M, Folkers E, Donnai D, Traupe H (2000) The Conradi-Hünermann-Happle syndrome (CDPX2) and emopamil binding protein: novel mutations, and somatic and gonadal mosaicism. Hum Mol Genet 9:1951–1955
103. Hasson A, Requena L, Arias D, Martin L, de Castro A (1991) Linear pemphigus vulgaris along a surgical scar. Dermatologica 182:191–192
104. Hauser C, Skaria A, Harms M, Saurat JH (1996) Morphoea following Blaschko's lines. Br J Dermatol 134:594–595
105. Herrick AL, Ennis H, Bhushan M, Silman AJ, Baildam EM (2010) Incidence of childhood linear scleroderma and systemic sclerosis in the UK and Ireland. Arthritis Care Res (Hoboken) 62:213–218
106. Hersh CS, Shwayder TA (1991) Unilateral progressive pigmentary purpura (Schamberg's disease) in a 15-year-old boy. J Am Acad Dermatol 24:651
107. Hladik F, Jurecka W, Hayek B, Stingl G, Volc-Platzer B (2005) Atopic dermatitis with increased severity along a line of Blaschko. J Am Acad Dermatol 53:S221–S224
108. Hofer T (2003) Lichen striatus in adults or 'adult blaschkitis'? There is no need for a new naming. Dermatology 207:89–92
109. Hofer T (2006) Does inflammatory linear verrucous epidermal nevus represent a segmental type 1/type 2 mosaic of psoriasis? Dermatology 212:103–107
110. Horio T, Komura J (1978) Linear unilateral basal cell nevus with comedo-like lesions. Arch Dermatol 114:95–97
111. Hughes BR, Cunliffe WJ (1987) An acne naevus. Clin Exp Dermatol 12:357–359
112. Hughes BR, Cunliffe WJ (1987) An acne naevus. Br J Dermatol 117:60–61
113. Indrieri A, van Rahden VA, Tiranti V, Morleo M, Iaconis D, Tammaro R, D'Amato I, Conte I, Maystadt I, Demuth S, Zvulunov A, Kutsche K, Zeviani M, Franco B (2012) Mutations in *COX7B* cause microphthalmia with linear skin lesions, an unconventional mitochondrial disease. Am J Hum Genet 91:942–949
114. Irgang S (1956) Pemphigoid: report of a zosteriform type. Br J Dermatol 68:132–134
115. Irgang S (1968) Zosteriform lichen planus. Cutis 4:1076–1078
116. Itin PH (2009) Happle-Tinschert syndrome: segmentally arranged basaloid follicular hamartomas, linear atrophoderma with hypo- and hyperpigmentation, enamel defects, ipsilateral hypertrichosis, and skeletal and cerebral anomalies. Dermatology 218: 221–225
117. Jaeckle Santos LJ, Xing C, Barnes RB, Ades LC, Megarbane A, Vidal C, Xuereb A, Tarpey PS, Smith R, Khazab M, Shoubridge C, Partington M, Futreal A, Stratton MR, Gecz J, Zinn AR (2008) Refined mapping of X-linked reticulate pigmentary disorder and sequencing of candidate genes. Hum Genet 123:469–476
118. Jaeger T, Andres C, Hein R, Ring J, Chen W (2011) Superimposed segmental manifestation of cherry angiomas. Eur J Dermatol 21:864–865
119. Jensen AL, Vanderhooft SL (2006) Pigmented purpura. In: Irvine AD, Hoeger PH, Yan AC (eds) Harper's textbook of pediatric dematology, vol 2, 3rd edn. Wiley-Blackwell, Chichester, pp 165.161–165.166
120. Jermann M, Eid K, Pfammatter T, Stahel R (2001) Maffucci's syndrome. Circulation 104:1693
121. Jue MS, Kim MH, Ko JY, Lee CW (2011) Digital image processing for the acquisition of graphic similarity of the distributional patterns between cutaneous lesions of linear scleroderma and Blaschko's lines. J Dermatol 38:778–783
122. Juern AM, Glick ZR, Drolet BA, Frieden IJ (2010) Nevus simplex: a reconsideration of nomenclature, sites of involvement, and disease associations. J Am Acad Dermatol 63:805–814
123. Julia M, Mascaró JM Jr, Guilabert A, Navarra E, Ferrando J, Herrero C (2008) Sclerodermiform linear lupus erythematosus: a distinct entity or coexistence of two autoimmune diseases? J Am Acad Dermatol 58:665–667
124. Kanoh H, Ichihashi N, Kamiya H, Seishima M, Akiyama T, Ichiki Y, Kitajima Y (1998) Linear pustular psoriasis that developed in a patient with generalized pustular psoriasis. J Am Acad Dermatol 39:635–637
125. Kanwar AJ, De D (2010) Lichen planus in childhood: report of 100 cases. Clin Exp Dermatol 35:257–562
126. Kaplan RP, Wang JT, Amron DM, Kaplan L (1993) Maffucci's syndrome: two case reports with a literature review. J Am Acad Dermatol 29:894–899
127. Kasap B, Soylu A, Türkmen M, Göktay Y, Sarioğlu S, Ozturk C, Kasapçopur O, Kavukcu S (2009) Juvenile linear scleroderma with unique forms of renal involvement. Pediatr Nephrol 24:2041–2045
128. Kawachi Y, Taguchi S, Fujisawa Y, Furuta J, Nakamura Y, Ishii Y, Otsuka F (2009) Superimposed segmental dermatitis with chronic prurigo. Eur J Dermatol 19:337–340

129. Kawachi Y, Taguchi S, Fujisawa Y, Furuta J, Nakamura Y, Ishii Y, Otsuka F (2011) Linear childhood discoid lupus erythematosus following the lines of Blaschko: successfully treated with topical tacrolimus. Pediatr Dermatol 28:205–207
130. Keegan BR, Kamino H, Fangman W, Shin HT, Orlow SJ, Schaffer JV (2007) "Pediatric blaschkitis": expanding the spectrum of childhood acquired Blaschko-linear dermatoses. Pediatr Dermatol 24:621–627
131. Kim BS, Seo SH, Jung HD, Kwon KS, Kim MB (2010) X-Linked reticulate pigmentary disorder in a female patient. Int J Dermatol 49:421–425
132. Kim JE, Won CH, Chang S, Lee MW, Choi JH, Moon KC (2012) Linear lichen planus pigmentosus of the forehead treated by neodymium:yttrium-aluminum-garnet laser and topical tacrolimus. J Dermatol 39:189–191
133. Kiorpelidou D, Stergiopoulou C, Zioga A, Bassukas ID (2008) Linear-agminated juvenile xanthogranulomas. Int J Dermatol 47:387–389
134. Kira M, Katayama I (2010) Superimposed linear psoriasis. J Dermatol 37:1063–1065
135. Kiritsi D, Diaz-Cascajo C, Hoffmann R, Happle R, Jakob T, Kern JS (2013) A band-like balding disorder. Lancet (Epub ahead of print) doi: 10.1016/S0140-6736(13)60804
136. Koga M, Tango T (1988) Clinical features and course of type A and type B vitiligo. Br J Dermatol 118:223–228
137. König A, Happle R (1999) Linear lesions reflecting lyonization in women heterozygous for IFAP syndrome (ichthyosis follicularis with atrichia and photophobia). Am J Med Genet 85:365–368
138. Küster W, Kind P, Hölzle E, Plewig G (1989) Linear lichen planopilaris of the face. J Am Acad Dermatol 21:131–132
139. Laxmisha C, Thappa DM, Mishra MM, Verma SK (2007) Linear syringocystadenoma papilliferum of the scalp. J Eur Acad Dermatol Venereol 21: 275–276
140. Lee DY, Kim PS, Lee JH (2009) Mixed vitiligo treated by suction blister epidermal grafting: long-term follow up. J Dermatol 36:672–673
141. Lee HE, Ko JY, Lee CW (2011) A case of lupus erythematosus panniculitis presenting as combined linear and lipoatrophic skin lesions. Int J Dermatol 50:203–205
142. Lee HW, Lee DK, Chang SE, Lee MW, Choi JH, Moon KC, Koh JK (2006) Segmental lichen aureus: combination therapy with pentoxifylline and prostacyclin. J Eur Acad Dermatol Venereol 20: 1378–1380
143. Lipsker D, Lenormand C (2009) Classification of polygenic inflammatory diseases distributed along the lines of Blaschko. Dermatology 219:99–101
144. Litani C, Engel G, Piette WW (2009) Tumescent liposuction in the treatment of hemihyperplasia multiple lipomatosis syndrome. Dermatol Surg 35: 1147–1151
145. Long CC, Finlay AY (1996) Multiple linear lichen planus in the lines of Blaschko. Br J Dermatol 135:275–276
146. López Jornet P (2008) White sponge nevus: presentation of a new family. Pediatr Dermatol 25:116–117
147. Lorette G, Toutain A, Barthes M, Saliba E, Georgesco G, Moraine C (1992) Menkes syndrome. An unusual pigmentation anomaly in a mother and three sisters. Ann Pediatr (Paris) 39:453–456
148. Luis-Montoya P, Domínguez-Soto L, Vega-Memije E (2005) Lichen planus in 24 children with review of the literature. Pediatr Dermatol 22:295–298
149. Ma HJ, Zhao G, Liu W, Dang YP, Li DG (2007) Unilateral linear capillaritis: two unusual Chinese cases. Eur J Dermatol 17:160–163
150. Macías-Flores MA, García-Cruz D, Rivera H, Escobar-Luján M, Melendrez-Vega A, Rivas-Campos D, Rodríguez-Collazo F, Moreno-Arellano I, Cantú JM (1984) A new form of hypertrichosis inherited as an X-linked dominant trait. Hum Genet 66:66–70
151. Margarot J, Lonjon M (1929) Psoriasis pustuleux zoniforme. Bull Soc Fr Dermatol Venereol 36:198
152. Margolis E (1962) A new hereditary syndrome–sex linked deafmutism associated with total albinism. Acta Genet Stat Med 12:12–19
153. Marrakchi S, Guigue P, Renshaw BR, Puel A, Pei XY, Fraitag S, Zribi J, Bal E, Cluzeau C, Chrabieh M, Towne JE, Douangpanya J, Pons C, Mansour S, Serre V, Makni H, Mahfoudh N, Fakhfakh F, Bodemer C, Feingold J, Hadj-Rabia S, Favre M, Genin E, Sahbatou M, Munnich A, Casanova JL, Sims JE, Turki H, Bachelez H, Smahi A (2011) Interleukin-36-receptor antagonist deficiency and generalized pustular psoriasis. N Engl J Med 365:620–628
154. Martorell-Calatayud A, Sanz-Motilva V, Garcia-Sales MA, Calatayud-Blas A (2011) Linear syringocystadenoma papilliferum: an uncommon event with a favorable prognosis. Dermatol Online J 17:5
155. Marzano AV, Tanzi C, Caputo R, Alessi E (2005) Sclerodermic linear lupus panniculitis: report of two cases. Dermatology 210:329–332
156. McCuaig CC, Russo P, Powell J, Pedneault L, Lebel P, Marcoux D (1996) Unilateral laterothoracic exanthem: a clinicopathologic study of forty-eight patients. J Am Acad Dermatol 34:979–984
157. McDow RA, Fields JP (1987) Linear granuloma annulare of the finger. Cutis 39:43–44
158. Megahed M, Reinauer S, Scharffetter-Kochanek K, Milde P, Hölzle E, Goerz G, Ruzicka T (1994) Acquired relapsing self-healing Blaschko dermatitis. J Am Acad Dermatol 31:849–852
159. Mensing H, Schmidt KU (1985) Diffuse fasciitis with eosinophilia associated with morphea and lichen sclerosus et atrophicus. Acta Derm Venereol 65:80–83
160. Micalizzi C, Farris A (2000) Erythema multiforme along Blaschko's lines. J Eur Acad Dermatol Venereol 14:203–204

161. Mitsuhashi Y, Kondo S (1996) Lichen striatus in an adult. J Dermatol 23:710–712
162. Moche J, Glassman S, Modi D, Grayson W (2011) Segmental lichen aureus: a report of two cases treated with methylprednisolone aceponate. Australas J Dermatol 52:e15–e18
163. Monteagudo B, Cabanillas M, Suarez-Amor O, Ramirez-Santos A, Alvarez JC, de Las HC (2010) Adult blaschkitis (lichen striatus) in a patient treated with adalimumab. Actas Dermosifiliogr 101: 891–892
164. Morice-Picard F, Boralevi F, Lepreux S, Labrèze C, Lacombe D, Taïeb A (2007) Severe linear form of granuloma annulare along Blaschko's lines preceding the onset of a classical form of granuloma annulare in a child. Br J Dermatol 157:1056–1058
165. Morice-Picard F, Kostrzewa E, Wolf C, Benlian P, Taïeb A, Lacombe D (2011) Evidence of postzygotic mosaicism in a transmitted form of Conradi-Hünermann-Happle syndrome associated with a novel *EBP* mutation. Arch Dermatol 147:1073–1076
166. Moulin G, Hill MP, Guillaud V, Barrut D, Chevallier J, Thomas L (1992) Acquired atrophic pigmented band-like lesions following Blaschko's lines. Ann Dermatol Venereol 119:729–736
167. Mrowietz U, Prinz JC (2012) Psoriasis. In: Plewig G, Landthaler M, Burgdorf WHC, Hertl M, Ruzicka T (eds) Braun-Falco's Dermatologie, Venerologie und Allergologie, vol 1, 6th edn. Springer Science+Business Media, Berlin, pp 625–648
168. Mulekar SV, Al Issa A, Asaad M, Ghwish B, Al Eisa A (2006) Mixed vitiligo. J Cutan Med Surg 10:104–107
169. Müller CS, Schmaltz R, Vogt T, Pföhler C (2011) Lichen striatus and blaschkitis: reappraisal of the concept of blaschkolinear dermatoses. Br J Dermatol 164:257–262
170. Mun JH, Park HJ, Kim HS, Kim SH, Ko HC, Kim BS, Kim MB (2012) Lichen striatus occurring after allogenic peripheral blood stem cell transplantation in an adult with aplastic anemia. Ann Dermatol 24:87–89
171. Nagai Y, Ishikawa O, Hattori T, Ogawa T (2003) Linear lupus erythematosus profundus on the scalp following the lines of Blaschko. Eur J Dermatol 13:294–296
172. Nagy G, Szabó E, Herpai S (1968) Lichen zosteriformis. Z Haut Geschlechtskr 43:509–514
173. Nanda A, Al-Ajmi HS, Al-Sabah H, Al-Hasawi F, Alsaleh QA (2001) Childhood lichen planus: a report of 23 cases. Pediatr Dermatol 18:1–4
174. Narang T, De D, Dogra S, Saikia UN, Handa S (2008) Linear papules and nodules on the neck: syringocystadenoma papilliferum (SP). Arch Dermatol 144:1509–1514
175. Neri I, Russo T, Piccolo V, Patrizi A (2012) Mixed vitiligo in childhood: a study on 13 Italian patients. J Eur Acad Dermatol Venereol 27:e140–e141
176. Ng SJ (2013) Segmental juvenile xanthogranuloma. Pediatr Dermatol 30 (Epub ahead of print) doi: 10.1111/pde.12199
177. Nnoruka EN (2007) Lichen planus in African children: a study of 13 patients. Pediatr Dermatol 24:495–498
178. Oeffner F, Fischer G, Happle R, König A, Betz RC, Bornholdt D, Neidel U, Boente Mdel C, Redler S, Romero-Gomez J, Salhi A, Vera-Casaño A, Weirich C, Grzeschik KH (2009) IFAP syndrome is caused by deficiency in *MBTPS2*, an intramembrane zinc metalloprotease essential for cholesterol homeostasis and ER stress response. Am J Hum Genet 84:459–467
179. Ogata T, Wakui K, Muroya K, Ohashi H, Matsuo N, Brown DM, Ishii T, Fukushima Y (1998) Microphthalmia with linear skin defects syndrome in a mosaic female infant with monosomy for the Xp22 region: molecular analysis of the Xp22 breakpoint and the X-inactivation pattern. Hum Genet 103:51–56
180. Oiso N, Tatsumi Y, Rai S, Matsumura I, Kawada A (2011) Superimposed linear graft-versus-host disease and secondary cutaneous involvement of anaplastic large cell lymphoma. Eur J Dermatol 21:636–638
181. Onder M, Adişen E, Happle R (2013) Superimposed linear lichen planus in a 4-year-old girl. Int J Dermatol (in press)
182. Online Mendelian Inheritance in Man (OMIM). Available from: http://www3.ncbi.nlm.nih.gov/Omim/. Accessed 6 Sep 2013
183. Özdemir M, Balevi A, Esen H (2012) An inflammatory verrucous epidermal nevus concomitant with psoriasis: treatment with adalimumab. Dermatol Online J 18:11
184. Özdemir M, Mevlitoğlu I, Balevi A (2009) Acitretin narrow-band TL-01 phototherapy but not etanercept treatment improves a localized inflammatory linear verrucous epidermal naevus with concomitant psoriasis. J Eur Acad Dermatol Venereol 23:1453–1454
185. Özkaya-Bayazit E, Akasya E, Buyukbabani N, Baykal C (2000) Pustular psoriasis with a striking linear pattern. J Am Acad Dermatol 42:329–331
186. Özkaya-Bayazit E, Baykal C (1997) Trimethoprim-induced linear fixed drug eruption. Br J Dermatol 137:1028–1029
187. Pahwa P, Kaushal S, Gupta S, Khaitan BK, Sharma VK, Sethuraman G (2011) Linear syringocystadenoma papilliferum: an unusual location. Pediatr Dermatol 28:61–62
188. Pansuriya TC, van Eijk R, d'Adamo P, van Ruler MA, Kuijjer ML, Oosting J, Cleton-Jansen AM, van Oosterwijk JG, Verbeke SL, Meijer D, van Wezel T, Nord KH, Sangiorgi L, Toker B, Liegl-Atzwanger B, San-Julian M, Sciot R, Limaye N, Kindblom LG, Daugaard S, Godfraind C, Boon LM, Vikkula M, Kurek KC, Szuhai K, French PJ, Bovée JV. Somatic mosaic IDH1 and IDH2 mutations are associated with enchondroma and spindle cell hemangioma in Ollier disease and Maffucci syndrome. Nat Genet 2011;43:1256-1261
189. Parsons JM (1996) Transient acantholytic dermatosis (Grover's disease): a global perspective. J Am Acad Dermatol 35:653–666, quiz 667–670

190. Partington MW, Marriott PJ, Prentice RS, Cavaglia A, Simpson NE (1981) Familial cutaneous amyloidosis with systemic manifestations in males. Am J Med Genet 10:65–75
191. Partington MW, Prentice RS (1989) X-linked cutaneous amyloidosis: further clinical and pathological observations. Am J Med Genet 32:115–119
192. Patrizi A, Neri I, Fiorentini C, Bonci A, Ricci G (2004) Lichen striatus: clinical and laboratory features of 115 children. Pediatr Dermatol 21:197–204
193. Patrizi A, Neri I, Orlandi C, Marini R (1996) Sacral medial telangiectatic vascular nevus: a study of 43 children. Dermatology 192:301–306
194. Patterson JW, Straka BF, Wick MR (2001) Linear syringocystadenoma papilliferum of the thigh. J Am Acad Dermatol 45:139–141
195. Perry D, Fazel N (2006) Zosteriform lichen planus. Dermatol Online J 12:3
196. Petrozzi JW, Shmunes E (1970) Linear lichen nitidus. Cutis 6:1109–1112
197. Pfau A, Landthaler M (1995) Recurrent inflammation in incontinentia pigmenti of a seven-year-old child. Dermatology 191:161–163
198. Prakash SK, Cormier TA, McCall AE, Garcia JJ, Sierra R, Haupt B, Zoghbi HY, Van den Veyver IB (2002) Loss of holocytochrome c-type synthetase causes the male lethality of X-linked dominant microphthalmia with linear skin defects (MLS) syndrome. Hum Mol Genet 11:3237–3248
199. Prigent F, Cavelier-Balloy B, Lemarchand-Venencie F, Civatte J (1989) Linear lichen nitidus. Ann Dermatol Venereol 116:814–815
200. Racette AJ, Adams AD, Kessler SE (2009) Simultaneous lichen striatus in siblings along the same Blaschko line. Pediatr Dermatol 26:50–54
201. Reiter H, Feldmann R, Breier F, Happle R, Gschnait F (2000) Lichen striatus or Blaschkitis in the adults: variations of the same entity? Hautarzt 51:770–773
202. Renner R, Colsman A, Sticherling M (2008) ILVEN: is it psoriasis? Debate based on successful treatment with etanercept. Acta Derm Venereol 88:631–632
203. Requena C, Torrelo A, de Prada I, Zambrano A (2002) Linear childhood cutaneous lupus erythematosus following Blaschko lines. J Eur Acad Dermatol Venereol 16:618–620
204. Rhee CH, Kim SM, Kim MH, Cinn YW, Ihm CW (2009) Two cases of linear alopecia on the occipital scalp. Ann Dermatol 21:159–163
205. Riou S, Morelon E, Guibaud L, Chotel F, Dijoud F, Marec-Berard P (2012) Efficacy of rapamycin for refractory hemangioendotheliomas in Maffucci's syndrome. J Clin Oncol 30:e213–e215
206. Röckmann H, Feller G, Schadendorf D, Goerdt S (2006) Subacute cutaneous lupus erythematosus on the lines of Blaschko. Eur J Dermatol 16: 302–306
207. Roholt NS, Lapière JC, Wang JI, Bernstein LJ, Woodley DT, Eramo LR (1995) Localized linear bullous eruption of systemic lupus erythematosus in a child. Pediatr Dermatol 12:138–144
208. Rompel R, Mischke AL, Langner C, Happle R (2000) Linear atrophoderma of Moulin. Eur J Dermatol 10:611–613
209. Rosser TL, Acosta MT, Packer RJ (2002) Aicardi syndrome: spectrum of disease and long-term prognosis in 77 females. Pediatr Neurol 27:343–346
210. Runne U, Fasshauer K (1977) Idiopathic and sclerodermic facial hemiatrophy with generalized myopathy: clinical, electromyographic and histologic examinations of six patients. Hautarzt 28:10–17
211. Russell B (1950) Psoriasis zoniformis. Br J Dermatol 62:314–316
212. Sabat M, Ribera M, Bielsa I, Mangas C, Fernandez-Chico N, Ferrandiz C (2006) Linear lupus erythematosus following the lines of Blaschko. J Eur Acad Dermatol Venereol 20:1005–1006
213. Şanli H, Anadolu R, Arat M, Ekmekci P, Birol A, Erdem C, Koç H (2003) Dermatomal lichenoid graft-versus-host disease within herpes zoster scars. Int J Dermatol 42:562–564
214. Saxena AK, Nigam PK (1988) Unilateral lichen planus. Cutis 42:142–143
215. Schulte TL, Liljenqvist U, Görgens H, Hackenberg L, Bullmann V, Tinschert S (2008) Hemihyperplasia-multiple lipomatosis syndrome (HHML): a challenge in spinal care. Acta Orthop Belg 74:714–719
216. Seitz CS, Bröcker EB, Trautmann A (2008) Linear variant of chronic cutaneous lupus erythematosus: a clue for the pathogenesis of chronic cutaneous lupus erythematosus? Lupus 17:1136–1139
217. Seitz CS, Garbaraviciene J, Bröcker EB, Hamm H (2009) Superimposed linear psoriasis: differential therapeutic response of linear and nonlinear lesions. Clin Exp Dermatol 34:e177–e179
218. Seo JK, Lee HJ, Lee D, Choi JH, Sung HS (2010) A case of linear lichen planus pigmentosus. Ann Dermatol 22:323–325
219. Sfia M, Roth-Mall B, Tortel MC, Guillaume JC, Cribier B (2009) Blaschko-linear psoriasis revealed by infliximab therapy. Ann Dermatol Venereol 136:898–903
220. Sharma R, Maheshwari V (1999) Childhood lichen planus: a report of fifty cases. Pediatr Dermatol 16:345–348
221. Shelley WB, Rawnsley HM, Beerman H (1969) Quadrant distribution of basal cell nevi. Arch Dermatol 100:741–743
222. Shepherd V, Godbolt A, Casey T (2005) Maffucci's syndrome with extensive gastrointestinal involvement. Australas J Dermatol 46:33–37
223. Shiloh Y, Litvak G, Ziv Y, Lehner T, Sandkuyl L, Hildesheimer M, Buchris V, Cremers FP, Szabo P, White BN et al (1990) Genetic mapping of X-linked albinism-deafness syndrome (ADFN) to Xq26.3-q27.I. Am J Hum Genet 47:20–27
224. Sigal-Nahum M, Konqui A, Gaulier A, Sigal S (1988) Linear fixed drug eruption. Br J Dermatol 118:849–851
225. Singh S, Kumar S (2009) Linear scleroderma with partial anonychia. Indian J Dermatol Venereol Leprol 75:623–625

226. Singh N, Noorunnisa N (2012) Linear psoriasis: a rare presentation. Indian Dermatol Online J 3:71–73
227. Smahi A, Courtois G, Vabres P, Yamaoka S, Heuertz S, Munnich A, Israel A, Heiss NS, Klauck SM, Kioschis P, Wiemann S, Poustka A, Esposito T, Bardaro T, Gianfrancesco F, Ciccodicola A, D'Urso M, Woffendin H, Jakins T, Donnai D, Stewart H, Kenwrick SJ, Aradhya S, Yamagata T, Levy M, Lewis RA, Nelson DL (2000) Genomic rearrangement in *NEMO* impairs NF-kappaB activation and is a cause of incontinentia pigmenti. The International Incontinentia Pigmenti (IP) Consortium. Nature 405:466–472
228. Soma Y, Kawakami T, Yamasaki E, Sasaki R, Mizoguchi M (2003) Linear scleroderma along Blaschko's lines in a patient with systematized morphea. Acta Derm Venereol 83:362–364
229. Staricco RG (1959) Lichen striatus: a study of fifteen new cases with special emphasis on the histopathological changes and a review of the literature. Arch Dermatol 79:311–324
230. Stojanović S, Jovanović M, Vučković N (2008) Lichen planus-like dermatosis with Blaschko line distribution: a case report. Acta Dermatovenerol Alp Panonica Adriat 17:137–138
231. Sutton VR, Hopkins BJ, Eble TN, Gambhir N, Lewis RA, Van den Veyver IB (2005) Facial and physical features of Aicardi syndrome: infants to teenagers. Am J Med Genet A 138A:254–258
232. Tada J, Arata J, Katayama H (1991) Linear lupus erythematosus profundus in a child. J Am Acad Dermatol 24:871–874
233. Tadin-Strapps M, Salas-Alanis JC, Moreno L, Warburton D, Martinez-Mir A, Christiano AM (2003) Congenital universal hypertrichosis with deafness and dental anomalies inherited as an X-linked trait. Clin Genet 63:418–422
234. Taïeb A (1994) Linear atopic dermatitis ('naevus atopicus'): a pathogenetic clue? Br J Dermatol 131:134–135
235. Taïeb A, el Youbi A, Grosshans E, Maleville J (1991) Lichen striatus: a Blaschko linear acquired inflammatory skin eruption. J Am Acad Dermatol 25:637–642
236. Taïeb A, Grosshans E (2006) Lichen striatus. In: Irvine AD, Hoeger PH, Yan AC (eds) Harper's textbook of dermatology, vol 1, 3rd edn. Wiley-Blackwell, Chichester, pp 86.81–86.87
237. Takahashi Y, Murota H, Tarutani M, Sano S, Okinaga T, Tominaga K, Yano T, Katayama I (2010) A case of juvenile dermatomyositis manifesting inflammatory epidermal nevus-like skin lesions: unrecognized cutaneous manifestation of blaschkitis? Allergol Int 59:425–428
238. Tamada Y, Arisawa S, Ikeya T, Yokoi T, Hara K, Matsumoto Y (1999) Linear lupus erythematosus profundus in a young man. Br J Dermatol 140:177–178
239. Tamiya H, Sowa J, Nakanishi T, Adachi M, Ishii M, Kobayashi H (2010) Linear lupus erythematosus profundus on the face, following the lines of Blaschko. Int J Dermatol 49:1459–1461
240. Tang L, Liang J, Wang W, Yu L, Yao Z (2011) A novel mutation in *MBTPS2* causes a broad phenotypic spectrum of ichthyosis follicularis, atrichia, and photophobia syndrome in a large Chinese family. J Am Acad Dermatol 64:716–722
241. Tejera-Vaquerizo A, Ruiz-Molina I, Solís-García E, Moreno-Giménez JC (2009) Adult blaschkitis (lichen striatus) successfully treated with topical tacrolimus. Actas Dermosifiliogr 100:631–632
242. Thind CK, Husain EA, Hewitt J (2009) A rare linear atrophic lesion on the face. Clin Exp Dermatol 34:e447–e448
243. Toda K, Okamoto H, Horio T (1986) Lichen striatus. Int J Dermatol 25:584–585
244. Torchia D (2011) Superimposed segmental dermatomyositis. Allergol Int 60:401
245. Torchia D, Schachner LA (2011) Segmental eczema: within the spectrum of mosaic spongiotic dermatitides. Eur J Pediatr Dermatol 21:199–203
246. Tsuruta D, Sowa J, Hiroyasu S, Ishii M, Kobayashi H (2011) Concomitant occurrence of patch granuloma annulare and classical granuloma annulare. J Dermatol 38:482–485
247. Tsuzaka S, Ishiguro N, Akashi R, Kawashima M (2012) A case of lupus erythematosus profundus with multiple arc-shaped erythematous plaques on the scalp and a review of the literature. Lupus 21:662–665
248. Turner RJ, Dahl MG, Shuster S, Rees JL (1998) Mosaic atopic eczema cured by autotransplantation? Lancet 352:961
249. Vachiramon V, Suchonwanit P, Thadanipon K (2010) Bilateral linear lichen planus pigmentosus associated with hepatitis C virus infection. Case Rep Dermatol 2:169–172
250. Van Baar H, Perret CM, Happle R (1989) Nuchal nevi flammei and alopecia areata. Dermatologica 179:52–53
251. van Geel N, De Lille S, Vandenhaute S, Gauthier Y, Mollet I, Brochez L, Lambert J (2011) Different phenotypes of segmental vitiligo based on a clinical observational study. J Eur Acad Dermatol Venereol 25:673–678
252. Verma SB, Wollina U (2012) Chronic disseminated discoid lupus erythematosus with linear lesions following Blaschko's lines on forearm and hand. J Dtsch Dermatol Ges 10:129–130
253. Volpintesta EJ (1974) Menkes kinky hair syndrome in a black infant. Am J Dis Child 128:244–246
254. Weibel L, Harper JI (2008) Linear morphoea follows Blaschko's lines. Br J Dermatol 159:175–181
255. Wimplinger I, Rauch A, Orth U, Schwarzer U, Trautmann U, Kutsche K (2007) Mother and daughter with a terminal Xp deletion: implication of chromosomal mosaicism and X-inactivation in the high clinical variability of the microphthalmia with linear skin defects (MLS) syndrome. Eur J Med Genet 50:421–431

256. Yaghoobi R, Zadeh SH, Zadeh AH (2009) Giant linear syringocystadenoma papilliferum on scalp. Indian J Dermatol Venereol Leprol 75:318–319
257. Yanaru E, Ueda M, Ichihashi M (2000) Linear lichen planopilaris of the face treated with low-dose cyclosporin A. Acta Derm Venereol 80:212
258. Yap FB, Lee BR, Baba R (2010) Syringocystadenoma papilliferum in an unusual location beyond the head and neck region: a case report and review of literature. Dermatol Online J 16:4
259. Yoon TY, Lee DY, Kim YJ, Lee JY, Kim MK (2013) Unilateral linear capillaropathy limited to the upper extremity in an infant. J Dermatol 40:71–75
260. Zhu H, Shang D, Sun M, Choi S, Liu Q, Hao J, Figuera LE, Zhang F, Choy KW, Ao Y, Liu Y, Zhang XL, Yue F, Wang MR, Jin L, Patel PI, Jing T, Zhang X (2011) X-linked congenital hypertrichosis syndrome is associated with interchromosomal insertions mediated by a human-specific palindrome near *SOX3*. Am J Hum Genet 88:819–826
261. Ziprkowski L, Krakowski A, Adam A, Costeff H, Sade J (1962) Partial albinism and deaf-mutism due to a recessive sex-linked gene. Arch Dermatol 86:530–539
262. Zulian F, Athreya BH, Laxer R, Nelson AM, Feitosa de Oliveira SK, Punaro MG, Cuttica R, Higgins GC, Van Suijlekom-Smit WA, Moore TL, Lindsley C, Garcia-Consuegra J, Esteves Hilário MO, Lepore L, Silva CA, Machado C, Garay SM, Uziel Y, Martini G, Foeldvari I, Peserico A, Woo P, Harper J (2006) Juvenile localized scleroderma: clinical and epidemiological features in 750 children: an international study. Rheumatology 45:614–620
263. Zvulunov A, Kachko L, Manor E, Shinwell E, Carmi R (1998) Reticulolinear aplasia cutis congenita of the face and neck: a distinctive cutaneous manifestation in several syndromes linked to Xp22. Br J Dermatol 138:1046–1052

A Note on Neoplastic Skin Lesions 13

Contents

At this point in time it seems reasonable to assume that most of the benign or malignant cutaneous neoplasias are mosaic lesions. As a rule, they reflect either LOH (see Sect. 10.1) or heterozygosity for a postzygotic new mutation.

By tradition, benign melanocytic proliferations are called nevi (see Sect. 7.2.1), whereas all other benign cutaneous neoplasias such as neurofibroma, lipoma, or syringoma do not represent nevi. For these tumors and for the malignant neoplasias such as basal cell carcinoma, the time-honored and simple two-hit model as first proposed by Armitage and Doll [1] and later elaborated by Knudson [10, 11], is probably not always true. Den Otter et al. [3–5] and Derkinderen et al. [6] have proposed a four-mutation model of tumorigenesis that appears to be more appropriate for many skin tumors. Recent molecular studies endorse this view [2, 8, 16, 17].

For the development of other malignant skin tumors such as squamous cell carcinoma or malignant melanoma, a multistep model of carcinogenesis can be taken as certain [7, 9, 12–15].

References

1. Armitage P, Doll R (1957) A two-stage theory of carcinogenesis in relation to the age distribution of human cancer. Br J Cancer 11:161–169
2. Christiaans I, Kenter SB, Brink HC, van Os TA, Baas F, van den Munckhof P, Kidd AM, Hulsebos TJ (2011) Germline *SMARCB1* mutation and somatic *NF2* mutations in familial multiple meningiomas. J Med Genet 48:93–97

R. Happle, *Mosaicism in Human Skin*,
DOI 10.1007/978-3-642-38765-4_13, © Springer-Verlag Berlin Heidelberg 2014

3. Den Otter W, Koten JW, Derkinderen DJ (1987) Carcinogenesis revisited. Cancer Invest 5:69–74
4. Den Otter W, Koten JW, Van der Vegt BJ, Beemer FA, Boxma OJ, De Graaf PW, Derkinderen DJ, Hill FW, Huber J, Klein WR et al (1990) Hereditary cancer and its clinical implications: a view. Anticancer Res 10:489–495
5. Den Otter W, Koten JW (1999) A 4-mutation model of carcinogenesis for tumour suppressor genes. Anticancer Res 19:4845–4852
6. Derkinderen DJ, Boxma OJ, Koten JW, Den Otter W (1990) Stochastic theory of oncogenesis. Anticancer Res 10:497–504
7. El-Naggar AK (2007) Pathobiology of head and neck squamous tumorigenesis. Curr Cancer Drug Targets 7:606–612
8. Hadfield KD, Newman WG, Bowers NL, Wallace A, Bolger C, Colley A, McCann E, Trump D, Prescott T, Evans DG (2008) Molecular characterisation of SMARCB1 and NF2 in familial and sporadic schwannomatosis. J Med Genet 45: 332–339
9. Hunt JL, Barnes L, Lewis JS Jr, Mahfouz ME, Slootweg PJ, Thompson LD, Cardesa A, Devaney KO, Gnepp DR, Westra WH, Rodrigo JP, Woolgar JA, Rinaldo A, Triantafyllou A, Takes RP, Ferlito A (2013) Molecular diagnostic alterations in squamous cell carcinoma of the head and neck and potential diagnostic applications. Eur Arch Otorhinolaryngol
10. Knudson AG Jr (1971) Mutation and cancer: statistical study of retinoblastoma. Proc Natl Acad Sci U S A 68:820–823
11. Knudson AG Jr, Hethcote HW, Brown BW (1975) Mutation and childhood cancer: a probabilistic model for the incidence of retinoblastoma. Proc Natl Acad Sci U S A 72:5116–5120
12. Koten JW, Neijt JP, Zonnenberg BA, Den Otter W (1993) The difference between benign and malignant tumours explained with the 4-mutation paradigm for carcinogenesis. Anticancer Res 13:1179–1182
13. Nambiar S, Mirmohammadsadegh A, Bar A, Bardenheuer W, Roeder G, Hengge UR (2004) Applications of array technology: melanoma research and diagnosis. Expert Rev Mol Diagn 4:549–557
14. Peeper DS (2011) Oncogene-induced senescence and melanoma: where do we stand? Pigment Cell Melanoma Res 24:1107–1111
15. Pfeifer GP, Besaratinia A (2012) UV wavelength-dependent DNA damage and human non-melanoma and melanoma skin cancer. Photochem Photobiol Sci 11:90–97
16. Reis AH, Vargas FR, Lemos B (2012) More epigenetic hits than meets the eye: microRNAs and genes associated with the tumorigenesis of retinoblastoma. Front Genet 3:284
17. Sestini R, Bacci C, Provenzano A, Genuardi M, Papi L (2008) Evidence of a four-hit mechanism involving SMARCB1 and NF2 in schwannomatosis-associated schwannomas. Hum Mutat 29:227–231

Index

R. Happle, *Mosaicism in Human Skin*,
DOI 10.1007/978-3-642-38765-4, © Springer-Verlag Berlin Heidelberg 2014

FSC
www.fsc.org
MIX
Papier | Fördert
gute Waldnutzung
FSC® C083411

Kolibri 360 GmbH
Ferdinand-Jühlke-Straße 7
99095 Erfurt, Deutschland
produktsicherheit@kolibri360.de